Springer
Berlin
Heidelberg
New York
Hong Kong
London
Milan
Paris
Tokyo

Guerino Mazzola · Gérard Milmeister ·
Jody Weissmann

Comprehensive Mathematics for Computer Scientists 1

Sets and Numbers, Graphs and Algebra,
Logic and Machines, Linear Geometry

With 82 Figures

Springer

Guerino Mazzola
Gérard Milmeister
Jody Weissmann

Department of Informatics
University of Zurich
Winterthurerstr. 190
8057 Zurich, Switzerland

The text has been created using LaTeX 2_e. The graphics were drawn using Dia, an open-source diagramming software. The main text has been set in the Y&Y Lucida Bright type family, the headings in Bitstream Zapf Humanist 601.

Library of Congress Control Number: 2004102307

Mathematics Subject Classification (1998): 00A06

ISBN 3-540-20835-6 Springer-Verlag Berlin Heidelberg New York

Springer-Verlag
is a part of Springer Science+Business Media
springer.de

© Springer-Verlag Berlin Heidelberg 2004
Printed in Germany

Cover Design: Erich Kirchner, Heidelberg
Typesetting: Camera ready by the authors
40/3142SR – 5 4 3 2 1 0 – Printed on acid-free paper

Preface

The need for better formal competence as it is generated by a sound mathematical education has been confirmed by recent investigations by professional associations, but also by IT opinion leaders such as Niklaus Wirth or Peter Wegner. It is rightly argued that programming skills are a necessary but by far not sufficient qualification for designing and controlling the conceptual architecture of valid software. Often, the deficiency in formal competence is compensated by trial and error programming. This strategy may lead to uncontrolled code which neither formally nor effectively meets the given objectives. According to the global view such bad engineering practice leads to massive quality breakdowns with corresponding economical consequences.

Improved formal competence is also urged by the object-oriented paradigm which progressively requires a programming style and a design strategy of high abstraction level in conceptual engineering. In this context, the arsenal of formal tools must apply to completely different problem situations. Moreover, the dynamics and life cycle of hard- and software projects enforce high flexibility of theorists and executives on all levels of the computer science academia and IT industry. This flexibility can only be guaranteed by a propaedeutical training in a number of typical styles of mathematical argumentation.

With this in mind, writing an introductory book on mathematics for computer scientists is a somewhat delicate task. On the one hand, computer science delves into the most basic machinery of human thought, such as it is traced in the theory of Turing machines, rewriting systems and grammars, languages, and formal logic. On the other hand, numerous applications of core mathematics, such as the theory of Galois fields (e.g., for coding theory), linear geometry (e.g., for computer graphics), or differential equations (e.g., for simulation of dynamic systems) arise in any relevant topic of computational science. In view of this wide field of mathematical subjects the common practice is to focus one's attention on a particular bundle of issues and to presuppose acquaintance with the background theory, or else to give a short summary thereof without any further details.

In this book, we have chosen a different presentation. The idea was to set forth and prove the entire core theory, from axiomatic set theory to numbers, graphs, algebraic and logical structures, linear geometry—in the present first volume, and then, in the second volume, topology and calculus, differential equations, and more specialized and current subjects such as neural networks, fractals, numerics, Fourier theory, wavelets, probability and statistics, manifolds, and categories.

There is a price to pay for this comprehensive journey through the overwhelmingly extended landscape of mathematics: We decided to omit any not absolutely necessary ramification in mathematical theorization. Rather it was essential to keep the global development in mind and to avoid an unnecessarily broad approach. We have therefore limited explicit proofs to a length which is reasonable for the non-mathematician. In the case of lengthy and more involved proofs, we refer to further readings. For a more profound reading we included a list of references to original publications. After all, the student should realize as early as possible in his or her career that science is vitally built upon a network of links to further knowledge resources.

We have, however, chosen a a modern presentation: We introduce the language of commutative diagrams, universal properties and intuitionistic logic as advanced by contemporary theoretical computer science in its topos-theoretic aspect. This presentation serves the economy and elegance of abstraction so urgently requested by opinion leaders in computer science. It also shows some original restatements of well-known facts, for example in the theory of graphs or automata. In addition, our presentation offers a glimpse of the unity of science: Machines, formal concept architectures, and mathematical structures are intimately related with each other.

Beyond a traditional "standalone" textbook, this text is part of a larger formal training project hosted by the Department of Informatics at the University of Zurich. The online counterpart of the text can be found on http://math.ifi.unizh.ch. It offers access to this material and includes interactive tools for examples and exercises implemented by Java

applets and script-based dynamic HTML. Moreover, the online presentation allows switching between textual navigation via classical links and a
quasi-geographical navigation on a "landscape of knowledge". In the latter, parts, chapters, axioms, definitions, and propositions are visualized
by continents, nations, cities, and paths. This surface structure describes
the top layer of a three-fold stratification (see the following screenshot of
some windows of the online version).

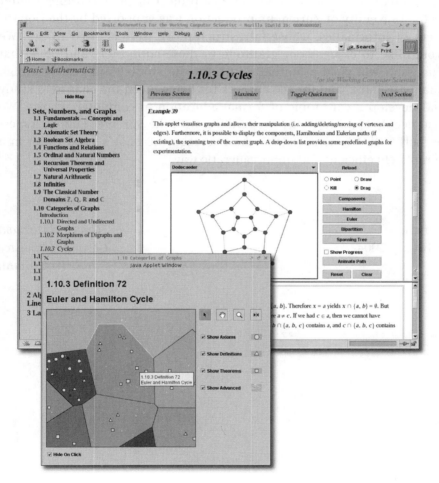

On top are the facts, below, in the middle layer, the user will find the
proofs, and in the third, deepest stratum, one may access the advanced
topics, such as floating point arithmetic, or coding theory. The online
counterpart of the book includes two important addenda: First, a list of

errata can be checked out. The reader is invited to submit any error encountered while reading the book or the online presentation. Second, the subject spectrum, be it in theory, examples, or exercises, is constantly updated and completed and, if appropriate, extended. It is therefore recommended and beneficial to work with both, the book and its online counterpart.

This book is a result of an educational project of the E-Learning Center of the University of Zurich. Its production was supported by the Department of Informatics, whose infrastructure we were allowed to use. We would like to express our gratitude to these supporters and hope that the result will yield a mutual profit: for the students in getting a high quality training, and for the authors for being given the chance to study and develop a core topic of formal education in computer science. We also deeply appreciate the cooperation with the Springer Publishers, especially with Clemens Heine, who managed the book's production in a completely efficient and unbureaucratic way.

Zurich,   *Guerino Mazzola*
February 2004   *Gérard Milmeister*
                                                                                    Jody Weissmann

Contents

Volume II

III   Topology and Calculus

Limits and Topology, Differentiability, Inverse and Implicit Functions, Integration, Fubini and Changing Variables, Vector Fields, Fixpoints, Main Theorem of ODEs

IV   Selected Higher Subjects

Numerics, Probability and Statistics, Splines, Fourier, Wavelets, Fractals, Neural Nets, Global Coordinates and Manifolds, Categories, Lambda Calculus

PART I

Sets, Numbers, and Graphs

CHAPTER 1

Fundamentals—Concepts and Logic

Die Welt ist alles, was der Fall ist.
Ludwig Wittgenstein

"The world is everything that is the case" — this is the first tractatus in Ludwig Wittgenstein's *Tractatus Logico-Philosophicus*.

In science, we want to know what is true, i.e., what is the case, and what is not. Propositions are the theorems of our language, they are to describe or denote what is the case. If they do, they are called true, otherwise they are called false. This sounds a bit clumsy, but actually it is pretty much what our common sense tells us about true and false statements. Perhaps an example would help to clarify things:

"This sentence contains five words" This proposition describes something which is the case, therefore it is a *true* statement.

"Every human being has three heads" Since I myself have only one head (and I assume this is the case with you as well), this proposition describes a situation which is not the case, therefore it is *false*.

In order to precisely handle propositions, science makes use of two fundamental tools of thought:

- Propositional Logic
- Architecture of Concepts

These tools aid a scientist to construct accurate concepts and to formulate new true propositions from old ones.

The following sections may appear quite diffuse to the reader; some things will seem to be obviously true, other things will perhaps not make much sense to start with. The problem is that we have to use our natural language for the task of defining things in a precise way. It is only by using these tools that we can define in a clear way what a set is, what numbers are, etc.

1.1 Propositional Logic

Propositional logic helps us to navigate in a world painted in black and white, a world in which there is only truth or falsehood, but nothing in between. It is a boiled down version of common sense reasoning. It is the essence of Sherlock Holmes' way of deducing that Professor Moriarty was the mastermind behind a criminal organization ("Elementary, my dear Watson"). Propositional logic builds on two propositions, which are declared to be true as basic principles (and they seem to make sense...):

Principle of contradiction (principium contradictionis) A proposition is never true and false at the same time.

Principle of the excluded third (tertium non datur) A proposition is either true or false—there is no third possibility.

In other words, in propositional logic we work with statements that are either true or false, no more and no less.

In propositional logic there are also some operations which are used to create new propositions from old ones:

Logical Negation The negation of a true proposition is a false proposition, the negation of a false proposition is a true proposition. This operation is also called 'NOT'.

Logical Conjunction The conjunction of two propositions is true if and only if both propositions are true. In all other cases it is false. This operation is also called 'AND'.

Logical Disjunction The disjunction of two propositions is true if at least one of the propositions is true. If both propositions are false, the disjunction is false, too. This operation is also known as 'OR'.

Logical Implication A proposition P_1 implies another proposition P_2 if P_2 is true whenever P_1 is true. This operation is also known as 'IMPLIES'.

Often one uses so-called truth tables to show the workings of these operations. In these tables, A stands for the possible truth values of a proposition $\mathcal{A}$, and B stands for the possible truth values of a proposition $\mathcal{B}$. The rows labeled "A AND B" and "A OR B" contain the truth value of the conjunction and disjunction of the propositions.

A	NOT A
F	T
T	F

A	B	A AND B
F	F	F
F	T	F
T	F	F
T	T	T

A	B	A OR B
F	F	F
F	T	T
T	F	T
T	T	T

A	B	A IMPLIES B
F	F	T
F	T	T
T	F	F
T	T	T

Let us look at a few examples.

1. Let proposition $\mathcal{A}$ be "The ball is red". The negation of $\mathcal{A}$, (i.e., NOT $\mathcal{A}$) is "It is not the case that the ball is red". So, if the ball is actually green, that means that $\mathcal{A}$ is false and that NOT $\mathcal{A}$ is true.

2. Let proposition $\mathcal{A}$ be "All balls are round" and proposition $\mathcal{B}$ "All balls are green". Then the conjunction $\mathcal{A}$ AND $\mathcal{B}$ of $\mathcal{A}$ and $\mathcal{B}$ is false, because there are balls that are not green.

3. Using the same propositions, the disjunction of $\mathcal{A}$ and $\mathcal{B}$, $\mathcal{A}$ OR $\mathcal{B}$ is true.

4. For any proposition $\mathcal{A}$, $\mathcal{A}$ AND (NOT $\mathcal{A}$) is always false.

5. For any proposition $\mathcal{A}$, $\mathcal{A}$ OR (NOT $\mathcal{A}$) is always true.

In practice it is cumbersome to say: "The proposition *It rains* is true". Instead, one just says: "It rains." Also, the formal combination of propositions by the above operators is often a formal overhead, we mostly use the common language denotation, such as: "2 = 3 is false" instead of "NOT (2 = 3)" or: "It's raining or/and I am tired." instead of "It's raining OR/AND I am tired", or: "If it's raining, then I am tired" instead of "It's raining IMPLIES I am tired." Moreover, we use the mathematical abbreviation "A iff B" for "(A IMPLIES B) AND (B IMPLIES A)". Observe that brackets $(\dots)$ are used in order to make the grouping of symbols clear if necessary.

1.2 Architecture of Concepts

In order to formulate unambiguous propositions, we need a way to describe the concepts we want to make statements about. An architecture of concepts deals with the question: "How does one build a concept?" Such an architecture defines ways to build new concepts from already existing concepts.

1. A concept has a *name*, for example, "Number" or "Set" are names of certain concepts.

2. Concepts have *components*, which are concepts, too.
   These components are used to construct a concept.

3. There are three fundamental principles of how to combine such components:

   * *Conceptual Selection:* requires one component

   * *Conceptual Conjunction:* requires one or two components

   * *Conceptual Disjunction:* requires two components

4. Concepts have *instances* (examples), which have the following properties:

   * Instances have a name

   * Instances have a value

The construction principles mentioned above are best described using instances:

The value of an instance of a concept constructed as a selection are the references to selected instances of the component.

The value of an instance of a concept constructed as a conjunction are the references to the instances of each component.

The value of an instance of a concept constructed as a disjunction is a reference to an instance of one of the components.

Perhaps some examples will clarify those three construction principles.

A selection is is really a selection in its common sense meaning: You point at a thing and say, "I select this", you point at another thing and say "I select this, too" and so on.

One example for a conjunction are persons' names which (at least in the western world) always consists of a first name and a family name. Another example are the points in the plane : every point is defined by an x- and a y-coordinate.

A disjunction is a simple kind of "addition": An instance of the disjunction of all fruits and all animals is either a fruit or an animal.

Notation

If we want to write about concepts and instances, we need an expressive and precise notation.

Concept: `ConceptName.ConstructionPrinciple(Component(s))`
> This means that we first write the concept's name followed by a dot. After the dot we put down the construction principle (Selection, Conjunction, or Disjunction) used to construct the concept. Finally we add the component or components which were used for the construction enclosed in brackets.

Instance: `InstanceName@ConceptName(Value)`
> In the case of a disjunction, a semicolon directly following the value denotes the first component, and a semicolon preceding the value denotes the second component. So in order to write down an instance, we write the instance's name followed by an '@'. After this, the name of the concept is added, followed by a value enclosed in brackets.

Very often it is not possible to write down the entire information needed to define a concept. In most cases one cannot write down components and values explicitly. Therefore, instead of writing the concept or instance, one only writes its name. Of course, this presupposes that these objects can be identified by a name, i.e., there are enough names to distinguish these objects from one another. Thus if two concepts have identical names, then they have identical construction principles and identical components. The same holds for instances: identical name means identical concept and identical value.

By identifying names with objects one can say "let X be a concept" or "let z be an instance", meaning that X and z are the names of such objects that refer to those objects in a unique way.

Here are some simple examples for concepts and instances:

CitrusFruits.Disjunction(Lemons, Oranges)

The concept **CitrusFruit** consists of the concepts **Lemons** and **Oranges**.

MyLemon@Citrusfruits(Lemon2;)

MyLemon is an instance of the concept **CitrusFruit**, and has the value **Lemon2** (which is itself an instance of the concept **Lemons**).

YourOrange@Citrusfruits(; Orange7)

YourOrange is an instance of the concept **CitrusFruits**, and has the value **Orange7** (which is itself an instance of the concept **Oranges**).

CompleteNames.Conjunction(FirstNames, FamilyNames)

The concept **CompleteNames** is a conjunction of the concept **FirstNames** and **FamilyNames**.

MyName@CompleteNames(Jody; Weissmann)

MyName is an instance of the concept **CompleteNames**, and has the value **Jody; Weissmann**.

SmallAnimals.Selection(Animals)

The concept **SmallAnimals** is a selection of the concept **Animals**.

SomeInsects@SmallAnimals(Ant, Ladybug, Grasshopper)

SomeInsects is an instance of the concept **SmallAnimals** and has the value **Ant, Ladybug, Grasshopper**.

Mathematics

The environment in which this large variety of concepts and propositions is handled, is Mathematics.

With the aid of set theory Mathematics is made conceptually precise and becomes the foundation for all formal tools. Especially formal logic is only possible on this foundation.

In Mathematics the existence of a concept means that it is conceivable without any contradiction. For instance, a set exists if it is conceivable without contradiction. Most of the useful sets exist (i.e., are conceivable without contradiction), but there are sets which don't exist. An example of such a set is the subject of the famous paradox advanced by Bertrand Russell: the set containing all sets that do not contain themselves—does this set contain itself, or not?

Set theory must be constructed successively to form an edifice of concepts which is conceivable without any contradictions.

In this section we will first show how one defines natural numbers using concepts and instances. After that, we go on to create set theory from "nothing".

Naive Natural Numbers

The natural numbers can be conceptualized as follows:

> **Number.Disjunction(Number, Terminator)**
> **Terminator.Conjunction(Terminator)**

The concept **Number** is defined as a disjunction of itself with a concept **Terminator**, the concept **Terminator** is defined as a conjunction of itself (and nothing else). There is a certain circularity involved here. The basic idea is to define a specific natural number as the successor of another natural number. This works out for 34, which is the successor of 33, and also for 786657, which is the successor of 786656. But what about 0? The number zero is not the successor of any other number. So in a way we use the **Terminator** concept as a starting point, and successively define each number (apart from 0) as the successor of the preceding number. The fact that the concept **Terminator** is defined as a conjunction of itself simply means: "**Terminator** is a thing". Now let us look at some instances of these concepts:

t@Terminator(t)
   In natural language: the value of the instance **t** of **Terminator** is itself.

0@Number(; t)
   The instance of **Number** which we call **0** has the value **t**;.

1@Number(0;)
   The instance of **Number** which we call **1** has the value **0**;.

2@Number(1;)
   The instance of **Number** which we call **2** has the value **1**;.

If we expand the values of the numbers which are neither **0** nor **t**, we get

* the value of **1** is **0**;

- the value of **2** is **1**; whose value is **0;;**
- the value of **3** is **2**; whose value is **0;;;**
- etc.

This could be interpreted by letting the semicolon stand for the operation "successor of", thus **3** is the successor of the successor of the successor of **0**.

Pure Sets

The pure sets are defined in the following circular way:

Set.Selection(Set)

Here, we say that a set is a selection of sets. Since one is allowed to select nothing in a conceptual selection, there is a starting point for this circularity. Let us look at some instances again:

∅@Set()
> Here we select nothing from the concept **Set**. We therefore end up with the empty set.

1@Set(∅)
> Since ∅ is a set we can select it from the concept **Set**. The value of *1* is a set consisting of one set.

2@Set(∅, 1)
> Here we select the two sets we have previously defined. The value of *2* is a set consisting of two sets.

Elements of the Mathematical Prose

In Mathematics, there is a "catechism" of true statements, which are named after their relevance in the development of the theory.

An *axiom* is a statement which is not proved to be true, but supposed to be so. In a second understanding, a theory is called *axiomatic* if its concepts are abstractions from examples which are put into generic definitions in order to develop a theory from a given type of concepts.

A *definition* is used for building—and mostly also for introducing a symbolic notation for—a concept which is described using already defined concepts and building rules.

A *lemma* is an auxiliary statement which is proved as a truth preliminary to some more important subsequent true statement. A *corollary* is a true statement which follows without significant effort from an already proved statement. Ideally, a corollary should be a straightforward consequence of a more difficult statement. A *sorite* is a true statement, which follows without significant effort from a given definition. A *proposition* is an important true statement, but less important than a *theorem*, which is the top spot in this nomenclature.

A mathematical *proof* is the logical deduction of a true statement $\mathcal{B}$ from another true statement C. Logical deduction means that the theorems of absolute logic are applied to establish the truth of $\mathcal{B}$, knowing the truth of C. The most frequent procedure is to use as the true statement C the truth of $\mathcal{A}$ and the truth of $\mathcal{A}$ IMPLIES $\mathcal{B}$, in short, the truth of $\mathcal{A}$ AND ($\mathcal{A}$ IMPLIES $\mathcal{B}$). Then $\mathcal{B}$ is true since the truth of the implication with the true antecedent $\mathcal{A}$ can only hold with $\mathcal{B}$ also being true. This is the so-called *modus ponens*. This scheme is also applied for *indirect proofs*, i.e., we use the true fact (NOT $\mathcal{B}$) IMPLIES (NOT $\mathcal{A}$), which is equivalent to $\mathcal{A}$ IMPLIES $\mathcal{B}$. Now, by the principle of the excluded third and the principle of contradiction, either $\mathcal{B}$ or NOT $\mathcal{B}$ will be true, but not both at the same time. Then the truth of NOT $\mathcal{B}$ enforces the truth of NOT $\mathcal{A}$. But by the principle of contradiction, $\mathcal{A}$ and NOT $\mathcal{A}$ cannot be both true, and since $\mathcal{A}$ is true, NOT $\mathcal{B}$ cannot hold, and therefore, by the principles of the excluded third and of contradiction, $\mathcal{B}$ is true. There are also more technical proof techniques, such as the proof by induction, but logically speaking, they are all special cases of the general scheme just described.

In this book, the end of a proof is marked by the symbol □.

CHAPTER 2

Axiomatic Set Theory

Axiomatic set theory is the theory of pure sets, i.e., it is built on a set concept which refers to nothing else but itself. One then states a number of axioms, i.e., propositions which are supposed to be true for sets (i.e., instances of the set concept). On this axiomatic basis, the whole mathematical concept framework is built, leading from elementary numbers to the most complex structures, such as differential equations or manifolds.

The concept of "pure sets" was already given in our introduction 1.2. It is given by the concept

$$Set.\textbf{Selection}(Set)$$

and the instance scheme

$$SetName@Set(Value)$$

where the value is described such that each reference is uniquely identified.

Notation 1 *If a set X has a set x amongst its values, one writes "$x \in X$" and one says "x is an element of X".*

If it is possible to write down the elements of a set explicitly, one uses curly brackets: Instead of "$A@Set(a, b, c, \ldots z)$" one writes "$A = \{a, b, c, \ldots z\}$". For example, the empty set is $\varnothing = \{\}$.

If there is an attribute $\mathcal{F}$ which characterizes the elements of a set A one writes "$A = \{x \mid \mathcal{F}(x)\}$", where "$\mathcal{F}(x)$" stands for "$x$ has attribute $\mathcal{F}$".

Definition 1 (Subsets and equality of sets) *Let A and B be sets. We say that A is a* subset *of B, and we write "A ⊂ B" if for every set x the proposition (x ∈ A* IMPLIES *x ∈ B) is true.*

One says that A equals *B and writes "A = B" if the proposition (A ⊂ B* AND *B ⊂ A) is true. If "A ⊂ B" is false, one writes "A ⊄ B". If "A = B" is false, one writes "A ≠ B". If A ⊂ B, but A ≠ B, one also writes "A ⊊ B".*

Example 1 Two empty sets $A = \{\}$ and $B = \{\}$ are equal.

It is impossible to decide whether two circular sets $I = \{I\}$ and $J = \{J\}$ are equal.

Axiomatic set theory is defined by two components: Its objects are pure sets, and the instances of such sets are required to satisfy a number of properties, the axioms, which cannot be deduced by logical reasoning, but must be claimed. It is hard to show that such axioms lead to mathematically existing sets. We circumvent this problem by stating a list of common axioms.

Fig. 2.1. In order to give the reader a better intuition about sets, we visualize them as bags, while their elements are shown as smaller bags— or as symbols for such elements—contained in larger ones. For example, ∅ is drawn as the empty bag. The set in this figure is $\{\emptyset, \{\emptyset\}\}$.

2.1 The Axioms

Axiom 1 (Axiom of Empty Set) *There is a set, denoted by ∅, which contains no element, i.e., for every set x, we have x ∉ ∅, or, differently formulated, ∅ = {}.*

Axiom 2 (Axiom of Equality) *If a, x, y are sets such that $x \in a$ and $x = y$, then $y \in a$.*

Axiom 3 (Axiom of Union) *If a is a set, then there is a set*

$$\{x \mid \text{there exists an element } b \in a \text{ such that } x \in b\}.$$

This set is denoted by $\bigcup a$ and is called the union *of a.*

Notation 2 *If $a = \{b, c\}$, or $a = \{b, c, d\}$, respectively, one also writes $b \cup c$, or $b \cup c \cup d$, respectively, instead of $\bigcup a$.*

Axiom 4 (Axiom of Pairs) *If a and b are two sets, then there is the* pair *set $c = \{a, b\}$.*

Notation 3 *If ϕ is a propositional attribute for sets, then, if $\phi(x)$ is true, we simply write "$\phi(x)$" to ease notation within formulas.*

Axiom 5 (Axiom of Subsets for Propositional Attributes) *If a is a set, and if ϕ is a propositional attribute for all elements of a, then there is the set $\{x \mid x \in a \text{ and } \phi(x)\}$; it is called the* subset *of a for ϕ, and is denoted by $a|\phi$.*

Axiom 6 (Axiom of Powersets) *If a is a set, then there is the* powerset *2^a, which is defined by $2^a = \{x \mid x \subset a\}$, i.e., the propositional attribute $\phi(x)$ iff $x \subset a$.*

Example 2 The powerset of $c = \{a, b\}$ is $2^c = \{\varnothing, \{a\}, \{b\}, \{a, b\}\}$. If the inclusion relation is drawn as an arrow from x to y if $x \subset y$ then the powerset of c can be illustrated as in figure 2.2.

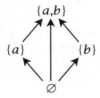

Fig. 2.2. The powerset of $\{a, b\}$.

For the next axiom, one needs the following proposition:

Lemma 1 *For any set a, there is the set $a^+ = a \cup \{a\}$. It is called the successor of a.*

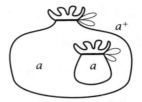

Exercise 1 Use the axioms 3 and 6 to give a proof of lemma 1.

Axiom 7 (Axiom of Infinity) *There is a set w with $\emptyset \in w$ and such that $x \in a$ implies $x^+ \in a$.*

Definition 2 *For two sets a and b, the set $\{x \mid x \in a \text{ and } x \in b\}$ is called the intersection of a and b and is denoted by $a \cap b$. If $a \cap b = \emptyset$, then a and b are called disjoint.*

Axiom 8 (Axiom of Choice) *Let a be a set whose elements are all non-empty, and such that any two different elements $x, y \in a$ are disjoint. Then there is a subset $c \subset \bigcup a$ such that for every non-empty $x \in a$, $x \cap c$ has exactly one element.*

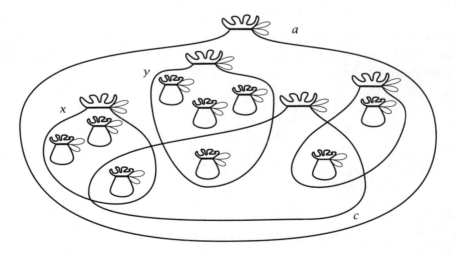

Fig. 2.3. Axiom of Choice: c is the choice set of the sets $x, y, \ldots \in a$.

We shall now develop the entire set theory upon these axioms. The beginning is quite hard, but we shall be rewarded by a beautiful result: all numbers from classical mathematics, all functions and all complex structures will emerge from this construction.

2.2 Basic Concepts and Results

Sorite 2 *For any three sets a, b, c, we have*

(i) $a \subset a$

(ii) *If $a \subset b$ and $b \subset a$, then $a = b$.*

(iii) *If $a \subset b$ and $b \subset c$, then $a \subset c$.*

Exercise 2 Give a proof of sorite 2.

Proposition 3 *For any sets a, b, c, d:*

(i) *(Commutativity of unions) the set $a \cup b$ exists and equals $b \cup a$,*

(ii) *(Associativity of unions) the sets $(a \cup b) \cup c$, $a \cup (b \cup c)$ exist and are equal, we then write $a \cup b \cup c$ instead,*

(iii) *$(a \cup b \cup c) \cup d$ and $a \cup (b \cup c \cup d)$ exist and are equal, we then write $a \cup b \cup c \cup d$ instead.*

Exercise 3 Give a proof of proposition 3 by applying axioms 4 and 3. Use statement (ii) to prove (iii).

Remark 1 The set whose elements are all sets x with $x \notin x$ does not exist, in fact both, the property $x \in x$, as well as $x \notin x$ lead to contradictions. Therefore, by axiom 5, the set of all sets does not exist.

Proposition 4 *If $a \neq \emptyset$, then the set $\{x \mid x \in z \text{ for all } z \in a\}$ exists, it is called the* intersection *of a and is denoted by $\bigcap a$. However, for $a = \emptyset$, the attribute "$x \in z$ for all $z \in a$" is fulfilled by every set z, and therefore $\bigcap \emptyset$ is inexistent, since it would be the non-existent set of all sets.*

Proof In fact, if $a \neq \emptyset$, and if $b \in a$ is one element satisfying the attribute, the required set is also defined by $\{x \mid x \in b \text{ and } x \in z \text{ for all } z \in a\}$. So this attribute selects a subset of b defined by the propositional attribute $\Phi(x) = (x \in z \text{ for all } z \in a)$, which is a legitimate set according to axiom 5. If $a = \emptyset$, then the attribute $\Phi(x)$ *alone* is true for every set x, which leads to the inexistent set of all sets. $\qquad\square$

Definition 3 *For two sets a and b, the* complement of a in b *or the dif-*
ference of b and a is the set $\{x \mid x \notin a \text{ and } x \in b\}$. It is denoted by
$b - a$.

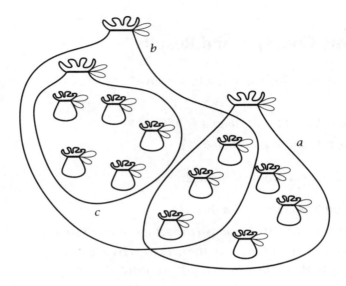

Fig. 2.4. The complement c of a in b, or $c = b - a$.

Sorite 5 *For any three sets a, b, c we have*

　(i) $c - a \subset c$,

　(ii) *If $a \subset c$, then $c - (c - a) = a$,*

　(iii) $c - \emptyset = c$,

　(iv) $c - c = \emptyset$,

　(v) $a \cap (c - a) = \emptyset$,

　(vi) *If $a \subset c$, then $a \cup (c - a) = c$,*

　(vii) $c - (a \cup b) = (c - a) \cap (c - b)$,

　(viii) $c - (a \cap b) = (c - a) \cup (c - b)$,

　(ix) $c \cap (a - b) = (c \cap a) - (c \cap b)$.

Exercise 4 Give a proof of sorite 5.

CHAPTER 3

Boolean Set Algebra

In this chapter, we shall give a more systematic account of the construction of sets by use of union, intersection and complement. The structures which emerge in this chapter are prototypes of algebraic structures which will appear throughout the entire course.

3.1 The Boolean Algebra of Subsets

Lemma 6 *For two sets a and b, the union $a \cup b$ is a* least upper bound, *i.e., $a, b \subset a \cup b$, and for every set c with $a, b \subset c$, we have $a \cup b \subset c$. This property uniquely determines the union.*

Dually, the intersection $a \cap b$ is a greatest lower bound, *i.e., $a \cap b \subset a, b$, and for every set c with $c \subset a, b$, we have $c \subset a \cap b$. This property uniquely determines the intersection.*

Proof Clearly, $a \cup b$ is a least upper bound. If x and y are any two least upper bounds of a and b, then by definition, we must have $x \subset y$ and $y \subset x$, therefore $x = y$. The dual statement is demonstrated by the analogous reasoning. □

Summarizing the previous properties of sets, we have the following important theorem, stating that the powerset 2^a of a set a is a *Boolean algebra*. We shall discuss this structure in a more systematic way in chapter 17.

Proposition 7 (Boolean Algebra of Subsets) *For a given set a, the powerset 2^a has the following properties. Let x, y, z be any elements of 2^a, i.e., subsets of a; also, denote $x' = a - x$. Then:*

(i) *(Reflexivity)* $x \subset x$,

(ii) *(Antisymmetry)* *if* $x \subset y$ *and* $y \subset x$, *then* $x = y$,

(iii) *(Transitivity)* *if* $x \subset y$ *and* $y \subset z$, *then* $x \subset z$,

(iv) *we have a "minimal" set* $\varnothing \in 2^a$ *and a "maximal" set* $a \in 2^a$, *and* $\varnothing \subset x \subset a$,

(v) *(Least upper bound)* *the union* $x \cup y$ *verifies* $x, y \subset x \cup y$, *and for every* z, *if* $x, y \subset z$, *then* $x \cup y \subset z$,

(vi) *(dually: Greatest lower bound)* *the intersection* $x \cap y$ *verifies* $x \cap y \subset x, y$, *and for every* z, *if* $z \subset x, y$, *then* $z \subset x \cap y$,

(vii) *(Distributivity)* $(x \cup y) \cap z = (x \cap z) \cup (y \cap z)$ *and (dually)* $(x \cap y) \cup z = (x \cup z) \cap (y \cup z)$,

(viii) *we have* $x \cup x' = a$, $x \cap x' = \varnothing$

Proof (i) is true for any set, see sorite 2.

(ii) is the very definition of equality of sets.

(iii) is immediate from the definition of subsets.

(iv) is clear.

(v) and (vi) were proved in lemma 6.

(vii) follows immediately from the definition of the complement x'. □

Example 3 Figure 3.1 uses Venn diagrams to illustrate distributivity of $\cap$ over $\cup$. In this intuitive representation, sets are drawn as surfaces of circles or parts thereof.

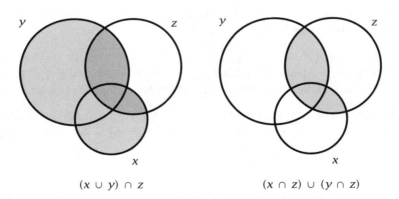

$(x \cup y) \cap z$ $(x \cap z) \cup (y \cap z)$

Fig. 3.1. Distributivity of $\cap$ over $\cup$.

Exercise 5 Given a set $a = \{r, s, t\}$ consisting of pairwise different sets, give a complete description of 2^a and the intersections or unions, respectively, of elements of 2^a.

Here is a second, also important structure on the powerset of a given set a:

Definition 4 If $x, y \in 2^a$, then we define $x + y = (x \cup y) - (x \cap y)$ (symmetric set difference). We further define $x \cdot y = x \cap y$. Both operations are illustrated using Venn diagrams in figure 3.2.

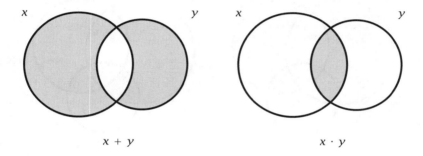

Fig. 3.2. Symmetric difference and intersection of sets.

Proposition 8 For a set a, and for any three elements $x, y, z \in 2^a$, we have:

(i) (Commutativity) $x + y = y + x$ and $x \cdot y = y \cdot x$,

(ii) (Associativity) $x + (y + z) = (x + y) + z$, $x \cdot (y \cdot z) = (x \cdot y) \cdot z$, we therefore also write $x + y + z$ and $x \cdot y \cdot z$, respectively,

(iii) (Neutral elements) we have $x + \varnothing = x$ and $x \cdot a = x$,

(iv) (Distributivity) $x \cdot (y + z) = x \cdot y + x \cdot z$,

(v) (Idempotency) $x \cdot x = x$,

(vi) (Involution) $x + x = \varnothing$,

(vii) the equation $x + y = z$ has exactly one solution w, i.e., there is exactly one set $w \subset a$ such that $w + y = z$.

Proof (i) follows from the commutativity of the union $a \cup b$ and the intersection $a \cap b$, see also lemma 3.

(ii) associativity also follows from associativity of union and intersection, see again lemma 3.

(iii) we have $x + \emptyset = (x \cup \emptyset) - (x \cap \emptyset) = x - \emptyset = x$, $x \cdot a = x \cap a = x$.

(iv) $x \cdot (y + z) = x \cap ((y \cup z) - (y \cap z)) = x \cap (y \cup z) - x \cap y \cap z = (x \cap y) \cup (x \cap z)) - x \cap y \cap z$, whereas $x \cdot (y + z) = x \cap (y + z) = x \cap ((y \cup z) - (y \cap z)) = (x \cap (y \cup z)) - x \cap y \cap z = (x \cap y) \cup (x \cap z)) - x \cap y \cap z$, and we are done.

(v) and (vi) are immediate from the definitions.

(vii) in view of (vi), $w = y + z$ is a solution. For any two solutions $w + y = w' + y$, one has $w = w + y + y = w' + y + y = w'$. □

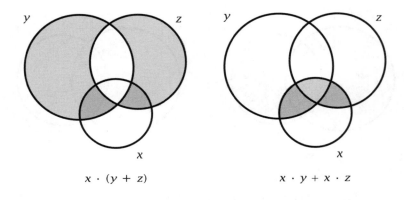

$$x \cdot (y + z) \qquad\qquad x \cdot y + x \cdot z$$

Fig. 3.3. Distributivity of · over +.

Remark 2 This structure will later be discussed as the crucial algebraic structure of a commutative ring, see chapter 15.

Exercise 6 Let $a = \{r, s, t\}$ as in exercise 5. Calculate the solution of $w + y = z$ within 2^a for $y = \{r, s\}$ and $z = \{s, t\}$.

Exercise 7 Let $a = \{\emptyset\}$. Calculate the complete tables of sums $x + y$ and products $x \cdot y$, respectively, for $x, y \in 2^a$, use the symbols $0 = \emptyset$ and $1 = a$. What do they remind you of?

CHAPTER 4

Functions and Relations

We have seen in chapter 1 that the conceptual architecture may be a selection, conjunction, or disjunction. Sets are built on the selection type. They are however suited for simulating the other types as well. More precisely, for the conjunction type, one needs to know the position of each of two conceptual coordinates: Which is the first, which is the second. In the selective type, no order between elements of a set is given, i.e., $\{x, y\} = \{y, x\}$. So far, we have no means for creating order among such elements. This chapter solves this problem in the framework of set theory.

4.1 Graphs and Functions

Definition 5 *If x and y are two sets, the* ordered pair (x, y) *is defined to be the following set:*

$$(x, y) = \{\{x\}, \{x, y\}\}$$

Observe that the set (x, y) always exists, it is a subset of the powerset of the pair set $\{x, y\}$, which exists according to axiom 4.

Here is the yoga of this definition:

Lemma 9 *For any four sets a, b, c, d, we have $(a, b) = (c, d)$ iff $a = c$ and $b = d$. Therefore one may speak of the* first *and* second, *respectively, coordinate a and b, respectively, of the ordered pair (a, b).*

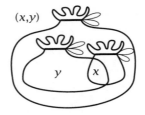

Fig. 4.1. The bag representation of the ordered pair (x, y).

Proof The ordered pair (x, y) has one single element $\{x\}$ iff $x = y$, it has different elements $\{x\} \neq \{x, y\}$ iff $x \neq y$. So, if $(a, b) = (c, d)$, then either $a = b$ and $c = d$, and then $\{\{a\}\} = (a, b) = (c, d) = \{\{c\}\}$, whence $a = c$. Or else, $a \neq b$ and $c \neq d$. But the the only element with one element in (a, b) is $\{a\}$, similarly the only element with one element in (c, d) is $\{d\}$. So $(a, b) = (c, d)$ implies $a = c$. Similarly, the other element $\{a, b\}$ of (a, b) must be equal to $\{c, d\}$. But since $a = c$ and $a \neq b$, we have $b = d$, and we are done. The converse implication is evident.   □

Exercise 8 Defining $(x, y, z) = ((x, y), z)$, show that $(x, y, z) = (u, v, w)$ iff $x = u$, $y = v$, and $z = w$.

Lemma 10 *Given two sets a and b, there is a set*

$$a \times b = \{(x, y) \mid x \in a \text{ and } y \in b\},$$

it is called the Cartesian product *of a and b.*

Proof We have the set $v = a \cup b$. Let $P = 2^{(2^v)}$ be the powerset of the powerset of v, which also exists. Then an ordered pair $(x, y) = \{\{x\}, \{x, y\}\}$, with $x \in a$ and $y \in b$ is evidently an element of P. Therefore $a \times b$ is the subset of those $p \in P$ defined by the propositional attribute $\Phi(p)$ iff there are $x \in a$ and $y \in b$ such that $p = (x, y)$.   □

Sorite 11 *Let a, b, c, d be sets. Then:*

(i) $a \times b = \varnothing$ *iff $a = \varnothing$ or $b = \varnothing$,*

(ii) *if $a \times b \neq \varnothing$, then $a \times b = c \times d$ iff $a = c$ and $b = d$.*

Proof The first claim is evident. As to the second, if $a \times b \neq \varnothing$, then we have $a \cup b = \bigcup(\bigcup(a \times b))$, as is clear from the definition of ordered pairs. Therefore we have the subset $a = \{x \mid x \in a \cup b, \text{ there is } z \in a \times b \text{ with } z = (x, y)\}$. Similarly for b, and therefore also $a = c$ and $b = d$.   □

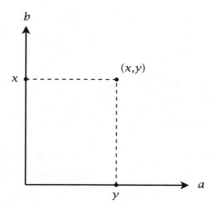

Fig. 4.2. It has become common practice to represent Cartesian products $a \times b$ intuitively by two axes, the horizontal axis representing a, i.e., $x \in a$ is drawn as a point on this axis, and the vertical axis representing the set b, i.e., $y \in b$ is drawn as a point on this axis. In traditional language, the horizontal axis is called the *abscissa*, while the vertical axis is called the *ordinate*. The element (x, y) is drawn as point on the plane, whose coordinates x and y are obtained by projections perpendicular to the respective axes.

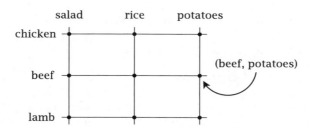

Fig. 4.3. The idea behind Cartesian products is to define sets which are composed from two given sets by simultaneous specification of two elements in order to define one of the Cartesian product. Here, to define a meal, we are given two intuitive sets: *meat dish* and accompanying *side dish*. In OOP, the meal class would have two instance variables, meat_dish and side_dish, both allowing appropriate character strings as values, "chicken", "beef", "lamb", and "salad", "rice", "potatoes", respectively. Logically, the information encoded in a Cartesian product is that of a conjunction: To know an object of a Cartesian product is to know its first coordinate AND its second coordinate.

Definition 6 *If a Cartesian product $a \times b$ is non-empty, we call the uniquely determined sets a and b, respectively, its* first *and* second projection, *respectively, and we write $a = pr_1(a \times b)$ and $b = pr_2(a \times b)$, respectively.*

The following concept of a graph is a formalization of the act of associating two objects out of two domains, such as the pairing of a man and a woman, or associating a human and its bank accounts.

Lemma 12 *The following statements about a set g are equivalent:*

(i) *The set g is a subset of a Cartesian product set $a \times b$.*

(ii) *Every element $x \in g$ is an ordered pair $x = (u, v)$.*

Proof Clearly, (i) implies (ii). Conversely, if g consists of ordered pairs, we may take $P = \bigcup(\bigcup(g))$, and then immediately see that $g \subset P \times P$. □

Definition 7 *A set which fulfills one of the equivalent properties of lemma 12 is called a* graph.

Example 4 For any set a, the *diagonal graph* is the graph

$$\Delta_a = \{(x, x) \mid x \in a\}.$$

Lemma 13 *For a graph g, there are two sets*

$$pr_1(g) = \{u \mid (u, v) \in g\} \ and \ pr_2(g) = \{v \mid (u, v) \in g\},$$

and we have $g \subset pr_1(g) \times pr_2(g)$.

Proof As in the previous proofs, we take the double union $d = \bigcup(\bigcup(g))$ and from d extract the subsets $pr_1(g)$ and $pr_2(g)$ as defined in this proposition. The statement $g \subset pr_1(g) \times pr_2(g)$ is then straightforward. □

Proposition 14 *If g is a graph, there is another graph, denoted by g^{-1} and called the* inverse graph *of g, which is defined by*

$$g^{-1} = \{(v, u) \mid (u, v) \in g\}.$$

We have $(g^{-1})^{-1} = g$.

Proof According to lemma 12, there are sets a and b such that $g \subset a \times b$. Then $g^{-1} \subset b \times a$ is the inverse graph. The statement about double inversion is immediate. □

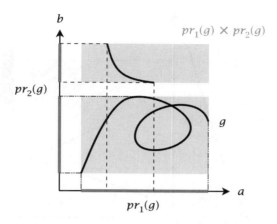

Fig. 4.4. The projections $pr_1(g)$ and $pr_2(g)$ (dark gray segments on the axes) of a graph g (black curves). Note that $g \subset pr_1(g) \times pr_2(g)$ (light gray rectangles).

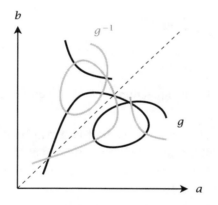

Fig. 4.5. The inverse g^{-1} of a graph g.

Exercise 9 Show that $g = \Delta_{pr_1(g)}$ implies $g = g^{-1}$; give counterexamples for the converse implication.

Definition 8 *If g and h are two graphs, there is a set $g \circ h$, the composition of g with h (attention: the order of the graphs is important here), which is defined by*

$$g \circ h = \{(v, w) \mid \text{there is a set } u \text{ such that } (v, u) \in h \text{ and } (u, w) \in g\}.$$

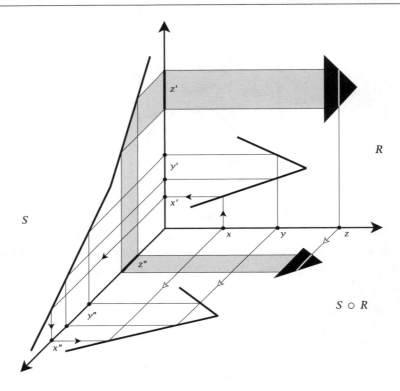

Fig. 4.6. The composition $S \circ R$ of two graphs R and S.

Figure 4.6 illustrates the composition of two graphs R and S. The point x is mapped by R to the point x', which in turn is mapped by S to the point x''. R maps the point y to *two* points both denoted by y', and these two points are mapped by S to *two* points y''. Last, z is mapped by S to a *segment* z'. This segment is mapped by S to a segment z''.

Sorite 15 *Let f, g, h be three graphs.*

(i) *(Associativity) We have $(f \circ g) \circ h = f \circ (g \circ h)$ and we denote this graph by $f \circ g \circ h$.*

(ii) *We have $\Delta_{pr_1(g)} \subset g^{-1} \circ g$.*

Proof These statements follow from a straightforward application of the definition.   □

Definition 9 *A graph g is called* functional *if $(u, v) \in g$ and $(u, w) \in g$ imply $v = w$.*

Exercise 10 Show that the composition $g \circ h$ of two functional graphs g and h is functional.

Example 5 For any sets a and b, the diagonal graph Δ_a is functional, whereas the Cartesian product $a \times b$ is not functional if $a \neq \varnothing$ and if there are sets $x, y \in b$ with $x \neq y$.

Definition 10 *A function is a triple* (a, f, b) *such that* f *is a functional graph, where* $a = pr_1(f)$ *and* $pr_2(f) \subset b$. *The set* a *is called the* domain *of the function, the set* b *is called its* codomain, *and the set* $pr_2(f)$ *is called the function's* image *and denoted by* $Im(f)$. *One usually denotes a function by a more graphical sign* $f : a \to b$. *For* $x \in a$, *the unique* y *such that* $(x, y) \in f$ *is denoted by* $f(x)$ *and called the* value *of the function at the argument* x. *Often, if the domain and codomain are clear, one identifies the function with the graph sign* f, *but this is not the valid definition. One then also notates* $a = dom(f)$ *and* $b = codom(f)$.

Example 6 For any set a, the *identity function (on a) Id_a* is defined by $Id_a = (a, \Delta_a, a)$.

Exercise 11 For the set $1 = \{\varnothing\}$ and for any set a, there is exactly one function $(a, f, 1)$. We denote this function by $! : a \to 1$. (The notation "1" is not quite arbitrary, we shall see the systematic background in chapter 5.) If $a = \varnothing$, and if b is any set, there is a unique function $(\varnothing, g, b)$, also denoted by $! : \varnothing \to b$.

Definition 11 *A function* $f : a \to b$ *is called* epi, *or "epimorphism" or* surjective *if* $Im(f) = codom(f)$.

It is called mono, *or "monomorphism" or* injective *if* $f(x) = f(y)$ *implies* $x = y$ *for all sets* $x, y \in dom(f)$.

The function is called iso, *or "isomorphism" or* bijective *if it is epi and mono. Isomorphisms are also denoted by special arrows, i.e.,* $f : a \overset{\sim}{\to} b$.

Example 7 Figure 4.7 illustrates three functions, $f : A \to B$, $g : B \to A$ and $h : B \to B$. An arrow from an element (point) $x \in X$ to an element $y \in Y$ indicates that (x, y) is in the graph $\kappa \subset X \times Y$ of a function $k = (X, \kappa, Y)$. The function f is epi, but not mono, g is mono, but not epi, and h is mono and epi, and thus, iso. The star-shaped points are the "culprits", i.e., the reasons that f, respectively g, is not mono, respectively epi.

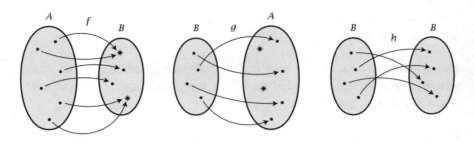

Fig. 4.7. Epimorphism f, monomorphism g and isomorphism h.

Exercise 12 The function $! : a \to 1$ is epi for $a \neq \varnothing$, the function $! : \varnothing \to b$ is always mono, and the identity function Id_a is always iso.

Definition 12 *Let $f : a \to b$ and $g : b \to c$ be functions, then their compo-sition is the function $g \circ f : a \to c$.*

When dealing with functions, one often uses a more graphical represen-tation of the involved arrows by so-called arrow diagrams. The domains and codomains are shown as symbols on the plane, which are connected by arrows representing the given functions. For example, the functions $f : a \to b$ and $g : b \to c$ and their composition $h = g \circ f$ are shown as a triangular diagram. This diagram is *commutative* in the sense that both "paths" $a \xrightarrow{f} b \xrightarrow{g} c$ and $a \xrightarrow{h} c$ define the same function $h = g \circ f$.

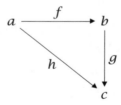

Sorite 16 *Let $f : a \to b$, $g : b \to c$ and $h : c \to d$ be functions.*

(i) *The compositions $(h \circ g) \circ f : a \to c$ and $h \circ (g \circ f) : a \to c$ are equal, we therefore denote them by $h \circ g \circ f : a \to c$.*

(ii) *The function $g : b \to c$ is mono iff the following condition holds: For any two functions $f, f' : a \to b$, $g \circ f = g \circ f'$ implies $f = f'$.*

(iii) *The function $f : a \to b$ is epi iff the following condition holds: For any two functions $g, g' : b \to c$, $g \circ f = g' \circ f$ implies $g = g'$.*

(iv) *If f and g are epi, mono, iso, respectively, then so is $g \circ f$.*

(v) *If f is mono and $a \neq \varnothing$, then there is a—necessarily epi—function $q : b \to a$ such that $q \circ f = Id_a$, such a function is called a* left inverse *or* retraction *of f.*

(vi) *If f is epi, then there is a—necessarily mono—function $p : b \to a$ such that $f \circ p = Id_b$, such a function is called a* right inverse *or* section *of f.*

(vii) *The function f is iso iff there is a (necessarily unique) inverse, denoted by $f^{-1} : b \to a$, such that $f^{-1} \circ f = Id_a$, and $f \circ f^{-1} = Id_b$.*

Proof (i) follows from the associativity of graph composition, see proposition 15.

(ii) If $g : b \to c$ is mono, then for $x \in a$, $g \circ f(x) = g \circ f'(x)$ means $g(f(x)) = g(f'(x))$, but since g is mono, $f(x) = f'(x)$, for all $x \in a$, whence $f = f'$. Conversely, take $u, v \in b$ such that $g(u) = g(v)$. Define two maps $f, f' : 1 \to b$ by $f(0) = u$ and $f'(0) = v$. Then $g \circ f = g \circ f'$. So $f = f'$, but this means $u = f(0) = f'(0) = v$, therefore g is injective.

(iii) If f is epi, then for every $y \in b$, there is $x \in a$ with $y = f(x)$. If $g \circ f = g' \circ f$, then $g(y) = g(f(x)) = g'(f(x)) = g'(y)$, whence $g = g'$. Conversely, if g is not epi, then let $z \notin Im(g)$. Define a function $f : b \to \{\varnothing, 1\}$ by $f(y) = 0$ for all $y \in b$. Define $f' : b \to \{\varnothing, 1\}$ by $f'(y) = 0$ for all $y \neq z$, and $f'(z) = 1$. We then have two different functions f and f' such that $f \circ g = f' \circ g$.

(iv) This property follows elegantly from the previous characterization: let f and g be epi, then for $h, h' : c \to d$, if $h \circ g \circ f = h' \circ g \circ f$, then, since f is epi, we conclude $h \circ g = h' \circ g$, and since g is epi, we have $h = h'$. The same formal argumentation works for mono. Since iso means mono and epi, we are done.

(v) If $f : a \to b$ is mono, then the inverse graph f^{-1} is also functional and $pr_2(f^{-1}) = a$. Take any element $y \in a$ and take the graph $g = f^{-1} \cup (\{y\} \times (b - Im(f)))$. This defines a retraction of f.

(vi) Let f be epi. For every $x \in b$, let $F(x) = \{y \mid y \in a, f(y) = x\}$. Since f is epi, no $F(x)$ is empty, and $F(x) \cap F(x') = \varnothing$ if $x \neq x'$. By the axiom of choice 8, there is a set $q \subset a$ such that $q \cap F(x) = \{q_x\}$ is a set with exactly one element q_x for every $x \in b$. Define $p(x) = q_x$. This defines the section $p : b \to a$ of f.

(vii) The case $a = \varnothing$ is trivial, so let us suppose $a \neq \varnothing$. Then characterizations (v) and (vi), together with the fact that 'mono + epi = iso' answer our problem. $\square$

Remark 3 The proof of statement (vi) in sorite 16 is a very strong one since it rests on the axiom of choice 8.

Definition 13 *Let $f : a \to b$ be a function, and let a' be a set. Then the* restriction *of f to a' is the function $f|_{a'} : a \cap a' \to b$, where the graph is $f|_{a'} = f \cap ((a \cap a') \times b)$.*

Definition 14 *Let $f : a \to b$ and $g : c \to d$ be two functions. Then the Cartesian product of f and g is the function $f \times g : a \times c \to b \times d$ with $(f \times g)(x, y) = (f(x), g(y))$.*

Sorite 17 *Let $f : a \to b$ and $g : c \to d$ be two functions. Then the Cartesian product $f \times g$ is injective (surjective, bijective) if f and g are so.*

Proof If one of the domains a or c is empty the claims are obvious. So suppose $a, c \neq \varnothing$. Let f and g be injective and take two elements $(x, y) \neq (x', y')$ in $a \times b$. Then either $x \neq x'$ or $y \neq y'$. in the first case, $(f \times g)(x, y) = (f(x), g(y)) \neq (f(x'), g(y)) = (f \times g)(x', y)$, the second case is analogous. A similar argument settles the cases of epi maps, and the case of iso maps is just the conjunction of the mono and epi cases. □

The next subject is the basis of the classification of sets. The question is: When are two sets are "essentially different"? This is the crucial definition:

Definition 15 *A set a is said to be* equipollent *to b or to have the same cardinality as b iff there is a bijection $f : a \xrightarrow{\sim} b$. We often just write $a \xrightarrow{\sim} b$ to indicate the fact that a and b are equipollent.*

Example 8 In figure 4.8, the set A of stars, the set B of crosses and the set C of plusses are equipollent. The functions $f : A \to B$ and $g : B \to C$ are both bijections. The composition of g and f is a bijection $h = g \circ f : A \to C$. The purpose of this example is to show that equipollence is a feature independent of the shape, or "structure" of the set. It only tells us that each element from the first set can be matched with an element from the second set, and vice-versa.

Proposition 18 *For every sets a, b and c, we have:*
  (i) *(Reflexivity) a is equipollent to a.*
  (ii) *(Symmetry) If a is equipollent to b, then b is equipollent to a.*
  (iii) *(Transitivity) If a is equipollent to b, and if b is equipollent to c, then a is equipollent to c.*

Proof Reflexivity follows from the fact that the identity Id_a is a bijection. Symmetry follows from statement (vii) in sorite 16. Transitivity follows from statement (iv) of sorite 16. □

Are there arbitrary large sets? Here are first answers:

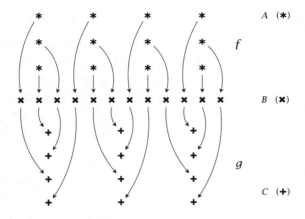

Fig. 4.8. The equipollence of sets A, B and C.

Exercise 13 Show that there is an injection $sing : a \to 2^a$ for each set, defined by $sing(x) = \{x\}$.

An injection in the reverse direction never exists. We have in fact this remarkable theorem.

Proposition 19 *For any set a, a and 2^a are never equipollent.*

Proof Suppose that we have a function $f : a \to 2^a$ We show that f cannot be surjective. Take the subset $g \subset a$ defined by $g = \{x \mid x \in a, x \notin f(x)\}$. We claim that $g \notin Im(f)$. If $z \in a$ with $f(z) = g$, then if $z \notin g$, then $z \in g$, contradiction. But if $z \in g$, then $z \notin f(z)$ by definition of g, again contradiction. So $f(z) = g$ is impossible.   □

To see why this contradicts the existence of an injection $2^a \to a$, we need some further results:

Sorite 20 *Let a, b, c, d be sets.*
  (i) *If $a \stackrel{\sim}{\to} b$ and $c \stackrel{\sim}{\to} d$, then $a \times c \stackrel{\sim}{\to} b \times d$.*
  (ii) *If $a \stackrel{\sim}{\to} b$ and $c \stackrel{\sim}{\to} d$, and if $a \cap c = b \cap d = \varnothing$, then $a \cup c \stackrel{\sim}{\to} b \cup d$.*

Proof (i) If $f : a \to b$ and $g : c \to d$ are bijections, then so is $f \times g$ by sorite 17.

(ii) If $f : a \to b$ and $g : c \to d$ are bijections, then the graph $f \cup g \subset (a \cup c) \times (c \cup d)$ clearly defines a bijection.   □

The following is a technical lemma:

Lemma 21 *If $p \cup q \cup r \overset{\sim}{\to} p$ and $p \cap q = \varnothing$, then $p \cup q \overset{\sim}{\to} p$.*

Proof This proof is more involved and should be read only by those who really are eager to learn about the innards of set theory. To begin with, fix a bijection $f : p \cup q \cup r \to p$. If $x \subset p \cup q \cup r$, we also write $f(x)$ for $\{f(z) \mid z \in x\}$.

Consider the following propositional attribute $\Psi(x)$ of elements $x \in 2^{p \cup q \cup r}$: We define $\Psi(x)$ iff $q \cup f(x) \subset x$. For example, $\Psi(p \cup q)$. We now show that if $\Psi(x)$, then also $\Psi(q \cup f(x))$. In fact, if $q \cup f(x) \subset x$, the $f(q \cup f(x)) \subset f(x)$, and a fortiori $f(q \cup f(x)) \subset q \cup f(x)$.

Next, let $b \subset 2^{p \cup q \cup r}$ with $\Psi(x)$ for all $x \in b$. Then we have $\Psi(\cap b)$. Denote $k = \cap b$. Since $k \subset x$ for all $x \in b$, we also have $f(k) \subset f(x)$ for all $x \in b$, and therefore $q \cup f(k) \subset q \cup f(x) \subset x$ for all $x \in b$. This implies $q \cup f(k) \subset k$.

Now let $e = \{x \mid x \in 2^{p \cup q \cup r}, \Psi(x)\}$; we know that e is non-empty. Let $d = \cap e$. Then $\Psi(d)$, i.e., $q \cup f(d) \subset d$. But by the first consideration, we also know that $\Psi(q \cup f(d))$, and therefore $d \subset q \cup f(d)$. This means that $q \cup f(d) \subset d \subset q \cup f(d)$, i.e., $q \cup f(d) = d$.

Moreover, $d \cap (p - f(d)) = \varnothing$. In fact, $d \cap (p - f(d)) = (q \cup f(d)) \cap (p - f(d)) = (q \cap (p - f(d))) \cup (f(d) \cap (p - f(d))) = \varnothing \cup \varnothing = \varnothing$ because we suppose $p \cap q = \varnothing$. So we have a disjoint union $q \cup p = d \cup (p - f(d))$ Now, we have a bijection $f : d \overset{\sim}{\to} f(d)$ and a bijection (the identity) $p - f(d) \overset{\sim}{\to} p - f(d)$. Therefore by sorite 20, we obtain the required bijection. □

This implies a famous theorem:

Proposition 22 (Bernstein-Schröder) *Let a, b, c be three sets such that there exist two injections $f : a \to b$ and $g : b \to c$. If a and c are equipollent, then all three sets are equipollent.*

Proof We apply lemma 21 as follows. Let $f : a \to b$ and $g : b \to c$ be injections and $h : a \to c$ a bijection. Then we may take the image sets $a' = g(f(a))$ and $b' = g(b)$ instead of the equipollent sets a and b, respectively, and show the theorem with the special situation of subsets $a \subset b \subset c$ such that a is equipollent to c. To apply our technical lemma we set $p = a$, $q = b - a$ and $r = c - b$. Therefore $c = p \cup q \cup r$ and $b = p \cup q$. In these terms, we are supposing that p is equipollent to $p \cup q \cup r$. Therefore the lemma yields that p is equipollent to $p \cup q$, i.e., a is equipollent to b. By transitivity of equipollence, b and c are also equipollent. □

In particular:

Corollary 23 *If $a \subset b \subset c$, and if a is equipollent to c, then all three sets are equipollent.*

Corollary 24 *For any set a, there is no injection $2^a \to a$.*

Proof If we had an injection $2^a \to a$, the existing reverse injection $a \to 2^a$ from exercise 13 and proposition 22 would yield a bijection $a \overset{\sim}{\to} 2^a$ which is impossible according to proposition 19. $\square$

4.2 Relations

Until now, we have not been able to deal with "relations" in a formal sense. For example, the properties of reflexivity, symmetry, and transitivity, as encountered in proposition 18, are only properties of single pairs of sets, but the whole set of all such pairs is not given. The theory of relations will deal with such problems.

Definition 16 *A binary relation on a set a is a subset $R \subset a \times a$; this is a special graph, where the domain and codomain coincide and are specified by the choice of a. Often, instead of "$(x, y) \in R$", one writes "xRy".*

Example 9 The *empty* relation $\varnothing \subset a \times a$, the *total* relation $R = a \times a$, and the diagonal graph Δ_a are relations on a. For each relation R on a, the *inverse* relation $R^{-1} = \{(y, x) \mid (x, y) \in R\}$ (the inverse graph) is a relation on a. If R and S are two relations on a, then the composed graph $R \circ S$ defines the *composed* relation on a. In particular, we have the second power $R^2 = R \circ R$ of relation R.

Notation 4 *Often, relation symbols are not letters, but special symbols such as $<, \le, \sim, \dots$. Their usage is completely context-sensitive and has no universal meaning.*

Definition 17 *Let $\le$ be a relation on a. The relation is called*

   (i) reflexive *iff $x \le x$ for all $x \in a$;*
   (ii) transitive *iff $x \le y$ and $y \le z$ implies $x \le z$ for all $x, y, z \in a$;*
   (iii) symmetric *iff $x \le y$ implies $y \le x$ for all $x, y \in a$;*
   (iv) antisymmetric *iff $x \le y$ and $x \ne y$ excludes $y \le x$ for all $x, y \in a$;*
   (v) total *iff for any two $x, y \in a$, either $x \le y$ or $y \le x$.*
   (vi) equivalence relation, *iff it is reflexive, symmetric, and transitive. In this case, the relation is usually denoted by $\sim$ instead of $\le$.*

Example 10 We shall illustrate these properties with examples from the real world. The relation "x is an ancestor of y" is transitive, but neither reflexive nor symmetric. The "subclass" relation of object-oriented programming is reflexive, transitive, antisymmetric, but not symmetric.

The relation "x lives within 10 kilometers from y" is reflexive, symmetric, but not transitive.

The relation "x is a sibling of y" is symmetric and transitive. It is not reflexive. None of these relations is total.

A total, transitive relation, is, for instance, "x is not taller than y".

For an equivalence relation, consider "x has the same gender as y". The set of equivalence classes of this relation partitions mankind into two sets, the set of males and the set of females (see the following proposition).

Definition 18 *Given a binary relation $R \subset X \times X$, we call the smallest set R_r, such that $R \subset R_r$ and R_r is reflexive, the* reflexive closure *of R. The smallest set R_s, such that $R \subset R_s$ and R_s is symmetric, is called the* symmetric closure *of R. The smallest set R_t, such that $R \subset R_t$ and R_t is transitive, is called the* transitive closure *of R. Finally, the smallest equivalence relation R_e containing R is called the* equivalence relation *generated by R.*

Proposition 25 *If $\sim$ is an equivalence relation on a, and if $s \in a$, then a subset*

$$[s] = \{r \mid r \in a \text{ and } s \sim r\}$$

is called an equivalence class *with respect to $\sim$. The set of equivalence classes—a subset of 2^a—is denoted by $a/\sim$. It defines a partition of a, i.e., for any two elements $s,t \in a$, either $[s] = [t]$ or $[s] \cap [t] = \varnothing$, and $a = \bigcup a/\sim$.*

Proof Since for every $s \in a$, $s \sim s$, we have $s \in [s]$, whence $a = \bigcup a/\sim$. If $u \in [s] \cap [t]$, then if $r \sim s$ we have $r \sim s \sim u \sim t$, whence $[s] \subset [t]$, the converse inclusion holds with the roles of s and t exchanged, so $[s] = [t]$.   □

Exercise 14 Show that the reflexive, symmetric, transitive closure $((R_r)_s)_t$ of a relation R is the smallest equivalence relation R_e containing R.

Example 11 A relation $R \subset A \times A$, where $A = \{a,b,c,d,e\}$ is shown in figure 4.9 in a graphical way. An element $(x,y) \in R$ is represented by a point at the intersection of the vertical line through x and the horizontal line through y. For the reflexive, symmetric and transitive closure, added elements are shown in gray.

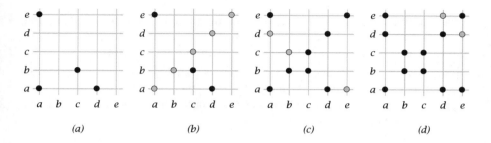

(a)   (b)   (c)   (d)

Fig. 4.9. A relation R *(a)*, its reflexive closure R_r *(b)*, the reflexive, symmetric closure $(R_r)_s$ *(c)*, and the reflexive, symmetric, transitive closure $R_e = ((R_r)_s)_t$ *(d)*.

Definition 19 *A binary relation $\leq$ on a set a is called a* partial ordering *iff it is reflexive, transitive and antisymmetric. A partial ordering is called* linear *iff it is total. A linear ordering is called a* well-ordering *iff every non-empty set $b \subset a$ has a minimal element m, i.e., $m \leq x$ for all $x \in b$.*

Example 12 We will later see that the set of natural numbers (0, 1, 2 ...) and the set of integers (... −2, −1, 0, 1, 2 ...) are both linearly ordered by "x is less than or equal to y". However, whereas the natural numbers are well-ordered with respect to this relation, the integers are not.

The aforementioned "subclass" relation is a partial ordering. It is not linear, because two classes can be completely unrelated, or derive from the same class in the hierarchy without one being a subclass of the other. Another example of a partial, but not linear, ordering is the inclusion relation on sets, which is also a well-ordering.

Lemma 26 *Let $\leq$ be a binary relation on a set a. Denoting $x < y$ iff $x \leq y$ and $x \neq y$, the following two statements are equivalent:*

(i) *The relation $\leq$ is a partial ordering.*

(ii) *The relation $\leq$ is reflexive, the relation $<$ is transitive, and for all $x, y \in S$, $x < y$ excludes $y < x$.*

If these equivalent properties hold, we have $x \leq y$ iff $x = y$ or else $x < y$. In particular, if we are given $<$ with the properties (ii), and if we define $x \leq y$ by the preceding condition, then the latter relation is a partial ordering.

Proof (i) implies (ii): If $\leq$ is a partial ordering, then it is reflexive by definition. The relations $a < b < c$ imply $a \leq c$, but $a = c$ is excluded since $\leq$ is antisymmetric. The last statement is a consequence of the transitivity of $<$.

(ii) implies (i): $\leq$ is reflexive by hypothesis. It is transitive, since $<$ is so, and the cases of equality are obvious. Finally, if $x < y$, then $y \leq x$ is impossible since equality is excluded by the exclusion of the simultaneous validity of $x < y$ and $y < x$, and inequality is by definition excluded by the same condition.   □

Definition 20 *If R is a relation on a, and if a' is any set, the induced relation $R|_{a'}$ is defined to be the relation $R \cap (a \cap a') \times (a \cap a')$ on $a \cap a'$.*

Exercise 15 Show that the induced relation $R|_{a'}$ is a partial ordering, a linear ordering, a well-ordering, if R is so.

Exercise 16 Given a relation R on a and a bijection $f : a \to b$, then we consider the image $f \circ R$ of the induced bijection $f \times f|_R$ in $b \times b$. This new relation is called "structural transport" of R. Show that $f \circ R$ inherits all properties of R, i.e., it is a partial ordering, a linear ordering, a well-ordering, iff R is so.

The strongest statement about relations on sets is this theorem (due to Ernst Zermelo):

Proposition 27 (Zermelo) *There is a well-ordering on every set.*

Proof We shall not prove this quite involved theorem, but see [46].   □

Remark 4 If every set admits a well-ordering, the axiom of choice is a consequence hereof. Conversely, the proposition 27 of Zermelo is proved by use of the axiom 8 of choice. In other words: Zermelo's theorem and the axiom of choice are equivalent axioms.

CHAPTER 5

Ordinal and Natural Numbers

Until now, our capabilities to produce concrete sets were quite limited. In particular, we were only capable of counting from zero to one: from the empty set $0 = \emptyset$ to the set $1 = \{0\}$. We are not even in state of saying something like: "For $n = 0, 1, \ldots$", since the dots have no sense up to now! This serious lack will be abolished in this chapter: We introduce the basic construction of natural numbers—together with one of the most powerful proof techniques in mathematics: proof by infinite induction.

5.1 Ordinal Numbers

We shall now construct the basic sets needed for every counting and number-theoretic task in mathematics (and all the sciences, which count on counting, be aware of that!)

Definition 21 *A set a is called* transitive *if $x \in a$ implies $x \subset a$.*

Example 13 The sets 0 and 1 are trivially transitive, and so is any set $J = \{J\}$ (if it exists).

Exercise 17 Show that if a and b are transitive, then so is $a \cap b$.

Definition 22 *A set a is called* alternative *if for any two elements $x, y \in a$, either $x = y$, or $x \in y$, or $y \in x$.*

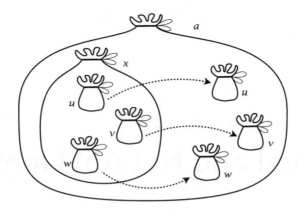

Fig. 5.1. Transitivity: Elements of an element $x \in a$ of a transitive set a are themselves elements of a.

Fig. 5.2. Alternativity: In an alternative set a, if $x \neq y$ are two elements of a, then either $x \in y$ or $y \in x$.

Exercise 18 If a is alternative and $b \subset a$, then b is alternative.

Definition 23 *A set a is called* founded *if for each non-empty $b \subset a$, there is $x \in b$ with $x \cap b = \varnothing$.*

Example 14 The sets 0 and 1 are founded, but a "circular" set $J = \{J\}$ (if it exists) is not.

The circular non-founded set is a special case of the following result:

Proposition 28 *Suppose that a is founded, then $x \in a$ implies $x \notin x$.*

Proof For $x \in a$, consider the subset $\{x\} \subset a$. Then $x \in x$ contradicts the fact that $\{x\} \cap x$ must be empty because a is founded.   □

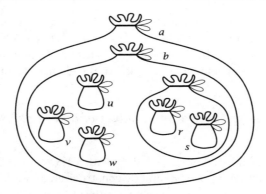

Fig. 5.3. Foundedness: If $b \in a$ is non-empty, and a is founded, then there is $x \in b$ such that $x \cap b = \varnothing$; here we have $x = \{r, s\}$, $b = \{x, u, v, w\}$ with $r \neq u, v, w$ and $s \neq u, v, w$.

Lemma 29 *If d is founded, and if $a, b, c \in d$ such that $a \in b$, $b \in c$, then $a \neq c$ and $c \notin a$.*

Proof Consider the subset $\{a, b\} \in d$. If $x = b$, $a \in x \cap \{a, b\}$. Therefore $x = a$ yields $x \cap \{a, b\} = \varnothing$. But $a = c$ enforces that b is in this intersection. Therefore $a \neq c$. If we had $c \in a$, then we cannot have $a \cap \{a, b, c\} = \varnothing$, by the hypothesis of the lemma, $b \cap \{a, b, c\}$ contains a, and $c \cap \{a, b, c\}$ contains b. Therefore $c \notin a$. □

Here is the fundamental concept for creating numbers:

Definition 24 *A set is called* ordinal *if it is transitive, alternative, and founded.*

The following series of results is destined to control the basic properties of ordinals.

Lemma 30 *Let d be an ordinal. If $a \subset d$ is non-empty, then there is $x_0 \in a$ such that whenever $x \in a$, then either $x_0 = x$ or $x_0 \in x$.*

Proof Since d is an ordinal, it is founded, and there is an element $x_0 \in a$ such that $x_0 \cap a = \varnothing$. Let $x \in a$ be any element different from x_0. Since d is alternative, either $x \in x_0$ or $x_0 \in x$. But $x \in x_0$ contradicts $x_0 \cap a = \varnothing$, and we are done. □

Proposition 31 *If d is ordinal, then $x \in d$ implies that x is ordinal.*

Proof Let $x \in d$. Then by transitivity of d, $x \subset d$. Hence x is alternative and founded. Let $b \in x$. Then $a \in b$ implies $a \in x$. In fact, $x \subset d$, therefore $b \in d$ and $a \in b \in d$. Since d is transitive, $a \in d$. So $a, x \in d$. Hence either $a = x$, or $a \in x$, or $x \in a$. But by lemma 29, applied to the element chain $a \in b \in x$, we must have $a \in x$. □

Proposition 32 *A proper transitive subset of an ordinal is an element of that ordinal.*

Proof Since d is founded, there is $y \in d - c$ with $y \cap (d - c) = \varnothing$. We claim that $c = y$. Since d is transitive, $y \subset d$, and by construction of y, $y \subset c$. Conversely, let $b \in c$, then $b \in d$. Since d is alternative, either $b \in y$ or $b = y$ or $y \in b$. But $b = y$ implies $b \in d - c$, a contradiction. Further, $y \in b$ and $b \in c$ imply $y \in c$ by transitivity of c. A contradiction. Therefore we have $b \in y$. □

Corollary 33 *If d is an ordinal, then a set a is an element of d iff it is an ordinal and $a \subsetneq d$.*

Proof Clearly, an element of an ordinal is a proper subset. The converse follows from proposition 32. □

Exercise 19 Show that if d is a non-empty ordinal, then $\varnothing \in d$.

Exercise 20 Show that if c is ordinal, then $a \in b, b \in c$ implies $a \in c$.

Exercise 21 Show that if a and b are ordinals, then $a \in b$, implies $b \notin a$.

Proposition 34 *If a and b are ordinals, then either $a \subset b$ or $b \subset a$.*

Proof Suppose both conclusions are wrong. Then the intersection $a \cap b$ is a proper subset of both a and b. But it is evidently transitive, hence an ordinal element of both, a and b, hence an element of itself, a contradiction. □

Corollary 35 *If a and b are ordinals, then exclusively either $a \in b$, or $a = b$, or $b \in a$.*

Proof Follows from propositions 32 and 34. □

Corollary 36 *If all elements $a \in b$ are ordinals, then b is alternative.*

Proof Follows from corollary 35. □

Proposition 37 *If every element $x \in u$ of a set u is ordinal, then there is exactly one element $x_0 \in u$ such that for every $x \in u$, we have either $x_0 \in x$ or $x_0 = x$.*

Proof Uniqueness is clear since all elements of u are ordinal.

Existence: If $u = \{a\}$, then take $x_0 = a$. Else, there are at least two different elements $c, c' \in u$, and either $c \in c'$ or $c' \in c$. So either $c \cap u$ or $c' \cap u$ is not empty. Suppose $c \cap u \neq \varnothing$. By lemma 30, there is an $x_0 \in c \cap u$ such that either $x_0 = x$ or $x_0 \in x$ for all $x \in c \cap u$. Take any $y \in u$. Then either $c \in y$; but $x_0 \in c$, and therefore $x_0 \in y$. Or $c = y$, whence $x_0 \in y$, or else $y \in c$, whence $y \in c \cap u$, and $x_0 = y$ or $x_0 \in y$ according to the construction of x_0. □

Corollary 38 *If all elements of a set u are ordinals, then u is founded.*

Proof Let $c \subset u$ be a non-empty subset. Take the element $x_0 \in c$ as guaranteed by proposition 37. Then clearly $x_0 \cap c = \varnothing$. □

Corollary 39 *A transitive set a is ordinal iff all its elements are so.*

Proof Follows immediately from the corollaries 36 and 38. □

Remark 5 There is no set *Allord* containing all ordinal sets. In fact, it would be transitive, and therefore ordinal, i.e., we would have the absurd situation *Allord* $\in$ *Allord*.

Recall the successor set a^+ defined in lemma 1.

Proposition 40 *For any set a, a^+ is non-empty, we have $a \in a^+$, and a is ordinal iff a^+ is so.*

Proof By definition, $a \in a^+$, a successor is never empty. If a^+ is ordinal, so is its element a. Conversely, all the elements of a^+ are ordinal. Moreover, a^+ is transitive, hence ordinal by corollary 39. □

Lemma 41 *If a and b are ordinals with $a \in b$, then*

(i) *either $a^+ \in b$ or $a^+ = b$;*

(ii) *$a^+ \in b^+$;*

(iii) *there is no x such that $a \in x$ and $x \in a^+$.*

Proof Since a^+ and b are ordinal, we have $a^+ \in b$ or $a^+ = b$ or $b \in a^+$. But the latter yields a contradiction to $a \in b$. The second statement follows from the first. The third follows from the two impossible alternatives $x = a$ or $x \in a$. □

Proposition 42 *If a and b are ordinals, then $a = b$ iff $a^+ = b^+$.*

Proof Clearly, $a = b$ implies $a^+ = b^+$. Conversely, $a^+ = b^+$ implies $b = a$ or $b \in a$, but the latter implies $b^+ \in a^+$, a contradiction. □

Corollary 43 *If two ordinals a and b are equipollent, then so are their successors.*

Proof Clear from the universal property (proposition 58, chapter 6) of coproducts since $a \cap \{a\} = b \cap \{b\} = \emptyset$. □

Proposition 44 *Let Φ be an attribute of sets such that whenever it holds for all elements $x \in a$ of an ordinal a, then it also holds for a. Then Φ holds for all ordinals.*

Proof Observe that in particular, Φ holds for $\emptyset$. Suppose that there is an ordinal b such that $\Phi(b)$ does not hold. Then the subset $b' = \{x \mid x \in b, \text{NOT } \Phi(x)\}$ of b is not empty (since NOT $\Phi(b)$) and a proper subset of b (since $\emptyset$ is not in b'). According to proposition 37, there is a minimal element $x_0 \in b'$, i.e., NOT $\Phi(x_0)$, but every element of x_0 is element of $b - b'$. This is a contradiction to the hypothesis about Φ. □

5.2 Natural Numbers

The natural numbers, known from school in the common writing $(0,)$ 1, 2, 3, 4, 5, ... are constructed from the ordinal sets as follows. This construction traditionally stems from an axiomatic setup, as proposed by Giuseppe Peano. In the set-theoretical framework, the Peano axioms appear as propositional statements. This is why proposition 45 and proposition 46 are named "Peano Axioms".

Definition 25 *A natural number is an ordinal set n which is either $\emptyset$ or a successor m^+ of an ordinal number m and such that every element x of n is either $x = \emptyset$ or a successor $x = y^+$ of an ordinal number y.*

The membership relation of sets can be illustrated by a diagram where an arrow from a set x to a set y means that x contains y as an element, or $y \in x$. The following picture, for example, represents the set $a = \{\{b, c, d\}, d, \emptyset\}$:

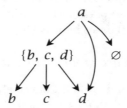

Using this method, the natural numbers $0 = \varnothing, 1 = \{\varnothing\}, 2 = \{\varnothing, \{\varnothing\}\}$, $3 = \{\varnothing, \{\varnothing\}, \{\varnothing, \{\varnothing\}\}\}, \ldots$ can be drawn:

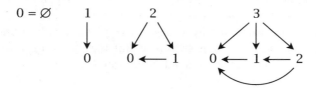

Proposition 45 (Peano Axioms 1 through 4) *We denote the empty set by* 0 *if we consider it as an ordinal.*

   (i) *The empty set* 0 *is a natural number.*

  (ii) *If* n *is natural, then so is* n^+.

 (iii) *The empty set* 0 *is not a successor.*

 (iv) *If* n *and* m *are natural, then* $n^+ = m^+$ *implies* $n = m$.

Proof This is straightforward from the propositions following the definition, lemma 1 in chapter 2, of a successor. □

Proposition 46 (Peano Axiom 5) *Let* Ψ *be an attribute of sets with these properties:*

   (i) *We have* $\Psi(0)$.

  (ii) *For every natural number* n, $\Psi(n)$ *implies* $\Psi(n^+)$.

Then Ψ *holds for every natural number.*

Proof Let m be a natural number with NOT $\Psi(m)$. Since $m \neq 0$, i.e., $m = n^+$, by hypothesis, the subset $\{x \mid x \in m, \text{NOT } \Psi(x)\}$ contains n. Take the minimum x_0 within this subset. This is a successor $x_0 = y^+$ such that $\Psi(y)$, a contradiction. □

Remark 6 Proposition 46 yields a proof scheme called "proof by induction". One is given an attribute Ψ of sets, usually only specified for natural numbers. For other sets, we tacitly set Ψ to be false, but that value is irrelevant for the given objective, which is to prove that Ψ holds for all natural numbers. To this end, one proves (i) and (ii) in proposition 46 and concludes as required.

The following proposition is the first result which follows from proposition 46, Peano's fifth axiom:

Proposition 47 *Let* n *be natural, and let* $a \subsetneq n$. *Then there is* $m \in n$ *such that* $a \overset{\sim}{\to} m$.

Proof We prove the proposition by induction: For any natural number n, let $\Psi(n)$ be the proposition that for every proper subset $y \subset n$, there is a natural m, which is equipollent to y. Clearly, $\Psi(0)$. Suppose that $n = m^+$ and $\Psi(m)$ holds. Let $y \subset n$. If $y \subset m$, either $y = m$ and we are done, or y is a proper subset of m and the induction hypothesis gives us the required bijection. Else, if $n \in y$, there is an element $u \in m - y$. But then $y' = (y - \{n\}) \cup \{u\}$ is equipollent to y, and we are redirected to the above case.   □

Proposition 48 *Let n be a natural number, and suppose that an ordinal a is equipollent to n. Then $a = n$. In particular, two natural numbers m and n are equipollent iff they are equal.*

Proof We prove the claim by induction and consider the property $\Psi(n)$ for natural numbers defined by $\Psi(n)$ iff for any ordinal a, the fact that a is equipollent to n implies $a = n$. Suppose that there is a counterexample m; evidently $m \neq 0$. Let $a \neq m$ be ordinal, but equipollent to m. Let $\overline{m} = \{x \mid x \in m, \Psi(x)\}$, which is not empty, since it contains 0.

1. $\overline{m} = m$. Since a and m are ordinal, either $a \in m$ or $m \in a$. In the first case, $\Psi(a)$, which contradicts to the choice of m. So $m \in a$ and $m \subset a$ is a proper subset. Let $f : a \to m$ be a bijection. Then $f(m)$ is a proper subset of m. By proposition 47, there is an element $n \in m$ which is equipollent to $f(m)$ and therefore also to m. But we know that $\Psi(n)$, whence $m = n$, which contradicts $n \in m$.

2. $m - \overline{m} \neq \varnothing$. Take the smallest x_0 in this difference set. There is an ordinal $b \neq x_0$, but equipollent to x_0. Then either $b \in x_0$ or $x_0 \in b$. In the first case, b is a natural number in m since $x_0 \subset m$. So $b = x_0$ by construction of x_0, a contradiction. If $x_0 \in b$, then since b is ordinal, $x_0 \subset b$ is a proper subset, and we may proceed as in the first case above. This concludes the proof.   □

Remark 7 This means that natural numbers describe in a unique way certain cardinalities of ordinals. Each natural number represents exactly one cardinality, and two different natural numbers are never equipollent.

Definition 26 *A set a is called* finite *if it is equipollent to a natural number. This number is then called* cardinality *of a and denoted by $card(a)$, by $\#(a)$, or by $|a|$, depending on the usage.*

This definition is justified by the fact that the cardinality of a finite set is a unique number which is equipollent to that set, i.e., which in our previous terminology has the same cardinality as the given set.

Corollary 49 *A subset of a finite set is finite.*

Proof We may suppose that we are dealing with a subset of a natural number, where the claim is clear. □

Corollary 50 *An ordinal set is finite iff it is natural.*

Proof This follows right from proposition 48. □

Corollary 51 *If an ordinal a is not finite, it contains all natural numbers.*

Proof For any natural number n, we have either $a \in n$ or $a = n$ or $n \in a$. The middle case is excluded by definition of a. The left alternative is excluded since every subset of a finite set is finite. So every n is an element of a. □

Corollary 52 *A finite set is not equipollent to a proper subset.*

Proof In fact, a proper subset of a natural number n is equipollent to an element of n, hence equal to this element, a contradiction. □

From the axiom 7 of infinity, we derive this result:

Proposition 53 *There is an ordinal $\mathbb{N}$ whose elements are precisely the natural numbers.*

Proof This follows immediately from the axiom 7. □

Notation 5 *The relation $n \in m$ for elements of $\mathbb{N}$ defines a well-ordering among natural numbers. We denote it by $n < m$ and say "n is smaller than m" or else "m is larger than n".*

Exercise 22 With the well-ordering $<$ among natural numbers, we have these facts: Every non-empty set of natural numbers has a (uniquely determined) minimal element. Let b be a limited non-empty set b of natural numbers, i.e., there is $x \in \mathbb{N}$ such that $y < x$ for all $y \in b$. Then there is a (uniquely determined) maximal element $z \in b$, i.e., $y < z$ for all $y \in b$.

CHAPTER 6

Recursion Theorem and Universal Properties

Before developing more specifically the arithmetic of natural numbers, we should summarize crucial properties of sets and their functions. These properties are the basis of a fundamental branch in mathematics, called topos theory, a branch which is of great importance to computer science too, since it provides a marriage of formal logic and geometry.

Proposition 54 *If a and b are two sets, then there is a set, denoted by $Set(a,b)$, whose elements are precisely the functions $f : a \to b$.*

Proof The elements of the asked for set $Set(a,b)$ are triples (a,b,f), where $f \subset a \times b$ is a graph. So $(a,b,f) = ((a,b),f) \in (2^a \times 2^b) \times 2^{a \times b}$, whence $Set(a,b) \subset (2^a \times 2^b) \times 2^{a \times b}$ is a subset selected by an obvious propositional attribute. □

Notation 6 *The set $Set(a,b)$ of functions is also denoted by b^a if we want to stress that it is a set, without emphasis on the specific nature, i.e., that its elements are the functions $f : a \to b$. This distinction may seem superfluous now, but we shall understand this point when we will deal with more general systems of objects and "functions" between such objects.*

Example 15 If $a = 0$ $(= \varnothing)$, then $Set(0,b) = \{! : \varnothing \to b\}$, if $b = 1$ $(= \{0\})$, then $Set(a,1) = \{! : a \to 1\}$. If $a \neq 0$, then $Set(a,0) = 0$.

6.1 Recursion Theorem

The set of functions allows a very important application of the fifth Peano
axiom (which is a theorem in our context!): construction by induction. To
this end, we need to look at functions $f : \mathbb{N} \to a$ for any non-empty set a.
If $n \in \mathbb{N}$, we have the restriction $f|_n : n \to a$ as declared in definition 13.
Observe that $f|_n$ is defined for all natural numbers strictly smaller than
n, but not for n. If $g : n^+ \to a$ is a function, then we denote by g^* the
function $g^* : \mathbb{N} \to a$ with $g^*(m) = g(n)$ for $n < m$ and $g^*(m) = g(m)$
for $m \leq n$. If g is $! : 0 \to a$, we pick an element $g_0 \in a$ and set $g^*(m) = g_0$
for all natural numbers m. Here is the general recursion theorem:

Proposition 55 (Recursion Theorem) *Let a be a set, and let $\Phi : a^{\mathbb{N}} \to a^{\mathbb{N}}$
be a function such that for every natural number n, if $f, g \in a^{\mathbb{N}}$ are such
that $f|_n = g|_n$, then $\Phi(f)(n) = \Phi(g)(n)$. Then Φ has a unique* fixpoint
$L_{\Phi} \in a^{\mathbb{N}}$, *which means that $\Phi(L_{\Phi}) = L_{\Phi}$. Consider the function $\Phi_n : a^n \to a$
which evaluates to $\Phi_n(g) = \Phi(g^*)(n)$. Then we have*

$$L_{\Phi}(0) = \Phi_0(! : 0 \to a)$$
$$L_{\Phi}(n^+) = \Phi_{n^+}(L_{\Phi}|_{n^+}).$$

Proof There is at most one such fixpoint. In fact, let L and M be two such
fixpoints, $\Phi(L) = L$ and $\Phi(M) = M$, and suppose that they are different. Then
there is a smallest value n_0 such that $L(n_0) \neq M(n_0)$. This means that $L|_{n_0} =
M|_{n_0}$. But then $\Phi(L)(n_0) = \Phi(M)(n_0)$, a contradiction. So there is at most one
such fixpoint. For every $n \in \mathbb{N}$, let $S(n) \subset a^n$ be the set of those functions
$f : n \to a$ such that for all $m \in n$, $f(m) = \Phi_m(f|_m)$. Clearly, either $S(n)$ is
empty or $S(n)$ contains precisely one function g_n. The set $S(0)$ is not empty. Let
N^+ be the smallest natural number such that $S(N^+)$ is empty. We may define a
function $h : N^+ \to a$ by $h|_N = g_N$ and $h(N) = \Phi_N(h|_N)$. But this is a function
in $S(N^+)$, so every $S(n)$ is non-empty. Now define $L(n) = g_{n^+}(n)$. Clearly, this
function is our candidate for a fixpoint: To begin with, if $n < m$, then evidently,
by the uniqueness of the elements of $S(n)$, $g(m)|_n = g(n)$. Therefore, $L|_n = g_n$
for all n. And L is a fixpoint, in fact: $L(n) = g_{n^+}(n) = \Phi_n(g_{n^+}|_n) = \Phi_n(g_n) =
\Phi(g_n^*)(n) = \Phi(L)(n)$ since $L|_n = g_n = g_n^*|_n$. The claimed formula then follows
by construction. $\square$

Remark 8 Very often, the above formal context will be treated in a rather
sloppy way. The common wording is this: One wants to define objects
(functions, sets of whatever nature) by the following data: you know
which object O_0 you have to start with, i.e., the natural number 0. Then

you suppose that you have already "constructed" the objects O_m, $m \leq n$ for a natural number n, and your construction of O_{n^+} for the successor n^+ is given from $O_0, O_1, \ldots O_n$ and some "formula" Φ. Then you have defined the objects O_n for every $n \in \mathbb{N}$.

6.2 Universal Properties

Definition 27 *A set b is called* final *iff for every set a, we have $|Set(a, b)| = 1$. A set a is called* initial *iff for every set b, $|Set(a, b)| = 1$.*

Proposition 56 (Existence of Final and Initial Sets) *A set b is final iff $|b| = 1$. A set a is initial iff $a = 0$.*

Proof This is immediate.   □

We shall usually pick the set 1 to represent the final sets.

Notation 7 *Given two sets a and b, denote by $pr_a : a \times b \rightarrow a$ and $pr_b : a \times b \rightarrow b$ the functions $pr_a(x, y) = x$ and $pr_b(x, y) = y$ for elements $(x, y) \in a \times b$.*

Proposition 57 (Universal Property of Cartesian Product) *Given two sets a and b and any set c, the function*

$$\beta : Set(c, a \times b) \xrightarrow{\sim} Set(c, a) \times Set(c, a)$$

defined by $\beta(u) = (pr_a \circ u, pr_b \circ u)$ is a bijection.

The following commutative diagram shows the situation:

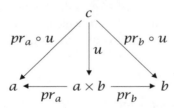

Proof If $u : c \rightarrow a \times b$, then for $x \in c$, $u(x) = (pr_a(u(x)), pr_b(u(x)))$, so β is injective. If we are given $v : c \rightarrow a$ and $w : c \rightarrow b$, then define $u(x) = (v(x), w(x))$. Then evidently, $\beta(u) = (v, w)$, so β is surjective.   □

Exercise 23 Suppose that a set q, together with two functions $p_a : q \rightarrow a$ and $p_b : q \rightarrow b$ has the property that

$$\beta : Set(c,q) \xrightarrow{\sim} Set(c,a) \times Set(c,b)$$

defined by $\beta(u) = (p_a \circ u, p_b \circ b)$ is a bijection. Show that there is a unique bijection $i : q \xrightarrow{\sim} a \times b$ such that $pr_a \circ i = p_a$ and $pr_b \circ i = p_b$.

Definition 28 *Given two sets a and b, the* disjoint sum *or* coproduct $a \sqcup b$ *of a and b is the set* $a \sqcup b = (1 \times a) \cup (\{1\} \times b)$, *together with the injections* $in_a : a \to a \sqcup b$, *and* $in_b : b \to a \sqcup b$, *where* $in_a(x) = (0, x)$ *and* $in_b(y) = (1, y)$ *for all* $x \in a$ *and* $y \in b$.

Evidently, the coproduct $a \sqcup b$ is the disjoint union of the two subsets $1 \times a$ and $\{1\} \times b$. Here is the "universal" property for the coproduct corresponding to the universal property of the Cartesian product proved in proposition 57:

Proposition 58 (Universal Property of Coproduct) *Given two sets a and b and any set c, then the function*

$$y : Set(a \sqcup b, c) \xrightarrow{\sim} Set(a, c) \times Set(b, c)$$

defined by $y(u) = (u \circ in_a, u \circ in_b)$ *is a bijection.*

The following commutative diagram shows the situation:

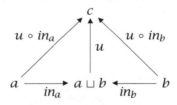

Proof Clearly, a map $u : a \sqcup b \to c$ is determined by its restrictions to its partitioning subsets $1 \times a$ and $\{1\} \times b$, which in turn is equivalent to the pair $u \circ in_a$ and $u \circ in_b$ of maps. So y is injective. Conversely, if $v : a \to c$ and $w : b \to c$ are any two sets, then we define $u((0, x)) = v(x)$ and $u((1, y)) = w(y)$, which shows the surjectivity of y.   □

Exercise 24 Suppose that a set q, together with two functions $i_a : a \to q$ and $i_b : b \to q$ has the property that

$$y : Set(q, c) \xrightarrow{\sim} Set(a, c) \times Set(b, c)$$

defined by $y(u) = (u \circ i_a, u \circ i_b)$ is a bijection. Show that there is a unique bijection $i : a \sqcup b \xrightarrow{\sim} q$ such that $i \circ in_a = i_a, i \circ in_b = i_b$.

Proposition 59 (Universal Property of Exponentials) *If a, b and c are sets, there is a bijection*

$$\delta : Set(a \times b, c) \xrightarrow{\sim} Set(a, c^b)$$

defined by

$$\delta(f)(\alpha)(\beta) = f(\alpha, \beta)$$

for all $\alpha \in a$ and $\beta \in b$, and $f \in Set(a \times b, c)$. This bijection is called the natural adjunction.

Proof The map δ is evidently injective. On the other hand, if $g : a \to c^b$, then we have the map $f : a \times b \to c$ defined by $f(\alpha, \beta) = g(\alpha)(\beta)$, and then $\delta(f) = g$. $\square$

For the next concepts we need to know what the fiber of a function is:

Definition 29 *If $f : a \to b$ is a function, and if $c \subset b$, then we call the set $\{x \mid x \in a \text{ and } f(x) \in c\}$ "fiber of f over c" and denote it by $f^{-1}(c)$. For a singleton $c = \{y\}$, we write $f^{-1}(y)$ instead of $f^{-1}(\{y\})$.*

We have this commutative diagram, where the horizontal arrows are the inclusions:

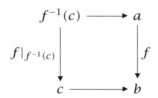

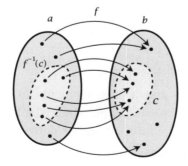

Fig. 6.1. The fiber $f^{-1}(c) \subset a$ of f over $c \subset b$.

We had introduced the somewhat strange notation 2^a for the powerset of a. Here is the resolution of this secret.

Proposition 60 (Subobject Classifier) *The natural number* $2 = \{0, 1\}$ *is a subobject classifier, i.e., for every set a, there is a bijection*

$$\chi : 2^a \xrightarrow{\sim} Set(a, 2)$$

defined by this prescription: If $b \subset a$ is an element of 2^a, then $\chi(b)(\alpha) = 0$ if $\alpha \in b$, and $\chi(b)(\alpha) = 1$ else. The function $\chi(b)$ is called the characteristic function of b. The inverse of χ is the zero fiber, i.e., $\chi^{-1}(x) = x^{-1}(0)$.

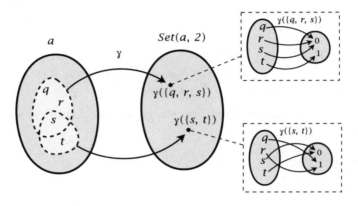

Fig. 6.2. Subobject classifier: The values of y for just two elements are shown. For example, the subset $\{q, r, s\}$ of $a = \{q, r, s, t\}$ is mapped by y to its characteristic function $y(\{q, r, s\})$ illustrated in the upper right of the figure. Note that $y^{-1}(\{q, r, s\}) = \{q, r, s\} = y(q, r, s)^{-1}(0)$, i.e., the zero fiber of the characteristic function.

Proof This result is immediate.   □

Observe that now the subsets of a, elements of 2^a, are identified with functions $a \to 2$, i.e., elements of $Set(a, 2)$ which we also denote by 2^a, and this is now perfectly legitimate by the above proposition.

A generalization of the Cartesian product is given by so-called families of sets.

Definition 30 *A family of sets is a surjective function $f : a \to b$. The images $f(x)$ are also denoted by f_x, and the function is also notated by*

$(f_x)_{x \in a}$ or by $(f_x)_a$. *This means that the elements of b are "indexed" by elements of a.*

If $c \subset a$, then the subfamily $(f_x)_{x \in c}$ *is just the restriction $f|_c$, together with the codomain being the image $Im(f|_c)$.*

The Cartesian product $\prod_{x \in a} f_x$ *of a family $(f_x)_{x \in a}$ of sets is the subset of $(\bigcup b)^a$ consisting of all functions $t : a \rightarrow \bigcup b$ such that $t(x) \in f_x$ for all $x \in a$. Such a function is also denoted by $(t_x)_{x \in a}$ and is called a* family of elements. *We shall always subsume that when a family of elements is given, that there is an evident family of sets backing this family of elements, often without mentioning these sets explicitly.*

For a given index x_0, we have the x_0-th projection $p_{x_0} : \prod_{x \in a} f_x \rightarrow f_{x_0}$ which sends $(t_x)_{x \in a}$ to t_{x_0}.

Example 16 Figure 6.3 shows a family of sets $f : a \rightarrow b$, with $a = \{x, y\}$ and $b = \{\{q, r, s\}, \{s, t\}\}$. It is defined as $f_x = \{q, r, s\}$ and $f_y = \{s, t\}$. The Cartesian product $\prod_{x \in a} f_x$ is given by the functions $t_i : a \rightarrow \{q, r, s, t\}$ with $t_1 = \{(x, q), (y, s)\}$, $t_2 = \{(x, r), (y, s)\}$, $t_3 = \{(x, s), (y, s)\}$, $t_4 = \{(x, q), (y, t)\}$, $t_5 = \{(x, r), (y, t)\}$, $t_2 = \{(x, s), (y, t)\}$, where the graphs of the functions have been used to describe them.

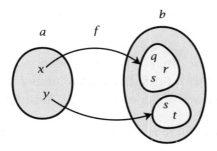

Fig. 6.3. A family of sets $f : a \rightarrow b$.

One defines the *union (respectively intersection) of the family $(f_x)_a$* by $\bigcup_a f_x = \bigcup Im(f)$ (respectively $\bigcap_a f_x = \bigcap Im(f)$), i.e., by the union (respectively intersection) of all its member sets. The intersection however exists only for a non-empty family.

Sorite 61 *Let $(f_x)_a$ be a non-empty family of sets, i.e., $a \neq \emptyset$.*

(i) *The Cartesian product $\prod_{x \in a} f_x$ is non-empty iff each f_x is non-empty.*

(ii) *If all sets f_x coincide and are equal to c, then $\prod_{x \in a} f_x = c^a$.*

(iii) *If $a = 2$, then $\prod_{x \in 2} f_x \xrightarrow{\sim} f_0 \times f_1$.*

(iv) *If $(u_x : d \to f_x)_{x \in a}$ is a family of functions, then there is a unique function $u : d \to \prod_{x \in a} f_x$ such that $u_x = p_x \circ u$ for all $x \in a$.*

Proof (i) By definition of the Cartesian product, this one is non-empty for $a \neq \emptyset$ iff each f_x is so.

(ii) This is an immediate consequence of the definition of the Cartesian product and of the fact that here, $c = \bigcup_{x \in a} c$.

(iii) The functions $g : 2 \to f_0 \cup f_1$, where $g(0) \in f_0$ and $g(1) \in f_1$, are in bijection with the pairs of their evaluations $(g(0), g(1)) \in f_0 \times f_1$.

(iv) The reader should think about the set where the family of functions is deduced. But if this is done, the statement is immediate.   □

Example 17 Families where the index set a is a natural number n or the set $\mathbb{N}$ of all natural numbers are called *sequences*. For $a = n$, we have the *finite* sequences $(t_i)_{i \in n}$, or, in an equivalent notation, $(t_i)_{i < n}$. One then often writes $t_i, i = 0, 1, \ldots n - 1$ instead, or also $t_0, t_1, \ldots t_{n-1}$. For $a = \mathbb{N}$, one writes also $(t_i)_{i=0,1,2,\ldots}$ or else $t_0, t_1, \ldots$. The *length of a sequence* $(t_i)_{i \in n}$ is the (uniquely determined) number n. One also calls such a sequence an *n-tuple*.

In computer science, one often calls sequences *lists*, and it is also agreed that list indexes start with 0, as do natural number indexes. The *empty list* is also that sequence with index set $a = 0$, the empty set.

Exercise 25 Prove that there is exactly one empty family, i.e., such that its index set is empty.

Cartesian products $\prod_{x \in a} f_x$ also admit linear orderings if their members do so. Here is the precise definition of this so-called "lexicographic ordering":

Definition 31 *Suppose that we are given a family $(f_x)_a$ of sets such that each f_x bears a linear ordering $<_x$, and such that the index set a is well-ordered by the relation $\prec$. Then, for two different families $(t_x)_a, (s_x)_a \in \prod_{x \in a} f_x$ the relation*

$(t_x)_a \prec (s_x)_a$ *iff the smallest index* y*, where* $t_y \neq s_y$*, has* $t_y <_y s_y$

is called the lexicographic *ordering on* $\prod_{x \in a} f_x$.

Lemma 62 *The lexicographic ordering is a linear ordering.*

Proof According to lemma 26, we show that $\prec$ is transitive, antisymmetric and total. Let $(t_x)_a \prec (s_x)_a \prec (u_x)_a$. If the smallest index y, where these three families differ, is the same, then transitivity follows from transitivity of the total ordering at this index. Else, one of the two smallest indexes is smaller than the other, let $y_1 < y_2$ for the index y_1 of the left pair $(t_x)_a \prec (s_x)_a$. Then the inequalities at this index are $t_{y_1} <_{y_1} t_{y_1} = u_{y_1}$, whence $t_{y_1} <_{y_1} u_{y_1}$, i.e., $(t_x)_a \prec (u_x)_a$; similarly for the other situation, namely, $y_2 < y_1$. The same argument works for antisymmetry. As to totality: Let $(t_x)_a$ and $(s_x)_a$ be any two families. If they are different, then the smallest index y where they differ has either $t_y <_y s_y$ or $s_y <_y t_y$ since $<_y$ is total. $\square$

Exercise 26 Show that the lexicographic ordering on $\prod_{n \in \mathbb{N}} f_n$ is a well-ordering iff each linear ordering $<_n$ on f_n is so. The same is true for a finite sequence of sets, i.e., for $\prod_{n < N} f_n$, where $N \in \mathbb{N}$.

Exercise 27 Suppose that we are given a finite alphabet set $\mathcal{A}$ of "letters". Suppose that a bijection $u : \mathcal{A} \overset{\sim}{\to} N$ with the natural number $N = card(\mathcal{A})$ is fixed, and consider the ordering of letters induced by this bijection, i.e., $X < Y$ iff $u(X) < u(Y)$. Suppose that an element $\_ \in \mathcal{A}$ is selected. Consider now the restriction of the lexicographic ordering on $\mathcal{A}^{\mathbb{N}}$ to the subset $\mathcal{A}^{(\mathbb{N})}$ consisting of all sequences $(\tau_n)_{\mathbb{N}}$ such that $\tau_n = \_$ for all but a finite number of indexes. Show that this set may be identified with the set of all finite words in the given alphabet. Show that the induced lexicographic ordering on $\mathcal{A}^{(\mathbb{N})}$ coincides with the usual lexicographic ordering in a dictionary of words from the alphabet $\mathcal{A}$; here the special sign $\_$ plays the role of the empty space unit.

The name "lexicographic" effectively originates in the use of such an ordering in compiling dictionaries. As an example we may consider words of length 4, i.e., the set $\mathcal{A}^4$. Let the ordering on $\mathcal{A}$ be $\_ < A < B \dots Z$. Then, writing the sequence $(t_i)_{i=0,1,2,3}$, $t_i \in \mathcal{A}$ as $t_0 t_1 t_2 t_3$, we have, for instance:

BALD $\prec$ BAR$\_$ $\prec$ BASH $\prec$ MAN$\_$ $\prec$ MANE $\prec$ MAT$\_$ $\prec$ SO$\_\_$ $\prec$ SORE

The minimal element of $\mathcal{A}^4$ is $\_\_\_\_$, the maximal element is ZZZZ.

Definition 32 *If $(f_x)_a$ is a family of sets, where each set f_x bears a binary relation r_x, then the Cartesian product $\prod_{x \in a} f_x$ bears the product relation $R = \prod_{x \in a} r_x$ which is defined "coordinatewise", i.e.,*

$$(t_x)_a R (s_x)_a \text{ iff } t_x \, r_x \, s_x \text{ for each } x \in a.$$

Attention! If each binary relation on the set f_x is a linear ordering, the *product relation* is not, in general, a linear ordering, so the lexicographic ordering is a remarkable construction since it "preserves" linear orderings.

Until now, we only considered binary relations. By use of Cartesian products of families of sets, one can now introduce the concept of an n-ary relation for $n \in \mathbb{N}$ as follows:

Definition 33 *If n is a natural number and a is a set, an n-ary relation on a is a subset R of the n-fold Cartesian product a^n, the binary relation being the special case of $n = 2$.*

Not every binary relation is an equivalence relation, but very often, one is interested in a kind of minimal equivalence relation which contains a given relation. Here is the precise setup:

Lemma 63 *If $(r_x)_a$ is a non-empty family of equivalence relations on a set b, then the intersection $\bigcap_a r_x$ is an equivalence relation. It is the largest equivalence relation (for the subset inclusion relation), which is contained in all relations $r_x, x \in a$.*

Proof This is straightforward to check.   □

Proposition 64 *Given a relation R on a set a, the smallest equivalence relation $\sim$ containing R consists of all pairs (x, y) such that either $x = y$ or there exists a finite sequence $x = x_0, x_1, \ldots x_{n^+} = y$ with $x_i R x_{i^+}$ or $x_{i^+} R x_i$ for all $i = 0, 1, \ldots n$.*

Proof Clearly, the smallest equivalence relation must contain these pairs. But these pairs visibly define an equivalence relation, and we are done.   □

Definition 34 *The equivalence relation $\sim$ defined in proposition 64 is called the* equivalence relation generated by the relation R.

6.3 Universal Properties in Relational Database Theory

Relational database theory serves a very useful exemplification of the universal properties of Cartesian product constructions. More than that: It even requires a construction which is slightly more general than the Cartesian product: the fiber product. Let us first introduce it, before we discuss a concrete database process as implemented for the relational database management system language SQL (Structured Query Language). SQL is an ANSI (American National Standards Institute) standard computer language for accessing and manipulating database systems. SQL statements are used to retrieve and update data in a database (see [24] for a reference to SQL and [18] for relational database theory).

Definition 35 (Universal property of fiber products) *Given three sets a, b and c and two maps $f : a \to c$ and $g : b \to c$, then a couple of maps $s_a : d \to a$ and $s_b : d \to b$ is called a* fiber product *with respect to f and g iff $f \circ s_a = g \circ s_b$, and if for every couple of maps $u : x \to a$ and $v : x \to b$ such that $u \circ s_a = v \circ s_b$, there is exactly one map $l : x \to d$ such that $s_a \circ l = u$ and $s_b \circ l = v$. Compare the following commutative diagram for this situation:*

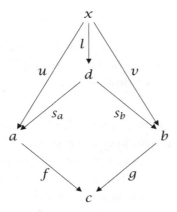

The existence and uniqueness of fiber products is easily shown, see the following proposition. But one special case is already at our hands: Suppose that $c = \{\varnothing\} = 1$. Then, by its universal property as a final set, there is always exactly one couple of maps $! : a \to 1$ and $! : b \to 1$. Further, the commutativity conditions $f \circ s_a = g \circ s_b$, $s_a \circ l = u$ and $s_b \circ l = v$ are also automatically fulfilled. So, if we set $d = a \times b$, this is a fiber product in this case. In other words, the Cartesian product is the special case, where $c = 1$ is the final set!

Proposition 65 *Given three sets a, b and c and two maps $f : a \to c$ and $g : b \to c$, there exists a fiber product $s_a : d \to a$ and $s_b : d \to b$ with respect to f and g, and for any fiber product $s'_a : d' \to a$ and $s'_b : d' \to b$ with respect to f and g, there is a unique bijection $t : d \overset{\sim}{\to} d'$ such that $s'_a \circ t = s_a$ and $s'_b \circ t = s_b$. The fiber product is denoted by $d = a \times_c b$, the two maps f and g being implicitly given.*

More explicitly, such a fiber product is constructed as follows: Take the Cartesian product $a \times b$, together with its projections pr_a and pr_b. Then consider the subspace $a \times_c b \subset a \times b$ consisting of all couples (x, y) such that $f(x) = g(y)$. On this set, take the restriction of the projections, i.e., $s_a = pr_a|_{a \times_c b}$ and $s_b = pr_b|_{a \times_c b}$.

Exercise 28 The easy proof is left to the reader.

Exercise 29 Given two subsets $a \subset c$ and $b \subset c$, show that $a \times_c b = a \cap b$.

Exercise 30 Given a map $f : a \to c$ and a subset $g : b \subset c$, prove that the fiber product of these two maps is the fiber $s_a : f^{-1}(c) \subset a$, with the second map $s_b = f|_{f^{-1}(c)}$.

With this small extension to theory, the relational database structure is easily described. We make an illustrative example and interpret its mechanisms in terms of the fiber product and other set-theoretical constructions.

To begin with, one is given domains from where values can be taken. However, these domains also have a name, not only values. We therefore consider sets with specific names X, but also isomorphisms $X \overset{\sim}{\to} V_X$ with given sets of values. Then, whenever we need to take elements $x \in X$, their values will be taken to lie in V_X, so that we may distinguish the elements, but nevertheless compare their values. In our example, we consider these sets of values: BIGINT $= \{1, 2, 3, 4, 5 \ldots\}$, this is a finite set of numbers (we assume that these are given, we shall discuss the precise definition of numbers later), whose size is defined by the standard implementation of numbers on a given operating system. Next, we are given a set

$$\text{TEXT} = \{\text{Apples, Cookies, Oranges, Donald Duck,}$$
$$\text{Mickey Mouse, Goofy, Bunny, Shrek, \dots}\}$$

of words, which also depends on the computer memory (again, words in a formal sense will be defined later). We now need the sets:

$$\text{ORDER\_ID} \overset{\sim}{\to} \text{BIGINT}$$
$$\text{CUSTOMER\_ID} \overset{\sim}{\to} \text{BIGINT}$$
$$\text{PRODUCT} \overset{\sim}{\to} \text{TEXT}$$
$$\text{NAME} \overset{\sim}{\to} \text{TEXT}$$
$$\text{ADDRESS} \overset{\sim}{\to} \text{TEXT}.$$

We consider two subsets ORDERS and CUSTOMERS, called *relations* in database theory,

$$\text{ORDERS} \subset \text{ORDER\_ID} \times \text{PRODUCT} \times \text{CUSTOMER\_ID}$$
$$\text{CUSTOMERS} \subset \text{CUSTOMER\_ID} \times \text{NAME} \times \text{ADDRESS}$$

which we specify as follows, to be concrete. The set ORDERS:

ORDER_ID	PRODUCT	CUSTOMER_ID
7	Apples	3
8	Apples	4
11	Oranges	3
13	Cookies	3
77	Oranges	7

and the set CUSTOMERS:

CUSTOMER_ID	NAME	ADDRESS
3	Donald Duck	Pond Ave.
4	Mickey Mouse	Cheeseway
5	Goofy	Dog Street
6	Bunny	Carrot Lane
7	Shrek	Swamp Alley

Each table has the single records as its rows. A first operation on such tables is their so-called *join*. This is not a mathematically correct terminology, but it is common in database theory. Mathematically speaking, it is a fiber product, which works as follows: Observe that the two relations ORDERS and CUSTOMERS are subsets of Cartesian products where factor spaces with common values appear, for example CUSTOMER_ID of ORDERS and CUSTOMER_ID of CUSTOMERS. In the join, we want to look for records having the same value for their CUSTOMER_ID coordinate. We therefore consider the composed maps

$$F : \text{ORDER\_ID} \times \text{PRODUCT} \times \text{CUSTOMER\_ID} \twoheadrightarrow \text{CUSTOMER\_ID} \overset{\sim}{\to} \text{BIGINT}$$

and

$$G : \text{CUSTOMER\_ID} \times \text{NAME} \times \text{ADDRESS} \twoheadrightarrow \text{CUSTOMER\_ID} \overset{\sim}{\to} \text{BIGINT}$$

derived from the first projections and the identification bijections for the values. Their restrictions $f = F|_{\text{ORDERS}}$ and $g = G|_{\text{CUSTOMERS}}$ yield the situation needed for a fiber product, and the join is exactly this construction:

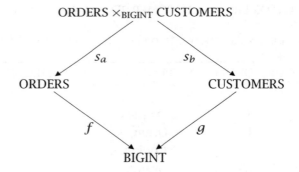

In the SQL syntax, the join is defined by the command

```
ORDERS JOIN CUSTOMERS
ON (ORDERS.CUSTOMER_ID = CUSTOMERS.CUSTOMER_ID).
```

The dot notation ORDERS.CUSTOMER_ID means the choice of the coordinate CUSTOMER_ID, i.e., this is the definition of the arrow f, whereas CUSTOMERS.CUSTOMER_ID defines g in the fiber product, and the equality sign "=" means that these two arrows f and g are taken for the fiber product construction.

We therefore obtain the following fiber product table:

ORDER_ID	PRODUCT	CUSTOMER_ID	NAME	ADDRESS
7	Apples	3	Donald Duck	Pond Ave.
11	Oranges	3	Donald Duck	Pond Ave.
13	Cookies	3	Donald Duck	Pond Ave.
8	Apples	4	Mickey Mouse	Cheeseway
77	Oranges	7	Shrek	Swamp Alley

For the next operation on this join we consider subsets thereof which are defined by the additional code WHERE... as in the following example:

```
ORDERS JOIN CUSTOMERS
ON (ORDERS.CUSTOMER_ID = CUSTOMERS.CUSTOMER_ID)
WHERE
        ORDERS.PRODUCT = 'Apples'
OR
        ORDERS.PRODUCT = 'Oranges'
```

This means that one chooses all elements in the join such that their projection to the coordinate ORDERS is either "Apples" or "Oranges". Mathematically, this again implies fiber products: We first consider the fiber of the projection to the factor PRODUCT:

$$p_{\text{PRODUCT}} : \text{ORDERS} \times_{\text{TEXT}} \text{CUSTOMERS} \to \text{PRODUCT} \xrightarrow{\sim} \text{TEXT}$$

including the identification with the value set TEXT. Then we consider the fiber of the singleton {"Apples"}:

$$
\begin{array}{ccc}
p_{\text{PRODUCT}}^{-1}(\text{"Apples"}) & \xrightarrow{\text{inclusion}} & \text{ORDERS} \times_{\text{TEXT}} \text{CUSTOMERS} \\
\downarrow & & \downarrow p_{\text{PRODUCT}} \\
\{\text{"Apples"}\} & \xrightarrow{\text{inclusion}} & \text{TEXT}
\end{array}
$$

This gives us all elements of the join which have the PRODUCT coordinate "Apples". The same is done with the "Oranges" coordinate:

$$
\begin{array}{ccc}
p_{\text{PRODUCT}}^{-1}(\text{"Oranges"}) & \xrightarrow{\text{inclusion}} & \text{ORDERS} \times_{\text{TEXT}} \text{CUSTOMERS} \\
\downarrow & & \downarrow p_{\text{PRODUCT}} \\
\{\text{"Oranges"}\} & \xrightarrow{\text{inclusion}} & \text{TEXT}
\end{array}
$$

So we have obtained two sets in the join set which we now may combine with the Boolean operation "OR", which amounts to taking the union $p_{\text{PRODUCT}}^{-1}(\text{"Apples"}) \cup p_{\text{PRODUCT}}^{-1}(\text{"Oranges"})$ of these two fibers.

Concluding this example, one then chooses a number of coordinates and omits the others in the union set by the prepended SELECT command

```
SELECT PRODUCT, NAME
FROM
ORDERS JOIN CUSTOMERS
ON (ORDERS.CUSTOMER_ID = CUSTOMERS.CUSTOMER_ID)
WHERE
        ORDERS.PRODUCT = 'Apples'
OR
        ORDERS.PRODUCT = 'Oranges'
```

Mathematically, we take the image

$$p_{\text{NAME,PRODUCT}}(p_{\text{PRODUCT}}^{-1}(\text{"Apples"}) \cup p_{\text{PRODUCT}}^{-1}(\text{"Oranges"}))$$

of the union under the projection

$$p_{\text{NAME,PRODUCT}} : \text{ORDERS} \times_{\text{TEXT}} \text{CUSTOMERS} \to \text{NAME} \times \text{PRODUCT}$$

which gives us this list:

PRODUCT	NAME
Apples	Donald Duck
Oranges	Donald Duck
Apples	Mickey Mouse
Oranges	Shrek

CHAPTER 7

Natural Arithmetic

This chapter is central insofar as the basic arithmetic operations, i.e., addition, multiplication, and exponentiation on natural numbers are introduced, operations which are the seed of the entire mathematical calculations.

7.1 Natural Operations

All these operations are defined by recursion, i.e., using the recursion theorem 55.

Definition 36 *Given a natural number a, addition to a is recursively defined as the function[1] $a + ? : \mathbb{N} \to \mathbb{N}$ which evaluates to*

$$a + 0 = a \quad and \quad a + b^+ = (a + b)^+.$$

Supposing that addition is defined, multiplication with a *is defined as the function $a \cdot ? : \mathbb{N} \to \mathbb{N}$ which evaluates to*

$$a \cdot 0 = 0 \quad and \quad a \cdot (b^+) = (a \cdot b) + a.$$

Supposing that addition and multiplication are defined, exponentiation of $a \neq 0$ is defined as the function $a^? : \mathbb{N} \to \mathbb{N}$ which evaluates to

$$a^0 = 1 \quad and \quad a^{(b^+)} = (a^b) \cdot a.$$

If $a = 0$, we define $0^0 = 1$ and $0^b = 0$ for $b \neq 0$.

---

[1] When defining function symbols, the question mark is used to indicate the position of the arguments.

Evidently, $a^+ = a + 1$, and from now on we identify these two expressions. The number $a + b$ is called the sum of a and b. The number $a \cdot b$ is called the product of a and b. These operations share the following properties. All these properties can be demonstrated by induction.

Sorite 66 *Let a, b, c be natural numbers. We have these laws:*

  (i) *(Additive neutral element)* $a + 0 = 0 + a = a$,

  (ii) *(Additive associativity)* $a + (b + c) = (a + b) + c$, *which is therefore written as* $a + b + c$,

  (iii) *(Additive commutativity)* $a + b = b + a$,

  (iv) *(Multiplicative neutral element)* $a \cdot 1 = 1 \cdot a = a$,

  (v) *(Multiplicative associativity)* $a \cdot (b \cdot c) = (a \cdot b) \cdot c$, *which is therefore written as* $a \cdot b \cdot c$,

  (vi) *(Multiplicative commutativity)* $a \cdot b = b \cdot a$,

  (vii) *(Multiplication distributivity)* $a \cdot (b + c) = a \cdot b + a \cdot c$,

  (viii) *(Exponential neutral element)* $a^1 = a$,

  (ix) *(Exponentiation $(+)$-distributivity)* $a^{b+c} = a^b \cdot a^c$,

  (x) *(Exponentiation $(\cdot)$-distributivity)* $(a \cdot b)^c = a^c \cdot b^c$,

  (xi) *(Additive monotony)* *if* $a < b$, *then* $a + c < b + c$,

  (xii) *(Multiplicative monotony)* *if* $c \neq 0$ *and* $a < b$, *then* $a \cdot c < b \cdot c$,

  (xiii) *(Exponential base monotony)* *if* $c \neq 0$ *and* $a < b$, *then* $a^c < b^c$,

  (xiv) *(Exponential exponent monotony)* *if* $c \neq 0, 1$ *and* $a < b$, *then* $c^a < c^b$,

  (xv) *(Ordering of operations)* *if* $a, b > 1$, *then* $a + b \leq a \cdot b \leq a^b$.

Proof (i) We have $a + 0 = a$ by definition, while $0 + (a^+) = (0 + a)^+ = a^+$ by recursion on a.

(ii) This is true for $c = 0$ by (i). By recursion on c, we have $a + (b + c^+) = a + ((b + c)^+) = (a + (b + c))^+ = ((a + b) + c)^+ = (a + b) + c^+$.

(iii)This is true for $b = 0$ by (i). We also have $a + 1 = 1 + a$, in fact, this is true for $a = 0$, and by recursion, $(a^+)^+ = (a + 1)^+ = (a + (1^+)) = a + (1 + 1) = (a + 1) + 1 = (a^+) + 1$. By recursion on b, we have $a + b^+ = (a + b)^+ = (b + a)^+ = b + (a^+) = b + (a + 1) = b + (1 + a) = (b + 1) + a = b^+ + a$.

(iv) We have $a \cdot 1 = (a \cdot 0) + a = 0 + a = a$. Therefore $1 \cdot 0 = 0 = 0 \cdot 1$, while $1 \cdot a^+ = (1 \cdot a) + 1 = a + 1 = a^+$.

(vii) We have $a \cdot (b + c) = a \cdot b + a \cdot c$ for $c = 0$. By recursion on c we have $a \cdot (b + c^+) = a \cdot ((b + c)^+) = a \cdot (b + c) + a = a \cdot b + a \cdot c + a = a \cdot b + a \cdot c^+$.

(v) For $c = 0, 1$ we have associativity by the previous results. By recursion on c we have $a \cdot (b \cdot c^+) = a \cdot (b \cdot c + b) = a \cdot (b \cdot c) + a \cdot b = (a \cdot b) \cdot c + (a \cdot b) \cdot 1 = (a \cdot b) \cdot c^+$.

(vi) Commutativity is known for $b = 1$. By recursion on b we have $a \cdot b^+ = a \cdot (b + 1) = a \cdot b + a = b \cdot a + a = 1 \cdot a + b \cdot a = (1 + b) \cdot a = b^+ \cdot a$.

(viii) We have $a^1 = a^0 \cdot a = 1 \cdot a = a$.

(ix) We have $a^{(b+0)} = a^b = a^b \cdot a^0$, and $a^{(b+c^+)} = a^{((b+c)^+)} = a^{(b+c)} \cdot a = a^b \cdot a^c \cdot a = a^b \cdot (a^c \cdot a) = a^b \cdot a^{(c^+)}$.

(x) We have $(a \cdot b)^0 = 1 = 1 \cdot 1 = a^0 \cdot b^0$, and $(a \cdot b)^{(c^+)} = (a \cdot b)^c \cdot (a \cdot b) = a^c \cdot b^c \cdot a \cdot b = (a^c \cdot a \cdot b^c \cdot b = a^{(c^+)} \cdot b^{(c^+)}$.

(xi) If $a < b$, then $a + 0 < b + 0$, and $a + c^+ = (a + c)^+ < (a + b)^+ = a + b^+$.

(xii) If $a < b$, then $a \cdot 1 < b \cdot 1$, and $a \cdot c^+ = a \cdot c + a < b \cdot c + a < b \cdot c + b = b \cdot c^+$.

(xiii) For $a = 0$ or $c = 1$ it is clear, suppose $a \neq 0$ and do recursion on c. Then $a^{(c^+)} = a^c \cdot a < b^c \cdot a < b^c \cdot b = b^{(c^+)}$.

(xiv) For $b = a^+$ it is clear, so suppose $b = d^+, a < d$. Then by recursion on b, $c^a < c^d$ and therefore $c^d < c^d \cdot c = c^b$.

(xv) To begin with denote $1^+ = 2$ (attention: we still do not know the usual notations for natural numbers!) and take $b = 2$. Then $a + 2 \leq a \cdot 2 \leq a^2$ is easily proved by induction on a, starting with the famous equality $2 + 2 = 2 \cdot 2 = 2^2$. We then prove the inequalities by induction on b, the details being left to be completed by the reader. □

Proposition 67 *If a and b are natural numbers such that $a \leq b$, then there is exactly one natural number c such that $a + c = b$.*

Proof We use induction on b. If $b = a$, then $c = 0$ solves the problem. If $b > a$, then $b = d + 1, d \leq a$. Then if $a + e = d$, set $c = e^+$, and we have by definition of addition $a + c = (a + e)^+ = d + 1 = b$. If c and c' are two different solutions, we must have $c < c'$, for example. Then monotony implies $a + c < a + c'$, i.e., $b < b$, an absurdity. □

If n is a natural number different from 0, we may look for the unique m, such that $m + 1 = n$. Clearly, it is the m such that $n = m^+$. This is the *predecessor of n*, which we denote by $n - 1$, a notation which will become clear later, when subtraction has been defined.

7.2 Euclid and the Normal Forms

The following theorem is Euclid's so-called "division theorem". It is a central tool for the common representation of natural, and also rational and real numbers in the well-known decimal format. Moreover, it is the central step in the calculation of the greatest common divisor of two natural numbers by the Euclidean algorithm[2], see chapter 16.

Proposition 68 (Division Theorem) *If a and b are natural numbers with $b \neq 0$, then there is a unique pair r and s of natural numbers with $s < b$ such that*

$$a = r \cdot b + s.$$

Proof Existence: Let t be the minimal natural number such that $a < t \cdot b$. For example, according to sorite 66, $a < a \cdot b$, so t exists and evidently is non-zero, $t = r^+$. This means that $r \cdot b \leq a$. So by proposition 67, there is s such that $a = r \cdot b + s$. If $b \leq s$ we have $s = b + q$, and by the choice of t, $a = r \cdot b + b + q = t \cdot b + q > a$, a contradiction. So the existence is proved.

Iff we have two representations $a = r \cdot b + s = r' \cdot b + s'$ with $r' \geq r^+$, then we have $a = r' \cdot b + s' \geq r \cdot b + b + s' > r \cdot b + s = a$, a contradiction. So $r = r'$, and equality of s and s' follows from proposition 67.   □

Proposition 69 *If a and b are natural numbers with $a \neq 0$ and $b \neq 0, 1$, then there is a unique triple c, s, r of natural numbers with $r < b^c$ and $s < b$ such that*

$$a = s \cdot b^c + r.$$

Proof Let t be the minimal natural number such that $a < b^t$, and clearly $t = w + 1$. Such a t exists since $a < b + a \leq b \cdot a \leq b^a$ by sorite 66. Therefore $a \geq b^w$. We now apply proposition 68 and have $a = r \cdot b^w + s, s < b^w$. Now, if $r = b + p$, then we also have $a = (b + p) \cdot b^w + s = b^t + p \cdot b^w + s$, a contradiction to the choice of t. So we have one such desired representation. Uniqueness follows by the usual contradiction from different s coefficients, and then form different r's for equal s coefficients.   □

In order to define the b-adic representation of a natural number, we need to define what is the sum of a finite sequence $(a_i)_{i<n}$ of length n of natural numbers a_i.

Definition 37 *Given a finite sequence $(a_i)_{i \leq n}$ of natural numbers, its sum is denoted by $\sum_{i \leq n} a_i$, by $a_0 + a_1 + \ldots a_n$, or by $\sum_{i=0,1,\ldots n} a_i$, and is defined by recursion on the sequence length as follows:*

---

[2] An *algorithm* is a detailed sequence of actions (steps), starting from a given input, to perform in order to accomplish some task, the output. It is named after Al-Khawarizmi, a Persian mathematician who wrote a book on arithmetic rules about A.D. 825.

$$n = 0 : \sum_{i \leq 0} a_i = a_0$$

$$n > 0 : \sum_{i \leq n} a_i = \left(\sum_{i \leq n-1} a_i \right) + a_n$$

Because of the associative law of addition, it is in fact not relevant how we group the sum from a_0 to a_n.

Proposition 70 (Adic Normal Form) *If a and b are non-zero natural numbers and $b \neq 1$, then there is a uniquely determined finite number n and a sequence $(s_i)_{i=0,...n}$, with $s_n \neq 0$ and $s_i < b$ for all i, such that*

$$a = \sum_{i=0,...n} s_i \cdot b^i. \tag{7.1}$$

Proof This immediately results from proposition 69 and by induction on the (unique!) remainder r in that proposition. $\square$

Definition 38 *Given non-zero natural numbers a and b, and $b \neq 1$ as in proposition 70, the number b which entails the representation (7.1), is called the base of the adic representation, and the representation is called the b-adic representation of a. It is denoted by*

$$a =_b s_n s_{n-1} \ldots s_1 s_0 \tag{7.2}$$

or, if the base is clear, by

$$a = s_n s_{n-1} \ldots s_1 s_0.$$

Remark 9 In computer science, the term *-adic* is usually replaced by the term *-ary*.

Example 18 For the basis $b = 2$, we have the 2-adic representation, which is also known as the *dual* or *binary* representation. Here, the representation $a =_b s_n s_{n-1} \ldots s_1 s_0$ from formula (7.2) reduces to a sequence of 1s and 0s.

With the well-known notations $3 = 2 + 1$, $4 = 3 + 1$, $5 = 4 + 1$, $6 = 5 + 1$, $7 = 6+1$, $8 = 7+1$, $9 = 8+1$, $Z = 9+1$, we have the *decadic* representation

$$a =_Z s_n s_{n-1} \ldots s_1 s_0 \qquad 0 \leq s_i \leq 9$$

with special cases $Z = 10$, $Z^2 = 100$, $Z^3 = 1000$, and so on.

For the *hexadecimal* base $H =_Z 16$, one introduces the symbols $1, 2, \ldots, 9$, $A =_Z 10$, $B =_Z 11$, $C =_Z 12$, $D =_Z 13$, $E =_Z 14$, $F =_Z 15$.

For example $x =_Z 41663$ becomes, in the hexadecimal base, $x =_H A2BF$, and, in the binary representation, $x =_2 1010001010111111$.

CHAPTER 8

Infinities

We already know that the powerset 2^a of any set a has larger cardinality than a itself, i.e., there is an injection $a \to 2^a$ but no injection in the other direction. This does not mean that constructions of larger sets from given ones always lead to strictly larger cardinalities.

8.1 The Diagonalization Procedure

We first have to reconsider the universal constructions of the Cartesian product and coproduct. Given a set a and a non-zero natural number n, we have the n-th power a^n, but we could also define recursively $a^{\times 1} = a$ and $a^{\times n+1} = a^{\times n} \times a$, and it is clear that $a^{\times n} \xrightarrow{\sim} a^n$. Dually, we define $a^{\sqcup 1} = a$ and $a^{\sqcup n+1} = a^{\sqcup n} \sqcup a$.

Proposition 71 *If a is a set that has the cardinality of $\mathbb{N}$ (in which case a is called* denumerable*), then for any positive natural number n, the sets a, $a^{\times n}$ and $a^{\sqcup n}$ have the same cardinality, i.e., are equipollent.*

Proof The proof of this proposition depends on the Bernstein-Schröder theorem 22, which we apply to the situation of pairs of injections $a \to a^{\times n} \to a$ and $a \to a^{\sqcup n} \to a$. Now, an injection $a \to a^{\sqcup n}$ is given by the known injection to the last cofactor defined in definition 28. An injection $a \to a^{\times n}$ is also given by the identity $a \to a$ on each factor. So we are left with the injections in the other direction. We may obviously suppose $n = 2$ and deduce from this the general case by induction on n. We also may suppose $a = \mathbb{N}$.

Now, a map $f : \mathbb{N} \sqcup \mathbb{N} \to \mathbb{N}$ is given as follows: for x in the left cofactor, we define $f(x) = x \cdot 2$, for x in the right cofactor, we set $f(x) = x \cdot 2 + 1$. By the uniqueness in proposition 68, this is an injection. To obtain an injection $g : \mathbb{N} \times \mathbb{N} \to \mathbb{N}$, we consider any pair $(x, y) \in \mathbb{N}^2$. We may then associate each pair (x, y) with the pair $(x, x + y) = (x, n) \in \mathbb{N}^2$ with $0 \leq x \leq n$, $n \in \mathbb{N}$. Call $\mathbb{N}^{<2}$ the set of these pairs. Then we have a bijection $u : \mathbb{N}^2 \xrightarrow{\sim} \mathbb{N}^{<2}$. Consider the function $f : \mathbb{N}^{<2} \to \mathbb{N}$ defined by $f(x, n) = 2^n + x$. We have $f(0, 0) = 1$, and for $0 < n, x \leq n < 2 \cdot n \leq 2^n$, so the uniqueness part of proposition 69 applies, and we have an injection (see figure 8.1).   □

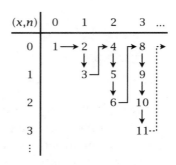

Fig. 8.1. The entries in the table are the values of the injection f from $\mathbb{N}^{<2}$ to $\mathbb{N}$. The arrows indicate the order on the natural numbers.

Remark 10 The proof of proposition 71 uses the so-called "diagonal procedure", which is a central tool in aligning n-tuples in a linear ordering. This procedure also works for non-denumerable infinite sets, i.e., the proposition is also true for any infinite sets, but this is not relevant for general computer science.

Remark 11 The equivalence of the axiom of choice and the proposition that every set can be well-ordered has a special meaning for finite or denumerable sets: If a set is denumerable, then it can be well-ordered by the ordering among natural numbers, and therefore, the axiom of choice easily follows from this well-ordering for denumerable sets. In other words, for denumerable sets, the axiom of choice is superfluous by the very definition of denumerability.

CHAPTER 9

The Classical Number Domains ℤ, ℚ, ℝ, and ℂ

This chapter is the recompensation for the abstract initialization of mathematics: it will give us all the central number domains as they are constructed from the natural numbers. In general, there are different construction methods for the same number domains. We have decided to present the fastest construction methods which also lead to effective representations of numbers in computer environments.

The construction first introduces the domain ℤ of integer numbers from natural numbers ℕ, it then yields rational numbers or fractions ℚ from integers, and real numbers (also called decimal numbers in a special representation) from rational numbers. The construction terminates by the building of complex numbers from real numbers. These examples are also very important for the understanding of subsequent chapters on algebra.

Basically, the construction of new number domains is motivated by the absence of many operations one would like to apply to numbers. For example, it is in general not possible to solve the simple equation $a + x = b$ or the multiplicative analog $a \cdot x = b$ on the domain of natural numbers. Moreover, we do not have any possibility to model most common geometric objects such as lines! Finally, non-linear equations such as $a^2 = b$ cannot be solved.

Mathematics must supply tools that help us understand the solution spaces for such problems. The set-theoretic approach has provided us with the construction of numbers, and we are now ready to find solutions to the aforementioned problems.

9.1 Integers $\mathbb{Z}$

We have seen that the equation $a + x = b$ for natural numbers has exactly one solution in the case $a \leq b$, call it $b - a$. But for $a > b$, such a solution is does not exist within the domain of natural numbers. Integer numbers, however, can solve this problem.

Lemma 72 *Consider the relation* $(a, b)R(p, q)$ *iff* $a + q = b + p$ *among pairs* $(a, b), (p, q) \in \mathbb{N}^2$.

(i) *The relation* R *is an equivalence relation.*

(ii) *A pair* $(a, b) \in \mathbb{N}^2$ *with* $a \geq b$ *is equivalent to* $(a - b, 0)$, *whereas a pair* $(a, b) \in \mathbb{N}^2$ *with* $a < b$ *is equivalent to* $(0, b - a)$.

(iii) *Two pairs* $(x, 0), (y, 0)$, *or* $(0, x), (0, y)$, *respectively, are equivalent iff* $x = y$. *Two pairs* $(x, 0), (0, y)$ *are equivalent iff* $x = y = 0$.

Proof The relation $(a, b)R(p, q)$ is reflexive since $(a, b)R(a, b)$ iff $a + b = b + a$ by the commutativity of addition. It is symmetric for the same reason. And if $(a, b)R(p, q)$ and $(p, q)R(r, s)$, then $a + q = b + p$ and $p + s = q + r$, whence $a + q + p + s = b + p + q + r$, but then we may cancel $p + q$ and obtain the desired equality $a + s = b + r$. The other claims follow immediately from equivalence of ordered pairs. $\square$

Definition 39 *The set* $\mathbb{N}^2/R$ *is denoted by* $\mathbb{Z}$, *its elements* $[a, b]$ *are called integers. Each integer is uniquely represented by either* $[a, 0]$ *for* $a \in \mathbb{N}$, *or by* $[0, b]$, $b \in \mathbb{N} - \{0\}$. *We identify the former class* $[a, 0]$ *with its unique associated natural number* a, *and the latter class* $[0, b]$ *with the uniquely determined natural number* b, *together with a minus sign, i.e., by* $-b$. *The numbers* $[a, 0]$ *with* $a \neq 0$ *are called the* positive *integers, while the numbers* $-b$, $b \neq 0$ *are called the* negative *integers. The meeting point between these number types is the* $0 = -0$ *(zero) number which is neither positive nor negative.*

The linear ordering on the natural numbers is extended to the integers by the rule that $[a, 0] < [c, 0]$ *iff* $a < c$, *and* $[0, a] < [0, c]$, *i.e.,* $-a < -c$ *iff* $a > c$. *Further, we set* $[a, 0] > [0, b]$, *i.e.,* $a > -b$ *for all natural* $b \neq 0$ *and natural* a. *We finally set* $|[a, 0]| = a$ *and* $|[0, b]| = b$ *and call this the* absolute value *of the integer.*

If $x = [a, b]$ is an integer, we further denote by $-x$ the integer $[b, a]$ and call it the additive inverse. *This definition evidently generalizes the convention $-c = [0, c]$. We also write $a - b$ for $a + (-b)$.*

This means that we have "embedded" the natural number set $\mathbb{N}$ in the larger set $\mathbb{Z}$ of integers as the non-negative integers, and that the ordering among natural numbers has been extended to a linear ordering on $\mathbb{Z}$. Observe however that the linear ordering on $\mathbb{Z}$ is not a well-ordering since there is no smallest integer!

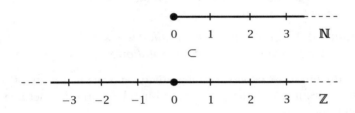

Fig. 9.1. The common representation of integers shows them as equidistant points on a straight line, with increasing values from left to right.

To calculate the "size" of $\mathbb{Z}$, observe that by lemma 72, $\mathbb{Z} = (\mathbb{N} - \{0\}) \sqcup -\mathbb{N}$. But we have two bijections $p : \mathbb{N} \overset{\sim}{\to} \mathbb{N} - \{0\} : n \mapsto n + 1$ and $q : \mathbb{N} \overset{\sim}{\to} -\mathbb{N} : n \mapsto -n$. Therefore $\mathbb{Z} \overset{\sim}{\to} \mathbb{N} \sqcup \mathbb{N}$, and $\mathbb{N} \overset{\sim}{\to} \mathbb{N} \sqcup \mathbb{N}$ by proposition 71, so $\mathbb{Z} \overset{\sim}{\to} \mathbb{N}$, i.e., $\mathbb{Z}$ and $\mathbb{N}$ have the same cardinality.

Next, we will create an arithmetic on $\mathbb{Z}$ which extends the arithmetic on the natural numbers. The following technique for defining addition of integers is a prototype of the definition of functions on equivalence classes: One defines these functions on elements (representatives) of such equivalence classes and then shows that the result is in fact not a function of the representative, but only of the class as a such. If a definition on representatives works in this sense, one says that the function is *well defined.*

Definition 40 *Given two integers $[a, b]$ and $[c, d]$, their sum is defined by $[a, b] + [c, d] = [a + c, b + d]$, i.e., "factor-wise". This function is well defined.*

Sorite 73 *Let $\mathbb{Z}$ be provided with the addition $+ : \mathbb{Z} \times \mathbb{Z} \to \mathbb{Z}$, and let a, b, c be any integers. Then we have these properties.*

(i) *(Associativity)* $(a + b) + c = a + (b + c) = a + b + c$,

(ii) *(Commutativity)* $a + b = b + a$,

(iii) *(Additive neutral element)* $a + 0 = a$,

(iv) *(Additive inverse element)* $a - a = 0$,

(v) *(Extension of natural arithmetic)* *If $a, b \in \mathbb{N}$, then $[a + b, 0] = [a, 0] + [b, 0]$, i.e., it amounts to the same if we add two natural numbers a and b or the corresponding non-negative integers, also denoted by a and b.*

(vi) *(Solution of equations)* *The equation $a + x = b$ in the "unknown" x has exactly one integer number solution x, i.e., $x = b - a$.*

Proof (i), (ii), (iii) Associativity, commutativity, and the neutrality of 0 immediately follows from associativity for natural numbers and the factorwise definition of addition.

(iv) For $a = [u, v]$, we have $a - a = a + (-a) = [u, v] + [v, u] = [u + v, u + v] = 0$. The rest is immediate from the definitions. $\qquad\square$

Definition 41 *Let $(a_i)_{i=0,\ldots n}$ be a sequence of integers. Then the sum of this sequence is defined by*

$$n = 0 \; : \; \sum_{i=0,\ldots n} a_i = a_0,$$

$$n > 0 \; : \; \sum_{i=0,\ldots n} a_i = \left(\sum_{i=0,\ldots n-1} a_i \right) + a_n.$$

It is also possible to extend the multiplication operation defined on $\mathbb{N}$ to the integers. The definition is again one by representatives of equivalence classes $[a, b]$. To understand the definition, we first observe that a class $[a, b]$ is equal to the difference $a - b$ of natural numbers with the above identification. In fact, $[a, b] = [a, 0] + [0, b] = a + (-b) = a - b$. So, if we want to extend the arithmetic on the naturals, we should try to observe the hoped for and given rules, and thereby get the extension. So we should have $[a, b] \cdot [c, d] = (a - b) \cdot (c - d) = ac + bd - ad - bc = [ac + bd, ad + bc]$. This motivates the following definition:

Definition 42 *Given two integers $[a, b]$ and $[c, d]$, their product is defined by $[a, b] \cdot [c, d] = [ac + bd, ad + bc]$. This function is well defined.*

Sorite 74 *Let a, b, c be three integers. We have these rules for their multiplication.*

(i) *(Associativity)* $(a \cdot b) \cdot c = a \cdot (b \cdot c) = a \cdot b \cdot c$,

(ii) *(Commutativity)* $a \cdot b = b \cdot a$,

(iii) *(Multiplicative neutral element)* *the element* $1 = [1, 0]$ *is neutral for multiplication*, $a \cdot 1 = a$,

(iv) *(Zero and negative multiplication)* $a \cdot 0 = 0$ *and* $a \cdot (-b) = -(a \cdot b)$,

(v) *(Distributivity)* $a \cdot (b + c) = a \cdot b + a \cdot c$,

(vi) *(Integrity)* *If* $a, b \neq 0$, *then* $a \cdot b \neq 0$,

(vii) *(Additive monotony)* *if* $a < b$, *then* $a + c < b + c$,

(viii) *(Multiplicative monotony)* *if* $a < b$ *and* $0 < c$, *then* $a \cdot c < b \cdot c$,

(ix) *(Extension of natural arithmetic)* *For two natural numbers a and b, we have $[a \cdot b, 0] = [a, 0] \cdot [b, 0]$. This allows complete identification of naturals as a subdomain of the integers, if we look at addition and multiplication.*

Proof Statements (i) through (v) and (ix) are straightforward and yield good exercises for the reader.

(vi) If $a = [r, s]$ and $b = [u, v]$, then the hypothesis means $r \neq s$ and $u \neq v$. Suppose that $r > s$ and $u > v$. Then $a = [r - s, 0]$ and $b = [u - v, 0]$, with the notation of differences of natural numbers as defined at the beginning of this section. But then $a \cdot b = [(r - s) \cdot (u - v), 0] \neq [0, 0]$. The other cases $r < s, u < v$, or $r < s, u > v$, or $r > s, u < v$ are similar.

(vii) First suppose that $r \geq s$ and $u \geq v$, and let $a = [r, s] = [r - s, 0]$ and $b = [u, v] = [u - v, 0]$. Then $a < b$ means $e = r - s < f = u - v$. So for $c = [g, h]$, we have $a + c = [e + g, h]$ and $b + c = [f + g, h]$. We may suppose that either h or g is zero. If $h = 0$, then $e + g < f + g$ implies $a + c = [e + g, 0] < [f + g, 0] = b + c$. Else we have $a + c = [e, h]$ and $b + c = [f, h]$. Suppose that $h \leq e$. Then $e - h < f - h$, whence $a + c = [e, h] = [e - h, 0] < [f - h, 0] = b + c$. If $e < h \leq f$, then $a + c = [e, h] = [0, h - e] < [f - h, 0] = b + c$. If $e, f < h$, then $h - e > h - f$, and then $a + c = [e, h] = [0, h - e] < [0, h - f]$. The other cases $r < s, u < v$, or $r < s, u > v$, or $r > s, u < v$ are similar.

(viii) If $0 \leq a < b$, we are done since this is the case of natural numbers, which we already know from sorite 66. Else if $a < b < 0$, then we have $0 < (-b) < (-a)$, and then by the previous case, $(-b) \cdot c < (-a) \cdot c$, whence, $-(-a) \cdot c < -(-b) \cdot c$, but $-(-a) \cdot c = a \cdot c, -(-b) \cdot c = b \cdot c$, whence the statement in this case. Else if $a \leq 0 < b$, then $a \cdot c \leq 0 < b \cdot c$. $\qquad \square$

We are now in state of adding and subtracting any two integers, and to solve an equation $a + x = b$. But the multiplication is still not in such a perfect state. For example, $2 \cdot x = 3$ has no integer solution.

Definition 43 *If a and b are two integers, we say that a divides b iff there is an integer c with $a \cdot c = b$. We then write $a|b$.*

Exercise 31 For any integer b, we have $b|b$, $1|b$, and $-1|b$.

Definition 44 *If b is such that it is divided only by itself, 1 or -1, then we call b a* prime *integer.*

We shall deal with the prime numbers in chapter 16. For the moment, we have the following exercise concerning prime decomposition of integers. For this we also need the product of a finite sequence of integers, i.e.,

Definition 45 *Let $(a_i)_{i=0,\ldots n}$ be a sequence of integers. Then the product of this sequence is defined by*

$$n = 0 \ : \ \prod_{i=0,\ldots n} a_i = a_0,$$

$$n > 0 \ : \ \prod_{i=0,\ldots n} a_i = \left(\prod_{i=0,\ldots n-1} a_i \right) \cdot a_n.$$

Exercise 32 Show that every non-zero integer $a \neq \pm 1$ has a multiplicative decomposition $a = \epsilon \cdot \prod_i p_i$ where $\prod_i p_i$ is a product of positive primes p_i and $\epsilon = \pm 1$.

Notation 8 *We shall henceforth also write ab instead of $a \cdot b$ if no confusion is possible.*

Proposition 75 (Triangle Inequality) *If a and b are two integers, then we have the* triangle inequality*:*

$$|a + b| \leq |a| + |b|.$$

Exercise 33 Give a proof of proposition 75 by distinction of all possible cases for non-negative or negative a, b.

9.2 Rationals $\mathbb{Q}$

The construction of the rational numbers is very similar to the procedure we have used for the construction of the integers. The main difference is that the underlying building principle is multiplication instead of addition. We denote by $\mathbb{Z}^*$ the set $\mathbb{Z} - \{0\}$.

Lemma 76 *On the set $\mathbb{Z} \times \mathbb{Z}^*$, the relation $(a,b)R(c,d)$ iff $ad = bc$ is an equivalence relation.*

Proof This is an exercise for the reader.   □

Definition 46 *The set $(\mathbb{Z} \times \mathbb{Z}^*)/R$ of equivalence classes for the relation R defined in lemma 76 is denoted by $\mathbb{Q}$. Its elements, the classes $[a,b]$, are called* rational numbers *and are denoted by $\frac{a}{b}$ or a/b. The (non uniquely determined) number a is called the* numerator, *whereas the (non uniquely determined) number b is called the* denominator *of the rational number $\frac{a}{b}$. Numerator and denominator are only defined relative to a selected representative of the rational number.*

Before we develop the arithmetic operations on $\mathbb{Q}$, let us verify that again, the integers are embedded in the rationals. In fact, we may identify an integer a with its fractional representation $\frac{a}{1}$, and easily verify that $\frac{a}{1} = \frac{b}{1}$ iff $a = b$.

Here is the arithmetic construction:

Definition 47 *Let $\frac{a}{b}$ and $\frac{c}{d}$ be two rationals. Then we define*

$$\frac{a}{b} + \frac{c}{d} = \frac{ad+bc}{bd}$$
$$\frac{a}{b} \cdot \frac{c}{d} = \frac{ac}{bd}.$$

We further set $-\frac{a}{b} = \frac{-a}{b}$ for the additive inverse *of $\frac{a}{b}$. If $a \neq 0$, we set $\left(\frac{a}{b}\right)^{-1} = \frac{b}{a}$, the latter being the* multiplicative inverse *of $\frac{a}{b}$. These operations are all well defined.*

In order to manage comparison between rational numbers, we need the following exercise.

Exercise 34 If $\frac{a}{b}$ and $\frac{c}{d}$ are rational numbers, then one may always find numerators and denominators such that $b = d$ (common denominator) and $0 < b$.

Definition 48 *If $\frac{a}{b}$ and $\frac{c}{d}$ are rational numbers, we have the (well defined) relation $\frac{a}{b} < \frac{c}{d}$ iff $a < c$, where we suppose that we have a common positive denominator $0 < b = d$.*

Sorite 77 *Let $\frac{a}{b}, \frac{c}{d}, \frac{e}{f}$ be rational numbers. Then these rules hold.*

(i) *(Additive associativity)* $(\frac{a}{b} + \frac{c}{d}) + \frac{e}{f} = \frac{a}{b} + (\frac{c}{d} + \frac{e}{f}) = \frac{a}{b} + \frac{c}{d} + \frac{e}{f}$

(ii) *(Additive commutativity)* $\frac{a}{b} + \frac{c}{d} = \frac{c}{d} + \frac{a}{b}$

(iii) *(Additive neutral element)* $\frac{a}{b} + \frac{0}{1} = \frac{a}{b}$

(iv) *(Additive inverse element)* $\frac{a}{b} + \frac{-a}{b} = \frac{0}{1}$

(v) *(Multiplicative associativity)* $(\frac{a}{b} \cdot \frac{c}{d}) \cdot \frac{e}{f} = \frac{a}{b} \cdot (\frac{c}{d} \cdot \frac{e}{f}) = \frac{a}{b} \cdot \frac{c}{d} \cdot \frac{e}{f}$

(vi) *(Multiplicative commutativity)* $\frac{a}{b} \cdot \frac{c}{d} = \frac{c}{d} \cdot \frac{a}{b}$

(vii) *(Multiplicative neutral element)* $\frac{a}{b} \cdot \frac{1}{1} = \frac{a}{b}$

(viii) *(Multiplicative inverse element)* *If $b \neq 0$, then* $\frac{a}{b} \cdot \frac{b}{a} = \frac{1}{1}$

(ix) *(Distributivity)* $\frac{a}{b} \cdot (\frac{c}{d} + \frac{e}{f}) = \frac{a}{b} \cdot \frac{c}{d} + \frac{a}{b} \cdot \frac{e}{f}$

(x) *(Linear ordering)* *The relation $<$ among rational numbers is a linear ordering. Its restriction to the integers $\frac{a}{1}$ induces the given linear ordering among integers.*

(xi) *(Additive monotony)* *If $\frac{a}{b} < \frac{c}{d}$, then $\frac{a}{b} + \frac{e}{f} < \frac{c}{d} + \frac{e}{f}$.*

(xii) *(Multiplicative monotony)* *If $\frac{a}{b} < \frac{c}{d}$ and $\frac{0}{1} < \frac{e}{f}$, then $\frac{a}{b} \cdot \frac{e}{f} < \frac{c}{d} \cdot \frac{e}{f}$.*

(xiii) *(Archimedean ordering)* *For any two positive rational numbers $\frac{a}{b}$ and $\frac{c}{d}$ there is a natural number n such that $\frac{n}{1} \cdot \frac{a}{b} > \frac{c}{d}$.*

Proof Everything here is straightforward once one has shown that in fact $\frac{a}{b} < \frac{c}{d}$ is well defined. For then we may calculate everything on (common) positive denominators and thereby reduce the problem to the integers, where we have already established these properties. So let $\frac{a}{b} < \frac{c}{b}$, and b be positive. Suppose that $\frac{a}{b} = \frac{a'}{b'}$ and $\frac{c}{b} = \frac{c'}{b'}$. We know $a < c$. Then $ab' = ba'$ and $cb' = c'b$, and therefore $a'b = b'a < b'c = c'b$, whence $a' < c'$, since b is positive. $\qquad\square$

Given the complete compatibility of original natural numbers within integers, and the complete compatibility of integers within rational numbers, we also "abuse" the injections

$$\mathbb{N} \to \mathbb{Z} \to \mathbb{Q}$$

and treat them as inclusions, in the sense that

- a natural number is denoted by n rather than by the equivalence class $[n, 0]$ in $\mathbb{Z}$,

- an integer is denoted by z rather than by the equivalence class $\frac{z}{1}$ in $\mathbb{Q}$.

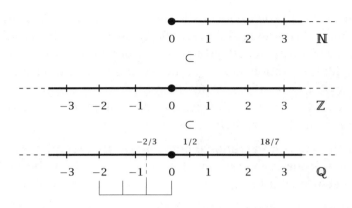

Fig. 9.2. The common representation of $\mathbb{N}$, $\mathbb{Z}$ and $\mathbb{Q}$ shows their sets as point-sets on a straight line, where the rational number $\frac{q}{p}$ is drawn as the equally divided line between the integer points 0 and q (supposing $p > 0$).

Definition 49 *If a rational number x is represented by $x = \frac{a}{b}$, then we define its absolute value $|x|$ by $|x| = \frac{|a|}{|b|}$. This is a well-defined definition.*

Exercise 35 Prove that $|x|$ is well defined. An alternative definition is this: $|x| = x$ if $0 \le x$, otherwise it is $|x| = -x$. Give a proof of this.

Proposition 78 (Triangle Inequality) *If a and b are rational numbers, then we have the triangle inequality:*

$$|a + b| \le |a| + |b|.$$

Exercise 36 Use proposition 75 to give a proof of proposition 78.

Since, in contrast to natural numbers and integers, there is a rational number between any two others, are there many more rationals than integers? The answer is negative. In fact, it can be shown that $\mathbb{Z}$ and $\mathbb{Q}$ have the same cardinality by a similar argument used to prove that $\mathbb{N}$ and $\mathbb{Z}$ have the same cardinality.

9.3 Real Numbers $\mathbb{R}$

We are now able to solve equations of the type $ax + b = c$ for any rational coefficients a, b, c provided that $a \neq 0$. We also have a linear ordering on $\mathbb{Q}$ and we are provided with arbitrary near numbers in the sense that for any two different rationals $x < y$, there is a rational number in-between, namely $x < \frac{1}{2}(x + y) < y$. But there are still many problems which cannot be solved in $\mathbb{Q}$. Intuitively, with rational numbers, there are now many points on the straight line, but there are still some gaps to be filled.

Some of the gaps on the rational number line are identified when we ask whether a so-called algebraic equation $x^2 = x \cdot x = 2$ can be solved in $\mathbb{Q}$. It will be shown later in chapter 16 that there is no such solution in $\mathbb{Q}$.

We recognize once again that in fact all our efforts in the construction of more numbers originate in the problem of solving special equations in a given number domain.

The real numbers are constructed from the rational numbers by an idea which results from the observation that the search for a solution of $x^2 = 2$ yields rational numbers y and z such that $y^2 < 2 < z^2$ and $|y - z| < 1/n$ for any non-zero natural number n, in other words, y and z are arbitrary near to each other. The point is that there is no rational number which gives us a precise solution, nonetheless, we would like to invent a number which solves the problem and in some sense is the "infinite approximation" suggested by the rational approximations.

In order to define the extension of rational numbers to real numbers, we need to formalize what we vaguely described by "infinite approximation". This definition goes back to the French mathematician Augustin Cauchy (1789–1857).

Definition 50 *A* Cauchy sequence *of rational numbers is a sequence* $(a_i)_{i \in \mathbb{N}}$ *of rational numbers* a_i *such that for any positive natural number* L, *there is an index number* N *such that whenever indexes* n, m *verify* $N < n, m$, *then* $|a_n - a_m| < \frac{1}{L}$.

The set of all Cauchy sequences is denoted by C.

Exercise 37 Every rational number r gives rise to the constant Cauchy sequence (r) which has $a_i = r$ for all indexes $0 \leq i$.

The sequence $(1/(i+1))_{i\in\mathbb{N}}$ is a Cauchy sequence. More generally, any sequence $(a + a_i)_i$ is a Cauchy sequence if a is a rational number and if $(a_i)_i$ is Cauchy.

A sequence of rational numbers $(a_i)_i$ is said to *converge* to a rational number a if for any positive natural number L, there is an index N such that $n > N$ implies $|a_n - a| < \frac{1}{L}$. Show that a convergent sequence is Cauchy, and that the rational number to which it converges is uniquely determined. We then write $\lim_{i\to\infty} a_i = a$. The sign ∞ means "infinity", but it has no precise meaning when used without a determined context, such as $\lim_{n\to\infty}$.

Definition 51 *A* zero sequence *is a Cauchy sequence* $(a_i)_i$ *which converges to* 0:

$$\lim_{i\to\infty} a_i = 0.$$

The zero sequences are those sequences which we would like to forget about. Let us make precise this option.

Definition 52 *If* $(a_i)_i$ *and* $(b_i)_i$ *are sequences of rational numbers, their sum is defined by*

$$(a_i)_i + (b_i)_i = (a_i + b_i)_i,$$

and their product is defined by

$$(a_i)_i \cdot (b_i)_i = (a_i \cdot b_i)_i.$$

Proposition 79 *The following properties hold for sequences of rational numbers:*

(i) *If* $(a_i)_i$ *and* $(b_i)_i$ *are Cauchy sequences, then so are their sum and product.*

(ii) *If* $(a_i)_i$ *and* $(b_i)_i$ *are zero sequences, then so is their sum.*

(iii) *If* $(a_i)_i$ *is a zero sequence and if* $(b_i)_i$ *is Cauchy, then their product is a zero sequence.*

Proof Let $(a_i)_i$ and $(b_i)_i$ be Cauchy sequences. Given any positive natural L, there is a common natural N such that $n, m > N$ implies $|a_n - a_m| < \frac{1}{L}$ and $|b_n - b_m| < \frac{1}{L}$ if $n, m > N$. Now, let N be such that $|a_n - a_m| < \frac{1}{2L}$ for $n, m > N$. Then

$$|(a_n + b_n) - (a_m + b_m)| = |(a_n - a_m) + (b_n - b_m)|$$
$$\leq |a_n - a_m| + |b_n - b_m|$$
$$< \frac{1}{2L} + \frac{1}{2L}$$
$$= \frac{1}{L}$$

by the triangle inequality. Further

$$|(a_n \cdot b_n) - (a_m \cdot b_m)| = |(a_n \cdot b_n) - (a_n \cdot b_m) + (a_n \cdot b_m) - (a_m \cdot b_m)|$$
$$= |(a_n \cdot (b_n - b_m) + (a_n - a_m) \cdot b_m)|$$
$$\leq |a_n \cdot (b_n - b_m)| + |(a_n - a_m) \cdot b_m|$$
$$= |a_n| \cdot |b_n - b_m| + |a_n - a_m| \cdot |b_m|$$
$$< (|a_n| + |b_m|)\frac{1}{L}$$

if $n, m > N$. Now, $|a_n| = |(a_n - a_{N+1}) + a_{N+1}| \leq \frac{1}{L} + |a_{N+1}|$ if $n > N$. Also $|b_m| \leq \frac{1}{L} + |b_{N+1}|$ if $m > N$. So, $|(a_n \cdot b_n) - (a_m \cdot b_m)| \leq k_N \cdot \frac{1}{L}$ if $n, m > N$, where k_N is a positive constant which is a function of N. Now, select N' such that $|a_n - a_m| < \frac{1}{k_N L}$ for $n, m > N'$ then we have $|(a_n \cdot b_n) - (a_m \cdot b_m)| \leq k_N \cdot \frac{1}{k_N L} = \frac{1}{L}$ for $n, m > N'$.

If $(a_i)_i$ and $(b_i)_i$ converge to 0, let N be such that $|a_n|, |b_n| < \frac{1}{2L}$ for $n > N$. Then $|a_n + b_n| \leq |a_n| + |b_n| < \frac{1}{2L} + \frac{1}{2L} = \frac{1}{L}$ for $n > N$.

Let $(a_i)_i$ and $(b_i)_i$ be two Cauchy sequences such that $(a_i)_i$ converges to 0. By the previous discussion we know that there is a positive constant k such that $|b_n| < k$ for all n. Now, let N be such that $|a_n| < \frac{1}{kL}$ for all $n > N$. Then $|a_n \cdot b_n| = |a_n| \cdot |b_n| < k \cdot \frac{1}{kL} = \frac{1}{L}$ for $n > N$. $\square$

The properties (ii) and (iii) of the set $\mathcal{O}$ of zero sequences make $\mathcal{O}$ into a so-called ideal. This is an important structure in algebra, we come back to its systematic discussion in chapter 15. We are now ready to define real numbers.

Lemma 80 *The binary relation R on C defined by $(a_i)_i R(b_i)_i$ iff $(a_i)_i - (b_i)_i = (a_i - b_i)_i$ is a zero sequence is an equivalence relation.*

Proof Clearly, R is reflexive and symmetric. Let $(a_i)_i, (b_i)_i, (c_i)_i$ be Cauchy sequences such that $(a_i)_i R(b_i)_i$ and $(b_i)_i R(c_i)_i$. Then $|a_n - c_n| = |a_n - b_n + b_n - c_n| \leq |a_n - b_n| + |b_n - c_n| < \frac{1}{L}$ for $n > N$ if N is such that $|a_n - b_n|, |b_n - c_n| < \frac{1}{2L}$ for $n > N$. $\square$

Definition 53 *The set C/R of equivalence classes under the relation R defined in lemma 80 is denoted by $\mathbb{R}$. Its elements are called* real numbers.

Lemma 81 *The equivalence class (the real number) of a Cauchy sequence $(a_i)_i$ is given by this rule: Take the "coset" of the ideal $\mathcal{O}$, i.e., $[(a_i)_i] = \{(a_i)_i + (c_i)_i \mid (c_i)_i \in \mathcal{O}\} = (a_i)_i + \mathcal{O}$.*

Proof If $(a_i)_i$ and $(b_i)_i$ are equivalent, then by definition $(a_i)_i = (b_i)_i + ((a_i)_i - (b_i)_i)$, and $(a_i)_i - (b_i)_i \in \mathcal{O}$. Conversely, if $(a_i)_i = (b_i)_i + (o_i)_i$, $(o_i)_i \in \mathcal{O}$, then $(a_i)_i R (b_i)_i$ by definition of R. $\quad\square$

Lemma 82 *We have an injection $e : \mathbb{Q} \to \mathbb{R}$ defined by $e(a) = (a)_i + \mathcal{O}$.*

Exercise 38 Give a proof of lemma 82.

On $\mathbb{R}$, we now want to develop the entire arithmetics, and we want to show that the purpose of this entire construction is effectively achieved.

Lemma 83 *Let $(a_i)_i, (b_i)_i, (c_i)_i$ be Cauchy sequences of rational numbers.*

   (i) *If $(a_i)_i$, $(b_i)_i$ are equivalent, then so are $(a_i)_i + (c_i)_i$, $(b_i)_i + (c_i)_i$.*

   (ii) *If $(a_i)_i$, $(b_i)_i$ are equivalent, then so are $(a_i)_i \cdot (c_i)_i$, $(b_i)_i \cdot (c_i)_i$.*

Proof (i) In fact, $((a_i)_i + (c_i)_i) - ((b_i)_i + (c_i)_i) = (a_i - b_i)_i$, which is a zero sequence.

(ii) Similarly, $((a_i)_i \cdot (c_i)_i) - ((b_i)_i \cdot (c_i)_i) = ((a_i - b_i)_i)(c_i)_i$, but by proposition 79 this is a zero sequence. $\quad\square$

This enables the definition of addition and multiplication of real numbers:

Definition 54 *If $(a_i)_i + \mathcal{O}$ and $(b_i)_i + \mathcal{O}$ are two real numbers, then we define*

$$((a_i)_i + \mathcal{O}) + ((b_i)_i + \mathcal{O}) = (a_i + b_i)_i + \mathcal{O}$$
$$((a_i)_i + \mathcal{O}) \cdot ((b_i)_i + \mathcal{O}) = (a_i \cdot b_i)_i + \mathcal{O}.$$

By lemma 83, this definition is independent of the representative Cauchy sequences.

Evidently, these operations, when restricted to the rational numbers r, embedded via $e(r)$ as above, yield exactly the operations on the rationals, i.e., $e(r + s) = e(r) + e(s)$ and $e(r \cdot s) = e(r) \cdot e(s)$. We therefore also use the rational numbers r instead of $e(r)$ when working in $\mathbb{R}$. If $x = (a_i)_i + \mathcal{O}$, we write $-x$ for $(-a_i)_i + \mathcal{O}$ and call it the *additive inverse* or *negative* of x.

The arithmetic properties of these operations on $\mathbb{R}$ are these:

Sorite 84 *Let x, y, z be real numbers.*

(i) *(Additive associativity)* $(x + y) + z = x + (y + z) = x + y + z$

(ii) *(Additive commutativity)* $x + y = y + x$

(iii) *(Additive neutral element) The rational zero 0 is also neutral on the reals, $x + 0 = x$.*

(iv) *(Additive inverse element)* $x + (-x) = 0$

(v) *(Multiplicative associativity)* $(x \cdot y) \cdot z = x \cdot (y \cdot z) = x \cdot y \cdot z$

(vi) *(Multiplicative commutativity)* $x \cdot y = y \cdot x$

(vii) *(Multiplicative neutral element) The rational unity 1 is also neutral on the reals, $x \cdot 1 = x$.*

(viii) *(Multiplicative inverse element) If $x \neq 0$, then there is exactly one multiplicative inverse x^{-1}, i.e., $x \cdot x^{-1} = 1$, more precisely, there exists in this case a Cauchy sequence $(a_i)_i$ representing x and such that $a_i \neq 0$ for all i, and we may represent x^{-1} by the Cauchy sequence $(a_i^{-1})_i$.*

(ix) *(Distributivity)* $x \cdot (y + z) = x \cdot y + x \cdot z$

Proof (i) through (vii) as well as (ix) are straightforward once we have learned that all the relevant operations are defined factorwise on the sequence members. As to (viii), since $(a_i)_i$ does not converge to zero, there is a positive natural L such that for every N there is $n > N$ with $|a_n| \geq \frac{1}{L}$. Choose N such that $|a_n - a_m| < \frac{1}{2L}$ for all $n, m > N$. Fix $n > N$ such that $|a_n| \geq \frac{1}{L}$ as above. Then $|a_m| \geq |a_n| - |a_n - a_m| \geq \frac{1}{L} - \frac{1}{2L} = \frac{1}{2L} > 0$ for $n, m > N$. Therefore $(a_i)_i$ is equivalent to a sequence $(a_i')_i$ without zero members, more precisely: there is I such that $a_i' = a_i$ for $i > I$. Then evidently the sequence $(1/a_i')_i$ is the inverse of $(a_i)_i$. The uniqueness of the inverse follows from the purely formal fact that $x \cdot y = x \cdot z = 1$ implies $y = 1 \cdot y = (y \cdot x) \cdot y = y \cdot (x \cdot y) = y \cdot (x \cdot z) = (y \cdot x) \cdot z = 1 \cdot z = z$. $\qquad \square$

Corollary 85 *If a, b, c are real numbers such that $a \neq 0$, then the equation $ax + b = c$ has exactly one solution x.*

This means that we have "saved" the algebraic properties of $\mathbb{Q}$ to $\mathbb{R}$. But we wanted more than that. Let us first look for the linear ordering structure on $\mathbb{R}$.

Definition 55 *A real number $x = (a_i)_i + \mathcal{O}$ is called positive iff there is a positive rational number $0 < \varepsilon$ such that $\varepsilon < a_i$ for all but a finite set of* *indexes. This property is well defined. We set $x < y$ for two real numbers x and y iff $y - x$ is positive. In particular, x is positive iff $0 < x$.*

Proposition 86 *The relation $<$ on $\mathbb{R}$ from definition 55 defines a linear ordering. The set $\mathbb{R}$ is the disjoint union of the subset $\mathbb{R}_+$ of positive real numbers, the subset $\mathbb{R}_- = -\mathbb{R}_+ = \{-x \mid x \in \mathbb{R}_+\}$ of negative real numbers, and the singleton set $\{0\}$. We have*

(i) $\mathbb{R}_+ + \mathbb{R}_+ = \{x + y \mid x, y \in \mathbb{R}_+\} = \mathbb{R}_+,$

(ii) $\mathbb{R}_+ \cdot \mathbb{R}_+ = \{x \cdot y \mid x, y \in \mathbb{R}_+\} = \mathbb{R}_+,$

(iii) $\mathbb{R}_- + \mathbb{R}_- = \{x + y \mid x, y \in \mathbb{R}_-\} = \mathbb{R}_-,$

(iv) $\mathbb{R}_- \cdot \mathbb{R}_- = \mathbb{R}_+,$

(v) $\mathbb{R}_+ + \mathbb{R}_- = \{x - y \mid x, y \in \mathbb{R}_+\} = \mathbb{R},$

(vi) $\mathbb{R}_+ \cdot \mathbb{R}_- = \{x \cdot (-y) \mid x, y \in \mathbb{R}_+\} = \mathbb{R}_-,$

(vii) *(Monotony of addition) if x, y, z are real numbers with $x < y$, then $x + z < y + z$,*

(viii) *(Monotony of multiplication) if x, y, z are real numbers with $x < y$ and $0 < z$, then $xz < yz$,*

(ix) *(Archimedean property) if x and y are positive real numbers, there is a natural number N such that $y < Nx$,*

(x) *(Density of rationals) if $0 < \varepsilon$ is a positive real number, then there is a rational number ρ with $0 < \rho < \varepsilon$.*

Proof Let us first show that $<$ is antisymmetric. If $x < y$, then $y - x$ is represented by a sequence $(a_i)_i$ with $a_i > \varepsilon$ for a positive rational ε. Then $x - y$ is represented by $(-a_i)_i$, and $-a_i < \varepsilon < 0$ for all but a finite set of indexes. If this were equivalent to a sequence $(b_i)_i$ with $b_i > \varepsilon' > 0$ except for a finite number of indexes, then $b_i - (-a_i) = b_i + a_i > \varepsilon + \varepsilon'$ could not be a zero sequence. Whence antisymmetry.

Also, if $y - x$ is represented by $(a_i)_i$ with $a_i > \varepsilon$ and $z - y$ is represented by $(b_i)_i$ with $a_i > \varepsilon'$ for all but a finite number of indexes, then $z - x = z - y + y - x$ is represented by $(a_i)_i + (b_i)_i = (a_i + b_i)_i$, and $a_i + b_i > \varepsilon + \varepsilon' > 0$ for all but a finite number of indexes, whence transitivity. Finally, if $x \neq y$, then $x - y \neq 0$. By the same argument as used in the previous proof, if $(d_i)_i$ represents $x - y$, then there is a positive rational ε and N, such that $|d_n| > \varepsilon$ for $n > N$. But since $(d_i)_i$ is a Cauchy sequence, too, the differences $|d_n - d_m|$ become arbitrary small for large n and m. So either d_n is positive or negative, but not both, for large n, and therefore $x - y$ is either positive or negative. This immediately entails statements (i) through (viii).

(ix) If x is represented by $(a_i)_i$ and y is represented by $(b_i)_i$, then there is a positive rational ε and an index N such that $a_n, b_n > \varepsilon$ for $n > M$. But since $(b_i)_i$ is Cauchy, there is also a positive δ such that $y_n < \delta$ for $n > M'$, and we may take the larger of M and M' and then suppose that $M = M'$ for our two conditions. Then, since $\mathbb{Q}$ has the Archimedean ordering property by sorite 77, there is a natural N such that $N\varepsilon > 2 \cdot \delta$. Then we have $N \cdot a_n > N \cdot \varepsilon > 2 \cdot \delta > \delta > b_n$ for $n > M$, whence $N \cdot a_n - b_n > \delta > 0$, whence the claim.

(x) If $\epsilon > 0$, and if this real number is represented by a Cauchy sequence $(e_i)_i$, by the very definition of positivity there is a positive rational δ such that $e_i > \delta$ for all but a finite number of indexes. But then $e_i - \frac{\delta}{2} > \frac{\delta}{2}$ for all but a finite number of indexes, and $\rho = \frac{\delta}{2}$ does the job. $\square$

Definition 56 *The absolute value $|a|$ of a real number a is a if it is non-negative, and $-a$ else.*

Proposition 87 (Triangle Inequality) *If a and b are two real numbers, then we have the* triangle inequality*:*

$$|a + b| \le |a| + |b|.$$

Proof Observe that, if x is represented by a Cauchy sequence $(a_i)_i$, then $|x|$ is represented by $(|a_i|)_i$. Therefore, if y is represented by $(b_i)_i$, then $|x + y|$ is represented by $(|a_i + b_i|)_i$, but by the triangle inequality for rationals, we have $|a_i| + |b_i| \ge |a_i + b_i|$, i.e., $|x| + |y| - |x + y|$ is represented by $(|a_i| + |b_i| - |a_i + b_i|)_i$, so it is not negative. $\square$

We now have a completely general convergence criterion on $\mathbb{R}$. We have to define convergence entirely along the lines of convergence for rational sequences.

Definition 57 *A sequence $(a_i)_i$ of real numbers is said to converge to a real number a iff for every real $\varepsilon > 0$, there is an index N such that $n > N$ implies $|a_n - a| < \varepsilon$. Clearly, convergence can only take place for one a, and therefore we denote convergence by $\lim_{i \to \infty} a_i = a$.*

The sequence $(a_i)_i$ is Cauchy, iff for every real number $\varepsilon > 0$, there is a natural number N such that $n, m > N$ implies that $|a_n - a_m| < \varepsilon$.

Proposition 88 (Convergence on $\mathbb{R}$) *A sequence $(a_i)_i$ of real numbers converges iff it is Cauchy.*

Proof We omit the detailed proof since it is quite technical. However, the idea of the proof is easily described: Let $(a_i)_i$ be a Cauchy sequence of real numbers.

For each $i > 0$, there is a rational number r_i such that $|a_i - r_i| < \frac{1}{i}$. This rational number can be found as follows: Represent a_i by a rational Cauchy sequence $(a_{ij})_j$. Then there is an index I_i such that $r, s > I_i$ implies $|a_{ir} - a_{is}| < \frac{1}{i}$. Then take $r_i = a_{ir}$ for any $r > I_i$. One then shows that $(r_i)_i$ is Cauchy and that $(a_i)_i$ converges to $(r_i)_i$. $\square$

This result entails a huge number of existence theorems of special numbers. We just mention one particularly important situation.

Corollary 89 (Existence of Suprema) *If A is a bounded, non-empty set, i.e., if there is an upper bound $b \in \mathbb{R}$ such that $b > a$ for all $a \in A$, in short $b > A$, then there is a uniquely determined supremum or least upper bound $s = sup(A)$, i.e., an upper bound $s \geq A$ such that for all $t < s$, there is $a \in A$ with $a > t$.*

Proof One first constructs a Cauchy sequence $(u_i)_i$ of upper bounds u_i of A as follows: Let u_0 be an existing upper bound and take $a_0 \in A$. Then consider the middle $v_1 = (a_0 + u_0)/2$. If v_1 is an upper bound, set $u_1 = v_1$, otherwise set $u_1 = u_0$. In the first case, set $a_1 = a_0$, in the second case, there is $a_1 > v_1$; then consider the pair a_1, u_1. In either case, their distance is half the first distance $|a_0 - u_0|$. We now again take the middle of this interval, i.e., $v_2 = (a_1 + u_1)/2$ and go on in the same way, i.e., defining a sequence of u_i by induction, and always taking the smallest possible alternative. This is a Cauchy sequence since the intervals are divided by 2 in each step. Further, the limit $u = \lim_{i \to \infty} u_i$, which exists according to proposition 88, is an upper bound, and there is no smaller one, as is easily seen from the construction. The details are left to the reader. $\square$

This corollary entails the following crucial fact:

Corollary 90 (Existence of n-th roots) *Let $a \geq 0$ be a non-negative real number and $n \geq 1$ a positive natural number, then there is exactly one number $b \geq 0$ such that $b^n = a$. This number is denoted by $\sqrt[n]{a}$ or by $a^{\frac{1}{n}}$, in case of $n = 2$ one simply writes $\sqrt[2]{a} = \sqrt{a}$.*

Proof The cases $n = 1$ or $a = 0$ are clear, so suppose $n > 1$ and $a > 0$. Let A be the set of q, $q^n < a$. This set is bounded since $a^n > a$. Take $b = sup(A)$ We claim that $b^n = a$. Clearly $b^n \leq a$. Suppose that $b^n < a$. Then consider this construction of a contradiction:

First show that for any $d > 0$ there is a natural number m such that $(1 + \frac{1}{m})^n < 1 + d$. We show this by induction on n. Suppose by induction that we found M such that $(1 + \frac{1}{m})^n < 1 + d/2$ for $m \geq M$. Then $(1 + \frac{1}{m})^n = (1 + \frac{1}{m})(1 + \frac{1}{m})^{n-1} < (1 + \frac{1}{m})(1 + d/2) = 1 + \frac{1}{m} + \frac{d}{2} + \frac{d}{2m}$. We require $1 + \frac{1}{m} + \frac{d}{2} + \frac{d}{2m} < 1 + d$, but this is true if $m > 1 + \frac{2}{d}$. Next, replace b by $b(1 + \frac{1}{m})$ and then we ask for a m such that $(b(1 + \frac{1}{m}))^n = b^n(1 + d) < a$. But we find such an m by the fact that for any positive d, we can find m such that $(1 + \frac{1}{m})^n < 1 + d$, and it suffices to take $d < a/b^n - 1$, this contradicts the supremum property of b, and we are done. $\square$

Exercise 39 Show that for two $a, b \geq 0$, we have $\sqrt[n]{a \cdot b} = \sqrt[n]{a} \cdot \sqrt[n]{b}$.

Definition 58 *One can now introduce more general rational powers x^q, $q \in \mathbb{Q}$ of a positive real number x as follows. First, one defines $x^0 = 1$, then for a negative integer $-n$, one puts $x^{-n} = 1/x^n$. For $q = n/m \in \mathbb{Q}$ with positive denominator m, one defines $x^{n/m} = (x^{1/m})^n$. One easily checks that this definition does not depend on the fractional representation of q.*

Exercise 40 Prove that for two rational numbers p and q and for two positive real numbers x and y, one has $x^{p+q} = x^p x^q$, $x^{pq} = (x^p)^q$, and $(xy)^p = x^p y^p$.

Exercise 41 Let $a > 1$ and $x > 0$ be real numbers. Show that the set

$$\{q \in \mathbb{Q} \mid \text{there are two integers } m, n, 0 < n,$$
$$\text{such that } a^m \leq x^n \text{ and } q = m/n\}$$

is non-empty and bounded from above. Its supremum is called the *logarithm of x for basis a*, and is denoted by $\log_a(x)$. It is the fundament of the construction of the exponential function and of the sine and cosine functions.

The general shape of real numbers as equivalence classes of rational Cauchy sequences is quite abstract and cannot, as such, be handled by humans when calculating concrete cases, and a fortiori cannot be handled by computers in hard- and software contexts. Therefore, one looks for more explicit and concrete representations of real numbers. Here is the construction, which generalizes the b-adic representation of natural numbers discussed in chapter 7. We consider a natural base number $1 < b$ and for an integer n the index set $n] = \{i \mid i \in \mathbb{Z} \text{ and } i \leq n\}$. Let $(a_i)_{i \in n]}$ be a sequence of natural numbers with $0 \leq a_i < b$ and $a_n \neq 0$. Consider the partial sums $\Sigma_j = \sum_{i=n,n-1,n-2,\dots j} a_i b^i$ for $j \in n]$.

Lemma 91 *The sequence $(\Sigma_j)_{n]}$ of partial sums as defined previously converges to a real number which we denote by $\sum_{i \in n]} a_i b^i$, and, when the base b is clear, by*

$$a_n a_{n-1} \ldots a_0.a_{-1}a_{-2}a_{-3} \ldots \qquad (9.1)$$

for non-negative n, or, if n < 0, by

$$0.00 \ldots a_n a_{n-1} \ldots, \qquad (9.2)$$

i.e., we put zeros on positions $-1, \ldots n + 1$ *until the first non-zero coefficient* a_n *appears on position n. The number zero is simply denoted by* 0.0 *or even* 0. *The dots are meant to represent the given coefficients. If the coefficients vanish after finite number of indexes, we can either stop the representation on the last non-vanishing coefficient* a_m: $a_n a_{n-1} \ldots a_0.a_{-1} \ldots a_{m+1}a_m$, *or append any number of zeros, such as* $a_n a_{n-1} \ldots a_{m+1} a_m 000$.

Proof Let us show that the differences $|\Sigma_k - \Sigma_j|$ of the rational partial sums become arbitrarily small for large k and j. We may suppose that $j > k$ and then have $\Sigma_k - \Sigma_j = a_{j-1}b^{j-1} + \ldots a_k b^k < b^j + \ldots b^{k+1} = b^j(1 + b^{-1} + \ldots b^{k+1-j})$. But we have this quite general formula which is immediately checked by evaluation: $1 + b^{-1} + \ldots b^{-r} = (1 - b^{-r-1})/(1 - b) < 1/(1 - b)$. And therefore $0 \le \Sigma_k - \Sigma_j < b^j \cdot 1/(1 - b)$. But the right side converges to zero as $j \to -\infty$, and we are done. $\square$

Definition 59 *The representation of a real number in the forms (9.1) or (9.2) is called the b-adic representation; in computer science, the synonymous term b-ary is more common. For* $b = 2$ *it is called the* binary representation, *for* $b = 10$, *the* decimal representation, *and for* $b = 16$, *it is called the* hexadecimal representation. *The extension of the b-adic representation to negative real numbers is defined by prepending the sign* $-$. *This extended representation is also called b-adic representation.*

Proposition 92 (Adic Representation) *Given a base number* $b \in \mathbb{N}$, $1 < b$, *every real number can be represented in the b-adic representation. The representation is unique up to the following cases: If the coefficients* a_i *are smaller than* $b - 1$ *until index m, and from* $m - 1$ *on equal to* $b - 1$ *until infinity, then this number is equal to the number whose coefficients are the old ones for* $i > m$, *while the coefficient at index i is* $a_m + 1$, *and all lower coefficients vanish.*

Proof We may evidently suppose that the real number to be represented is positive, the other cases are deduced from this. We construct the representation by induction as follows: Observe that we have $b^i > b^j > 0$ whenever $i > j$, i and j being integers. Moreover, b^i converges to 0 as $i \to -\infty$, and b^i becomes arbitrarily large if $i \to \infty$. Therefore, there is a unique j such that $b^j \le x < b^{j+1}$. Within this interval of b powers there is a unique natural a_j with $0 < a_j < b$

such that $a_j b^j \leq x < (a_j + 1)b^j$. Consider the difference $x' = x - a_j b^j$, then we have $0 \leq x' < b^j$ and there is a unique natural $0 \leq a_{j-1} < b$ such that $a_{j-1} b^{j-1} \leq x' < (a_{j-1} + 1)b^{j-1}$. Then $x'' = x' - a_{j-1}b^{j-1} = x - a_j b^j - a_{j-1}b^{j-1}$ has $0 \leq x'' < b^{j-1}$, and we may go on in this way, defining a b-adic number. It is immediate that this number converges to x. The failure of uniqueness in the case where one has $b - 1$ until infinity is left as an exercise to the reader. □

Example 19 In the decadic representation, the number $0.999\ldots$ with a non-terminating sequence of 9s is equal to 1.0. Often, one writes $\ldots a_{m+1}\overline{a}_m$ in order to indicate that the coefficient is constant and equal to a_m for all indexes $m, m - 1, m - 2, \ldots$. Thus, in the binary representation, $0.\overline{1}$ equals 1.0.

Example 20 Show that for every real number, there is exactly one representation (without the exceptional ambiguity, i.e., avoiding the typical decimal $0.999\ldots$ representation) of a real number r in the form

$$r = \pm(a_0.a_{-1}a_{-2}\ldots) \cdot b^e$$

with $a_0 \neq 0$ for $r \neq 0$.

This is the so-called *floating point representation*, which for $b = 2$ has been standardized by the IEEE society to computerized representations. See chapter 14 on the first advanced topic for this subject.

At this point we are able to tackle the question whether the cardinalities of $\mathbb{N}$ and $\mathbb{R}$ are equal. This is not the case, in fact, there are many more reals than natural numbers. More precisely, there is an injection $\mathbb{N} \to \mathbb{R}$, which is induced by the chain $\mathbb{N} \to \mathbb{Z} \to \mathbb{Q} \to \mathbb{R}$ of injections, but there is no bijection $\mathbb{N} \xrightarrow{\sim} \mathbb{R}$. To show this, let us represent the reals by decimal numbers $x = n(x) + 0.x_0 x_1 \ldots$, where $n(x) \in \mathbb{Z}$ and $0 \leq 0.x_0 x_1 \ldots < 1$. Suppose that we take the unique representation $x = n(x) + 0.x_0 x_1 \ldots x_t \overline{0}$ instead of $x = n(x) + 0.x_0 x_1 \ldots (x_t - 1)\overline{9}$, for the case of an ambiguous representation. Now, suppose we are given a bijection $f : \mathbb{N} \xrightarrow{\sim} \mathbb{R}$. The f-image of $m \in \mathbb{N}$ is then $f(m) = n(f(m)) + 0.f(m)_0 f(m)_1 \ldots$. Let $a \in \mathbb{R}$ be the following "antidiagonal" element $a = 0.a_0 a_1 \ldots a_m \ldots$. We define $a_m = 2$ if $f(m)_m = 1$ and $a_m = 1$ else. This is a decimal representation of a number a which must occur in our bijection. Suppose that $a = f(m_0)$. Then by construction of a, the digit of $f(m_0)$ at position m_0 after the dot is different from the digit of a at position m_0 after the dot, so a cannot occur, which is a contradiction to the claimed bijection f.

$f(0)$	92736.$\underline{2}$82109927835...
$f(1)$	2.8$\underline{1}$4189264762...
$f(2)$	1623.10$\underline{9}$473637637...
$\vdots$	$\vdots$
$f(m_0) = a$?	**0.121**...

Fig. 9.3. A tentative "bijection" $f : \mathbb{N} \to \mathbb{R}$.

9.4 Complex Numbers $\mathbb{C}$

The last number domain which we need in the general mathematical environment are the complex numbers. The theory that we have developed so far enables us to solve equations such as $ax + b = c$, and we have convergence of all Cauchy sequences, including the standard adic representations. But there is still a strong deficiency: General equations cannot be solved in $\mathbb{R}$. More precisely, one can easily show that an equation of form $ax^3 + bx^2 + cx + d = 0$ with real coefficients a, b, c, d and $a \neq 0$ always has a solution (a "root") in $\mathbb{R}$. But an equation with an even maximal power of the unknown, such as $ax^4 + bx^3 + cx^2 + dx + e = 0$, cannot be solved in general. Two hundred years ago, mathematicians were searching for a domain of numbers where the special equation $x^2 + 1 = 0$ has a solution. Evidently, such a solution does not exist in $\mathbb{R}$ since in $\mathbb{R}$ any square is non-negative, and therefore $x^2 + 1 \geq 1$.

The solution was rigorously conceptualized by Carl Friedrich Gauss. Instead of working in $\mathbb{R}$, he considers the plane $\mathbb{R}^2$ of pairs of real numbers. His trick is to give this set an arithmetic structure, i.e., addition and multiplication, such that the existing $\mathbb{R}$ arithmetic and the entire Cauchy sequence structure are embedded and such that we can effectively solve the critical equation $x^2 + 1 = 0$. But Gauss' invention is much deeper: It can be shown that in his construction, every equation $a_n x^n + a_{n-1} x^{n-1} + \ldots + a_1 x + a_0 = 0$ has a solution, this is the fundamental theorem of algebra, see chapter 15. Therefore, the solvability of that single equation $x^2 + 1 = 0$ implies that we are done once for all with such a type of equation solving. The equation type is called *polynomial equation*, we shall come back to this structure in the second part of the course (see chapter 15).

Gauss' complex numbers work as follows. We consider the Cartesian product $\mathbb{R}^2$, i.e., the set of ordered pairs (x, y) of real numbers. When thinking of the arithmetical structure (addition, multiplication) on $\mathbb{R}^2$, we denote this domain by $\mathbb{C}$ and call it the domain of *complex numbers*. Here are the two fundamental operations, addition and multiplication of complex numbers:

Definition 60 *Given two complex numbers* (x, y) *and* $(u, v) \in \mathbb{C}$, *we define the sum*

$$(x, y) + (u, v) = (x + u, y + v),$$

while the product is defined by

$$(x, y) \cdot (u, v) = (xu - yv, xv + yu).$$

Here is the sorite for this arithmetic structure:

Sorite 93 *Let* x, y, z *be complex numbers, and denote* $0 = (0, 0)$ *and* $1 = (1, 0)$. *Then:*

(i) *(Additive associativity) We have* $(x + y) + z = x + (y + z)$ *and denote this number by* $x + y + z$.

(ii) *(Multiplicative associativity) We have* $(x \cdot y) \cdot z = x \cdot (y \cdot z)$ *and denote this number by* $x \cdot y \cdot z$, *or also* xyz, *if no confusion is likely.*

(iii) *(Commutativity) We have* $x + y = y + x$ *and* $x \cdot y = y \cdot x$.

(iv) *(Distributivity) We have* $x \cdot (y + z) = x \cdot y + x \cdot z$.

(v) *(Additive and multiplicative neutral elements) We have* $0 + x = x$ *and* $1 \cdot x = x$.

(vi) *If* $a \neq 0$, *then every equation* $a \cdot x = b$ *has a unique solution; in particular, the solution of* $a \cdot x = 1$, *the multiplicative inverse of* a, *is denoted by* a^{-1}. *The solution of* $a + x = 0$, *the additive inverse (or negative) of* a, *is denoted by* $-a$.

Proof The statements (i) through (v) are straightforward.

(vi) Let $a = (x, y) \neq (0, 0)$, Then $x^2 + y^2 > 0$. But then $(\frac{1}{x^2 + y^2}, 0) \cdot (x, -y) \cdot (x, y) = (1, 0)$, so $a^{-1} = (\frac{1}{x^2 + y^2}, 0) \cdot (x, -y)$ is an inverse of a. It is unique by the common argument: If w and z are two inverses of a, then $w = 1 \cdot w = (w \cdot a) \cdot w = w \cdot (a \cdot w) = w \cdot 1 = w \cdot (a \cdot z) = (w \cdot a) \cdot z = z$. □

The geometric view of Gauss is this: We have an injection $\mathbb{R} \to \mathbb{C}$ which sends a real number a to the complex number $(a, 0)$. Like with the embedding $\mathbb{Q} \to \mathbb{R}$ discussed above, all arithmetic operations, addition and multiplication, "commute" with this embedding, i.e., $(a + b, 0) = (a, 0) + (b, 0)$

and $(a \cdot b, 0) = (a, 0) \cdot (b, 0)$. We therefore identify real numbers a with their image $(a, 0)$ in $\mathbb{C}$. With this convention, denote the complex number $(0, 1)$ by i, and call it the *imaginary unit*. Evidently, $i^2 = -1$. This means that in $\mathbb{C}$, the equation $x^2 + 1 = 0$ now has a solution, namely $x = i$! Further, for a complex number $x = (a, b)$, we write $Re(x) = a$ and call it the *real part of x*, similarly we write $Im(x) = b$ and call it the *imaginary part of x*; complex numbers of the shape $(0, b)$ are called *imaginary*. Clearly, x is uniquely determined by its real and imaginary parts, in fact:

$$x = (Re(x), Im(x)).$$

We then have this crucial result, which justifies the geometric point of view:

Proposition 94 *For any complex number x, we have a unique representation*
$$x = Re(x) + i \cdot Im(x)$$
as a sum of a real number (i.e., $Re(x)$) and an imaginary number (i.e., $i \cdot Im(x)$).

Proof This is obvious. □

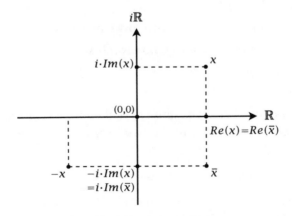

Fig. 9.4. The representation of complex numbers in the plane as introduced by the German mathematician Carl Friedrich Gauss (1777–1855). In the Gauss-plane, the conjugate $\overline{x}$ of x is the point obtained by reflecting x at the abscissa, while $-x$ is the point obtained from x by a rotation of $180°$ around the origin $(0, 0)$ of the Gauss-plane.

Exercise 42 Using the representation in 94, show that we have these arithmetical rules:

1. $(a + i \cdot b) + (c + i \cdot d) = (a + c) + i \cdot (b + d)$,

2. $(a + i \cdot b) \cdot (c + i \cdot d) = (ac - bd) + i \cdot (ad + bc)$.

The complex numbers have a rich inner structure which is related to the so-called conjugation.

Definition 61 *The* conjugation *is a map* $\mathbb{C} \to \mathbb{C} : x \mapsto \overline{x}$ *defined by* $\overline{x} = Re(x) - i \cdot Im(x)$, *i.e.,* $Re(\overline{x}) = Re(x)$ *and* $Im(\overline{x}) = -Im(x)$.

The norm *of a complex number* x *is defined by* $|x| = \sqrt{x \cdot \overline{x}}$, *which is real, since* $x \cdot \overline{x} = Re(x)^2 + Im(x)^2 \geq 0$.

Observe that the norm of a complex number $x = a + i \cdot b$ is the Euclidean length of the vector $(a, b) \in \mathbb{R}^2$ known from high school.

Sorite 95 *Let* $x, y \in \mathbb{C}$. *Then*

(i) $x = \overline{x}$ *iff* $x \in \mathbb{R}$, *and* $\overline{x} = -x$ *iff* x *is imaginary,*

(ii) $|x| = 0$ *iff* $x = 0$,

(iii) $Re(x) = \frac{x + \overline{x}}{2}$ *and* $Im(x) = \frac{x - \overline{x}}{2i}$,

(iv) *if* $x \neq 0$, *then the multiplicative inverse of* x *is* $x^{-1} = |x|^{-2} \cdot \overline{x}$,

(v) $\overline{\overline{x}} = x$; *in particular, conjugation is a bijection,*

(vi) $\overline{x + y} = \overline{x} + \overline{y}$,

(vii) $\overline{x \cdot y} = \overline{x} \cdot \overline{y}$,

(viii) *if* x *is real, then* $|x|$ *in the sense of real numbers coincides with* $|x|$ *in the sense of complex numbers, which justifies the common notation,*

(ix) $|x \cdot y| = |x| \cdot |y|$,

(x) *(Triangle inequality)* $|x + y| \leq |x| + |y|$.

Proof The only non-trivial statement is the triangle inequality. It suffices to show that $|x + y|^2 \leq (|x| + |y|)^2$. This gives us the inequality $y\overline{x} + x\overline{y} \leq 2|x||y|$, and then, by putting $a = x\overline{y}$, we get inequality $a + \overline{a} \leq |a| + |\overline{a}|$ which is obvious by simple explication of the coordinates of the complex number a. $\square$

CHAPTER 10

Categories of Graphs

In this chapter, we introduce the concept of a graph. However, this is homonymous with but really different from the already known concept of a graph *relation*. Please do observe this historically grown ambiguity. Of course, both concepts are related by the fact that they allude to something being drawn: The graph of a function is just what in nice cases will be drawn as a graphical representation of that function, whereas the other meaning is related to the graphical representation of assignments between nodes of a processual assembly. So the concrete situations are completely different.

As a preliminary construction we need this setup: Given a set V, always finite in this context, we have the Cartesian product $V^2 = V \times V$. In addition we define the *edge set as* $^2V = \{a \subset V \mid 1 \leq card(a) \leq 2\}$. It has this name because it parametrizes the set of all undirected lines, i.e., edges, including single points ("loop at x"), between any two elements of V. We have the evident surjection $|?| : V^2 \to {}^2V : (x, y) \mapsto \{x, y\}$, which has a number of sections which we (somewhat ambiguously) denote by $\overrightarrow{?} : {}^2V \to V^2$, i.e., $|?| \circ \overrightarrow{?} = Id_{{}^2V}$.

Exercise 43 Give reasons for the existence of sections $\overrightarrow{?}$ of $|?|$.

We further denote by $?^* : V^2 \to V^2 : (x, y) \mapsto (y, x)$ the exchange bijection, and note that $(x, y)^{**} = (x, y)$.

10.1 Directed and Undirected Graphs

Definition 62 *A directed graph or* digraph *is a map* $\Gamma : A \to V^2$ *between finite sets. The elements of the set A are called* arrows, *the elements of V are called* vertexes *of the directed graph. By the universal property of the Cartesian product (see proposition 57), these data are equivalent to the data of two maps, $head_\Gamma : A \to V$ and $tail_\Gamma : A \to V$; more precisely, we set $tail_\Gamma = pr_1 \circ \Gamma$ and $head_\Gamma = pr_2 \circ \Gamma$. For $a \in A$ and $h = head_\Gamma(a)$ and $t = tail_\Gamma(a)$, we also write $a : t \to h$ or $t \xrightarrow{a} h$.*

The intuitive meaning of a directed graph is that we are given a set of objects some of which are connected by arrows, and that there may exist several "parallel" arrows between a pair of given tail and head objects.

Example 21 The graph Γ consists of the set of vertexes $V = \{B, C, D, F\}$ and the set of arrows $A = \{a, b, c, d, e, g\}$

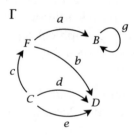

Fig. 10.1. The graph $\Gamma : A \to V^2$.

If the simplified notation is used, the map Γ is given by:

$$F \xrightarrow{a} B, F \xrightarrow{b} D, C \xrightarrow{c} F, C \xrightarrow{d} D, C \xrightarrow{e} D, B \xrightarrow{g} B$$

Taking the arrow a as an example, we have

$$tail_\Gamma(a) = F \text{ and } head_\Gamma(a) = B.$$

Example 22 For every directed graph $\Gamma : A \to V^2$, we have the *dual graph* $\Gamma^* = ?^* \circ \Gamma$, evidently $\Gamma^{**} = \Gamma$.

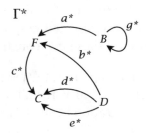

Fig. 10.2. The dual graph Γ^* of the graph Γ from example 21.

Example 23 Given a binary relation A in a set V, we have the associated directed graph defined by $\Gamma = A \subset V^2$, the inclusion of A in V^2. So the arrows here identify with the associated ordered pairs. In particular, the *complete* directed graph $CompDi(V)$ over a set V is defined by the identity on the Cartesian product $A = V^2$. Clearly, $CompDi(V)^* = CompDi(V)$. The *discrete* directed graph $DiDi(V)$ over the set V is the one defined by the empty set of arrows.

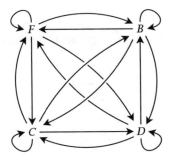

Fig. 10.3. The complete digraph $CompDi(V)$ over the vertex set $V = \{B, C, D, F\}$.

Example 24 In process theory, a labeled transition system (LTS) is a subset $T \subset S \times Act \times S$ of the Cartesian product of a *state space S*, a set *Act* of *labels*, together with a selected start state $s_0 \in S$. For each state $s \in S$, there is a number of transitions, e.g., the triples $(s, l, t) \in T$, parameterizing "transitions from state s to state t via the transition type l". Defining the directed graph $\Gamma = T \to S^2$ via $head_\Gamma = pr_3$ and $tail_\Gamma = pr_1$, we associate a directed graph with the LTS, together with a distinct vertex s_0. Show that conversely, every directed graph, together with a distinct vertex defines an LTS. How are these two constructions related?

Example 25 If for a directed graph $\Gamma = A \subset V^2$, the vertex set is a disjoint union $V = V_1 \cup V_2$ of subsets, and if for all arrows a, we have $\Gamma(a) \in V_1 \times V_2 \cup V_2 \times V_1$, then the graph is called *bipartite (with respect to V_1 and V_2)*. For any partition $V = V_1 \cup V_2$ of a finite set V, one has the *complete bipartite digraph BipDi(V_1, V_2)*, defined by the inclusion $V_1 \times V_2 \cup V_2 \times V_1 \subset V^2$. Clearly, $BipDi(V_1, V_2) = BipDi(V_2, V_1)$ and $BipDi(V_1, V_2)^* = BipDi(V_1, V_2)$.

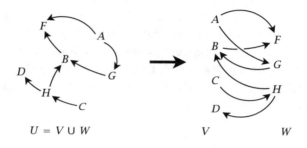

Fig. 10.4. A bipartite graph with vertexes $U = V \cup W$.

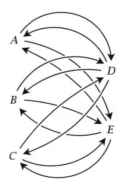

Fig. 10.5. A complete bipartite graph with vertexes $U = \{A, B, C\} \cup \{D, E\}$.

Example 26 In automata theory, a special type of directed graphs are called *Petri nets*, they were introduced by Carl Adam Petri in 1962. A

Petri net has two sets: the set P of *places*, and the set *Tr of transitions*. It is supposed that some places are related "by input arcs" to transitions as *input places*, whereas some places are related "by output arcs" to transitions as *output places*. It is also assumed that every transition has at least one input and one output place, and that an input place cannot be an output place of the same transition. This means that we can view a Petri net as an LTS (except that no initial state s_0 is specified) where the labels are the transitions, and where the triples $(p, t, q) \in P \times Tr \times P$ are the elements of the ternary state space relation T in example 24. The axiom eliminating the possibility "input = output" means that the directed graph of the Petri net has no *loop*, i.e., no arrow of type $p \xrightarrow{a} p$.

The next subject relates set theory to graph theory.

Definition 63 *For a set x and for $n \in \mathbb{N}$, we define inductively $\bigcup^0 x = x$ and $\bigcup^{n+1} x = \bigcup(\bigcup^n x)$. The set x is called* totally finite *iff there is a natural number m such that $\bigcup^m x = \varnothing$. Call the minimal such m the level $lev(x)$ of x.*

Example 27 Let $x = \{\{\{\{\varnothing\}, \varnothing\}, \varnothing, \{\varnothing\}\}, \{\{\varnothing\}\}\}$. Then $x_0 = \bigcup^0 x = x$, $x_1 = \bigcup^1 x = \bigcup x_0 = \{\{\{\varnothing\}, \varnothing\}, \varnothing, \{\varnothing\}\}$, $x_2 = \bigcup^2 x = \bigcup x_1 = \{\{\varnothing\}, \varnothing\}$, $x_3 = \bigcup^3 x = \bigcup x_2 = \{\varnothing\}$, and, finally, $x_4 = \bigcup^4 x = \bigcup x_3 = \varnothing$. Thus x is totally finite and $lev(x) = 4$.

The set $\mathbb{N}$ of natural numbers is not totally finite. The set $a = \{a\}$ is finite, but not totally finite, as can be easily verified.

Clearly, if $y \in x$, then y is also totally finite and $lev(y) < lev(x)$. We now associate the notation $Fi(x)$ to any totally finite set as follows (Paul Finsler (1894–1970) was a mathematician at the University of Zurich).

Definition 64 *The vertex set V of the* Finsler digraph $Fi(x)$ *of a totally finite set x is the union $V = \{x\} \cup \bigcup_{i=0,\dots lev(x)-1} \bigcup^i x$. Observe that all pairs $\{x\}, \bigcup^i x$ of sets are mutually disjoint, otherwise, x would not be totally finite! The arrow set is the set $\{(r, s) \mid s \in r\} \subset V^2$, i.e., the arrows $r \xrightarrow{a} s$ of $Fi(x)$ correspond to the element $(\in)$ relation.*

Example 28 With x defined as in example 27, the vertex set of $Fi(x)$ is $V = \{x\} \cup \bigcup_{i=0,1,2,3} \bigcup^i x = \{x\} \cup \bigcup_{i=0,1,2,3} x_i = \{x\} \cup x_0 \cup x_1 \cup x_2 \cup x_3$. Denoting $a = \{\{\{\varnothing\}, \varnothing\}, \varnothing, \{\varnothing\}\}$, $b = \{\{\varnothing\}\}$, $c = \{\{\varnothing\}, \varnothing\}$, and $d = \{\varnothing\}$, then $V = \{x\} \cup \{a,b\} \cup \{c,d,\varnothing\} \cup \{d,\varnothing\} \cup \{\varnothing\} = \{x,a,b,c,d,\varnothing\}$. The arrow set is $\{(x,a),(x,b),(a,c),(a,d),(a,\varnothing),(b,d),(c,d),(c,\varnothing),(d,\varnothing)\}$. The resulting Finsler digraph is shown in figure 10.6.

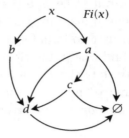

Fig. 10.6. The Finsler digraph $Fi(x)$ of the set x from example 28.

The Finsler digraphs characterize totally finite sets, in other words, we may redefine such sets starting from graphs. This is a new branch of theoretical computer science used in parallel processor theory, formal ontologies, and artificial intelligence (search for the keyword "hyperset" on the Web for more details.) Important contributions to this branch of computer science, mathematics, and formal logic have been made by Jon Barwise, Lawrence Moss [3], and Peter Aczel [1].

Example 29 In an object-oriented language, for example Java, we consider a class library which we suppose being given as a set L, the elements C of which represent the library's classes, including (for simplicity) the primitive type classes of integers, floating point numbers, strings, and booleans. For each class C we have a number of fields (instance variables) defined by their names F and class types $T \in L$. This defines a digraph Λ_L, the vertex set being L, and the arrows being either the triples (C,F,T) where F is the name of a field of class C, and where T is the type class of F, or the pairs (C,S), where S is the direct superclass of C; we set $head_\Lambda(C,F,T) = T$ and $tail_\Lambda(C,F,T) = C$, or, for superclass arrows, $head_\Lambda(C,S) = T$ and $tail_\Lambda(C,S) = S$.

If one forgets about the direction in a digraph, the remaining structure is that of an "undirected" graph, or simply "graph". *We shall henceforth always write digraph for a directed graph, and graph for an undirected graph, if a confusion is unlikely.*

Definition 65 *A(n undirected) graph is a map* $\Gamma : A \to {}^2V$ *between finite sets. The elements of the set A are called* edges, *the elements of V are called* vertexes *of the graph. For $a \in A$ and $A(a) = \{x, y\}$, we also write* $x \xrightarrow{\ a\ } y$, *which is the same as* $y \xrightarrow{\ a\ } x$.

Example 30 For each directed graph Γ, one generates the *associated graph* $|\Gamma| = |?| \circ \Gamma$, and for any given (undirected) graph Γ, one generates an *associated directed graph* $\vec{\Gamma} = \vec{?} \circ \Gamma$, the latter construction supposing that a section $\vec{?}$ for the graph's vertex set is given.

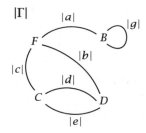

Fig. 10.7. The associated graph $|\Gamma|$ to the graph from example 21.

Example 31 A graph $\Gamma : A \to {}^2V$ is *complete* iff Γ is a bijection onto the subset ${}^2V - V$. The complete graph *Comp(V)* of a set V is the inclusion ${}^2V - \{\{x\} \mid x \in V\} \subset {}^2V$, i.e., intuitively the set of all lines between any two different points in V. A graph $\Gamma : A \to {}^2V$ is *bipartite*, iff there is a partition $V = V_1 \cup V_2$ such that for all edges a, $\Gamma(a) = \{x, y\}$ with $x \in V_1$ and $y \in V_2$. The complete bipartite graph *Bip(V₁, V₂)* for two disjoint sets V_1 and V_2 is the embedding $\{\{x, y\} \mid x \in V_1, y \in V_2\} \subset {}^2V$. The discrete (undirected) graph over the vertex set V is denoted by *Di(V)*.

10.2 Morphisms of Digraphs and Graphs

Evidently, there are many (directed or undirected) graphs which look much the same, such as with sets which are essentially the same, in the sense that they are equipollent, i.e., they have same cardinality. In order to control this phenomenon, we need a means to compare graphs in much the same way as we had to learn how to compare sets by use of functions.

Definition 66 *Let* $\Gamma : A \to V^2$ *and* $\Delta : B \to W^2$ *be two digraphs. A morphism* $f : \Gamma \to \Delta$ *of digraphs is a pair* $f = (u, v)$ *of maps* $u : A \to B$ *and* $v : V \to W$ *such that* $v^2 \circ \Gamma = \Delta \circ u$, *in other words, for any arrow* $t \xrightarrow{a} h$ *in* Γ, *we have* $v(t) \xrightarrow{u(a)} v(h)$. *This means that we have the following commutative diagram:*

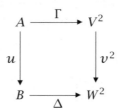

In particular, the identity morphism $Id_\Gamma = (Id_A, Id_V)$ *is a morphism, and, if* $f = (u, v) : \Gamma \to \Delta$ *and* $g = (u', v') : \Delta \to \Theta$ *are two morphisms, then their* composition *is a morphism* $g \circ f = (u' \circ u, v' \circ v) : \Gamma \to \Theta$.

We denote the set of digraph morphisms $f : \Gamma \to \Delta$ *by* $Digraph(\Gamma, \Delta)$.

Example 32 Figure 10.8 shows two digraphs Γ and Δ and a morphism $f = (u, v)$. The map v on the vertexes is drawn with light gray arrows, the map u on edges is drawn with dark gray arrows.

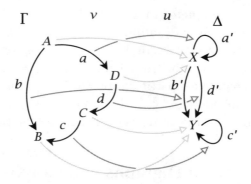

Fig. 10.8. A morphism of digraphs.

Sorite 96 *Call the two maps u and v defining a digraph morphism $f = (u, v)$ its components.*

   (i) *If $f : \Gamma \to \Delta, g : \Delta \to \Theta, h : \Theta \to \Psi$ are morphisms of digraphs, then $h \circ (g \circ f) = (h \circ g) \circ f$, which we denote (as usual) by $h \circ g \circ f$.*

   (ii) *There is a morphism $k : \Delta \to \Gamma$ such that $k \circ f = Id_\Gamma$ and $f \circ k = Id_\Delta$ iff the components u and v of $f = (u, v)$ are bijections. In this case, f is called an* isomorphism *of digraphs.*

Proof (i) Associativity of digraph morphisms follows directly from the associativity of set maps which define digraph morphisms.

(ii) Since the composition of two digraph morphisms is defined by the set-theoretic composition of their two components, the claim is immediate from the synonymous facts for set maps. □

Remark 12 Informally speaking, the system *Digraph* of digraphs $\Gamma, \Delta, \ldots$ together with their sets *Digraph*(Γ, Δ) of morphisms, including their associative composition (sorite 96) is called the *category* of digraphs. A formal definition of a category will be given in volume II of this book. Evidently, we already have encountered the structure of category for sets and their functions. Categories constantly appear in all of mathematics, they are the most powerful unifying structure of modern mathematics and computer science.

Example 33 An immediate class of morphisms is defined by subgraphs. More precisely, given a digraph $\Gamma : A \to V^2$, if inclusions $u : A' \subset A$ and $v : V' \subset V$ of subsets A' and V' are such that $\Gamma(A') \subset (V')^2$, then we have an *induced digraph* $\Gamma' : A' \to (V')^2$, and a subgraph inclusion morphism $(u, v) : \Gamma' \subset \Gamma$. In particular, if for a subset $V' \subset V$, we take $A' = \Gamma^{-1}(V')$, we obtain the subgraph $\Gamma|_{V'}$ *induced on* V'.

The set cardinality classifies sets up to bijections. In particular, for finite sets, the natural number $card(a)$ is classifying. Similarly, for digraphs we also want to have prototypes of digraphs such that each digraph is isomorphic to such a prototype. To begin with we have this:

Exercise 44 Show that each digraph $\Gamma : A \to V^2$ is isomorphic to a digraph $\Gamma' : A \to card(V)^2$.

For a digraph $\Gamma : A \to V^2$ and $e, f \in V$, we consider the sets $A_{e,f} = \Gamma^{-1}(e, f) = \{a \mid a \in A, tail(a) = e, head(a) = f\}$.

Definition 67 *Let* $\Gamma : A \rightarrow V^2$ *be a digraph, and fix a bijection* $c :$ *card*$(V) \rightarrow V$. *The* adjacency matrix *of* Γ (*with respect to* c) *is the function* $Adj_c(\Gamma) : card(V)^2 \rightarrow \mathbb{N} : (i, j) \mapsto card(A_{c(i),c(j)})$.

We shall deal extensively with matrixes in chapter 21. But we will already now show the standard representation of a matrix, and in particular of the adjacency matrix. It is a function of a finite number of pairs (i, j). Such a function is usually represented in a graphical form as a tabular field with n rows and n columns, and for each row number i and column number j, we have an entry showing the function value $Adj_c(i, j)$, i.e.,

$$Adj_c = \begin{pmatrix} Adj_c(0,0) & Adj_c(0,1) & \ldots & Adj_c(0, n-1) \\ Adj_c(1,0) & Adj_c(1,1) & \ldots & Adj_c(1, n-1) \\ \vdots & \vdots & & \vdots \\ Adj_c(n-1,0) & Adj_c(n-1,1) & \ldots & Adj_c(n-1, n-1) \end{pmatrix}$$

Observe however that here, the row and column indexes start at 0, whereas in usual matrix theory, these indexes start at 1, i.e., for usual matrix notation, the entry at position (i, j) is our position $(i - 1, j - 1)$.

Example 34 For simplicity's sake, the vertexes of the digraph Γ are all the natural numbers in 6. Thus the bijection c is the identity. The adjacency matrix of Γ is shown below the digraph in figure 10.9.

The following numerical criterion for isomorphic digraphs is very important for the representation of digraphs in computerized representations:

Proposition 97 *If* Γ *and* Δ *are two digraphs such that* $Adj(\Gamma) = Adj(\Delta)$, *then they are isomorphic, i.e., there is an isomorphism* $\Gamma \xrightarrow{\sim} \Delta$.

Proof Suppose that for two digraphs $\Gamma : A \rightarrow V^2$ and $\Delta : B \rightarrow W^2$, $Adj(\Gamma) = Adj(\Delta)$. Then the number of rows of the two matrixes is the same, and therefore we have a bijection $v : V \rightarrow W$ such that the matrix index (i, j) in $Adj(\Gamma)$ corresponds to the same index of $Adj(\Delta)$. Moreover, we have $A = \bigsqcup_{x \in V^2} \Gamma^{-1}(x)$ and $B = \bigsqcup_{y \in W^2} \Delta^{-1}(y)$. But since by hypothesis $card(\Gamma^{-1}(x)) = card(\Delta^{-1}(v^2(x))$ for all $x \in V$, setting $v^2 = v \times v$, we have bijections $u_x : \Gamma^{-1}(x) \rightarrow \Delta^{-1}(v^2(x))$ and their disjoint union yields a bijection $u : A \rightarrow B$ which defines the desired isomorphism (u, v). $\square$

So the adjacency matrix of a digraph describes the digraph "up to isomorphisms". The converse is not true, but it can easily be said what is missing: Essentially, we have to take into account the bijection c, which labels the vertexes. We come back to this subject in chapter 15.

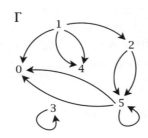

$$Adj_c(\Gamma) = \begin{pmatrix} 0 & 0 & 0 & 0 & 0 & 0 \\ 1 & 0 & 1 & 0 & 2 & 0 \\ 0 & 0 & 0 & 0 & 0 & 2 \\ 0 & 0 & 0 & 1 & 0 & 0 \\ 0 & 0 & 0 & 0 & 0 & 0 \\ 2 & 0 & 0 & 0 & 0 & 1 \end{pmatrix}$$

Fig. 10.9. Adjacency matrix of a digraph.

Example 35 The bipartite digraphs can now be denoted by simple numbers, i.e., if n and m are natural numbers, we have the bipartite digraph $BipDi(n, m)$, and every bipartite digraph $BipDi(V_1, V_2)$ with $card(V_1) = n$ and $card(V_2) = m$ is isomorphic to $BipDi(n, m)$. The same situation holds for complete or discrete digraphs. Given a natural number n, the complete digraph $CompDi(n)$ is isomorphic to any $CompDi(V)$ such that $card(V) = n$. And the discrete digraph $DiDi(n)$ is isomorphic to any $DiDi(V)$ if $card(V) = n$.

An important type of morphisms with special digraphs, namely chains, as domains is defined as follows:

Definition 68 *Given* $n \in \mathbb{N}$, *the* directed chain $[n]$ *of length* n *is the digraph on the vertex set* $V = n + 1$ *with the* n *arrows* $(i, i + 1), i \in n$. *In particular, if* $n = 0$, *then* $[0] = DiDi(1)$ *(discrete with one vertex), whereas in general, the leftmost arrow is* $0 \longrightarrow 1$, *followed by* $1 \longrightarrow 2$, *etc., up to* $(n - 1) \longrightarrow n$.

Definition 69 *Given a digraph* Γ *and* $n \in \mathbb{N}$, *a* path *of length* n *in* Γ *is a morphism* $p : [n] \to \Gamma$, *write* $l(p) = n$ *for the length. (Equivalently,* p *may be described as a sequence of arrows* $(a_i)_{i=1,...n}$ *in* Γ *such that for every* $i < n$, *head*$(a_i) = $ *tail*(a_{i+1}).) *If* $p(0) = v$ *and* $p(n) = w$, *one also* *says that p is a path* from *v* to *w*. *A path of length 0—just one vertex v in Γ—is called the* lazy path at *v*, *and also denoted by v*. *A non-lazy path p* : [*n*] → Γ *such that p(0) = p(n) is called a* cycle in Γ. *A cycle of length 1 at a vertex v is called a* loop at *v*.

Directed chain [0] of length 0

    0

Directed chain [1] of length 1

    0 ⟶ 1

Directed chain [2] of length 2

    0 ⟶ 1 ⟶ 2

Directed chain [4] of length 4

    0 ⟶ 1 ⟶ 2 ⟶ 3 ⟶ 4

Fig. 10.10. Directed chains.

Example 36 In the digraph of figure 10.11, the light gray arrows form a path from *A* to *B*, the dark gray ones a cycle from *A* to *A*, and the dashed arrow is a loop on *D*.

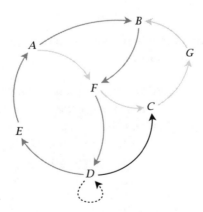

Fig. 10.11. Paths and cycles in a digraph.

Definition 70 *A vertex v in a digraph Γ such that for every vertex w in Γ, there is a path from v to w, is called a* root *or* source *of Γ, a root in the dual digraph is called a* co-root *or* sink *of Γ. A digraph without directed cycles and with a (necessarily unique) root is called a* directed tree. *A vertex v in a directed tree, which is not the tail of an arrow, is called a* leaf *of the tree.*

Example 37 The roots in the figure are $\{A, C, E\}$ and the co-roots are $\{D, B\}$. Note that if a root lies on a cycle, all vertexes on the cycle are roots. The analogue statement is valid for co-roots.

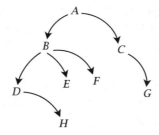

The following figure shows a tree with root A and no co-root, with leafs $\{E, F, G, H\}$.

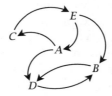

Exercise 45 Given two paths p and q of lengths n and m, respectively, in a digraph Γ such that p ends where q begins, we have an evident *composition qp of paths* of length $l(qp) = n + m = l(q) + l(p)$, defined by joining the two arrow sequences in the connecting vertex. Composition of paths is associative, and the composition with a lazy path at v, if it is defined, yields $vp = p$, or $pv = p$.

This allows us to characterize totally finite sets by their Finsler digraphs (see definition 64). To this end, we consider the following subgraph construction: If Γ is a graph and v a vertex, we define $v\rangle$ to be the set of all vertexes w such that there is a path from v to w. The directed subgraph induced on $v\rangle$ is also denoted by $v\rangle$ and is called the *sub(di)graph generated by v.*

Proposition 98 *A digraph Γ is isomorphic to a Finsler digraph $Fi(x)$ of a totally finite set x iff it is a directed tree such that for any two vertexes v_1 and v_2, if $v_1\rangle$ is isomorphic to $v_2\rangle$, then $v_1 = v_2$.*

Proof Let x be totally finite. Then in the Finsler digraph $Fi(x)$, every $y \in x$ is reached from the vertex x. And if $z \in \bigcup^{n+1} x = \bigcup\bigcup^n x$, then it is reached from a selected element in $\bigcup^n x$. So x is the root of $Fi(x)$. If y is a vertex in $Fi(x)$, then clearly $y\rangle = Fi(y)$. Let us show by induction on $lev(y)$ that an isomorphism $Fi(y) \overset{\sim}{\to} Fi(z)$ implies $y = z$. In fact, if $lev(y) = 0$, then $Fi(y), Fi(z)$ are the one-point digraph, and both, y and z must be the empty set. In general, if $Fi(y) \overset{\sim}{\to} Fi(z)$, then the roots must also correspond under the given isomorphism, and therefore there is a bijection between the vertexes y_i reached from y by an arrow, and the vertexes z_i reached from z by an arrow. If z_i corresponds to y_i, then also $z_i\rangle \overset{\sim}{\to} y_i\rangle$, and therefore, by recursion, $y_i = z_i$, which implies $y = z$. Conversely, if a directed tree $F : V \to A^2$ is such that any two vertexes v_1 and v_2 with $v_1\rangle \overset{\sim}{\to} v_2\rangle$ must be equal, then it follows by induction on the length of the maximal path from a given vertex that F is isomorphic to the Finsler digraph of a totally finite set. In fact, if a directed tree has maximal path length 0 from the root, it is the Finsler digraph of the empty set. In general, the vertexes $y_i, i = 1, \dots k$ reached from the root y by an arrow have a shorter maximal path than the root and are also directed trees with the supposed conditions. So by recursion, they are Finsler digraphs $Fi(t_i)$ of totally finite, mutually different sets t_i. Then evidently, y is isomorphic to the Finsler digraph of the set $\{t_i \mid i = 1, \dots k\}$. $\square$

Corresponding to morphisms between directed graphs, we have morphisms between undirected graphs. We need the undirected variant of the square map $v^2 : V^2 \to W^2$ associated with a map $v : V \to W$, i.e., $^2v : {}^2V \to {}^2W : \{x, y\} \mapsto \{v(x), v(y)\}$.

Definition 71 *Let $\Gamma : A \to {}^2V$ and $\Delta : B \to {}^2W$ be two graphs. A morphism $f : \Gamma \to \Delta$ of graphs is a pair $f = (u, v)$ of maps $u : A \to B$ and $v : V \to W$ such that $^2v \circ \Gamma = \Delta \circ u$, in other words, for any edge $t \overset{a}{\text{———}} h$ in Γ, we have $v(t) \overset{u(a)}{\text{———}} v(h)$. This means we have the commutative diagram*

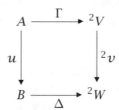

In particular, the identity morphism $Id_\Gamma = (Id_A, Id_V)$ is a morphism, and, if $f = (u, v) : \Gamma \to \Delta$ and $g = (u', v') : \Delta \to \Theta$ are two morphisms, then their composition *is a morphism $g \circ f = (u' \circ u, v' \circ v) : \Gamma \to \Theta$.*

Exercise 46 Show that the mapping $\Gamma \mapsto |\Gamma|$ can be extended to morphisms, i.e., if $f = (u, v) : \Gamma \to \Delta$, is a morphism, then so is $|f| = f : |\Gamma| \to |\Delta|$. Also, $|Id_\Gamma| = Id_{|\Gamma|}$, and $|f \circ g| = |f| \circ |g|$.

The following sorite looks exactly like its corresponding version for digraphs:

Sorite 99 *Call the two maps u and v defining a graph morphism $f = (u, v)$ its components.*

(i) *If $f : \Gamma \to \Delta$, $g : \Delta \to \Theta$, and $h : \Theta \to \Psi$ are morphisms of graphs, then $h \circ (g \circ f) = (h \circ g) \circ f$, which we (as usual) denote by $g \circ h \circ f$.*

(ii) *There is a morphism $k : \Delta \to \Gamma$ such that $k \circ f = Id_\Gamma$ and $f \circ k = Id_\Delta$ iff the components u and v of $f = (u, v)$ are bijections. In this case, f is called an isomorphism of graphs.*

Proof This results from the corresponding set-theoretic facts much as it did for the digraph sorite 96. □

Example 38 In complete analogy with exercise 33 on directed subgraphs, we have subgraphs for undirected graphs, i.e., we require that for a graph $\Gamma : A \to {}^2V$ and two subsets $u : A' \subset A$ and $v : V' \subset V$, Γ restricts to $\Gamma' : A' \to {}^2V'$, and in particular, if $A' = \Gamma^{-1}(V')$, we get the subgraph $\Gamma|_{V'}$ induced on V'.

As for digraphs, clearly every graph with vertex set V is isomorphic to a graph the vertexes of which are elements of the natural number $n = card(V)$. If $\{e, f\} \in {}^2V$, we denote $A_{\{e,f\}} = \Gamma^{-1}(\{e, f\})$. Then

Definition 72 *Let $\Gamma : A \to {}^2V$ be a graph, and fix a bijection $c : card(V) \to V$. The* adjacency matrix *of Γ (with respect to c) is the function $Adj_c(\Gamma) : card(V)^2 \to \mathbb{N} : (i, j) \mapsto card(A_{\{c(i),c(j)\}})$.*

Example 39 The adjacency matrix for $|\Gamma|$ (where Γ is the same as in example 34) is:

$$Adj_c(|\Gamma|) = \begin{pmatrix} 0 & 1 & 0 & 0 & 0 & 2 \\ 1 & 0 & 1 & 0 & 2 & 0 \\ 0 & 1 & 0 & 0 & 0 & 2 \\ 0 & 0 & 0 & 1 & 0 & 0 \\ 0 & 2 & 0 & 0 & 0 & 0 \\ 2 & 0 & 2 & 0 & 0 & 1 \end{pmatrix}$$

Obviously the adjacency matrix of an undirected graph is symmetrical, i.e., $Adj_c(\Gamma)(i,j) = Adj_c(\Gamma)(j,i)$ for all pairs i, j.

Proposition 100 *If Γ and Δ are two graphs such that $Adj(\Gamma) = Adj(\Delta)$, then they are isomorphic.*

Proof The proof of this proposition is completely analogous to the proof of the corresponding proposition 97 for digraphs. $\square$

Definition 73 *Given $n \in \mathbb{N}$, the* chain $|n|$ *of length n is the graph on the vertex set $V = n + 1$ such that we have the n edges $\{i, i + 1\}, i \in n$. In particular, if $n = 0$, then $|0| = Di(1)$ (discrete with one vertex), whereas in general, the leftmost arrow is $0 \underline{\hspace{1cm}} 1$, followed by $1 \underline{\hspace{1cm}} 2$, etc., up to $(n - 1) \underline{\hspace{1cm}} n$.*

Example 40 For any natural number n, we have the complete graph $Comp(n)$, which in literature is often denoted by K_n, it is isomorphic to any complete graph $Comp(V)$ with $card(V) = n$. For two natural numbers n and m, we have the complete bipartite graph $Bip(n, m)$, often denoted by $K_{n,m}$, which is isomorphic to any complete bipartite graph $Bip(V_1, V_2)$ such that $card(V_1) = n$ and $card(V_2) = m$.

Definition 74 *Given a graph Γ and $n \in \mathbb{N}$, a* walk *of length n in Γ is a morphism $p : |n| \to \Gamma$, we write $l(p) = n$ for the length. If $p(0) = v$ and $p(n) = w$, one also says that p is a walk from v to w. A walk of length 0—just one vertex v in Γ—is called the* lazy *walk at v, and also denoted by v. A non-lazy walk $p : |n| \to \Gamma$ such that $p(0) = p(n)$ is called a* cycle *in Γ. A cycle of length 1 at a vertex v is called a* loop *at v.*

A graph is said to be connected *if any two vertexes can be joined by a walk. A connected graph without (undirected) cycles is called a* tree.

A directed graph Γ is called connected *if $|\Gamma|$ is so.*

As with directed graphs, we may also compose walks, more precisely, if p is a walk from v to w, and q is one from w to z, then we have an evident walk qp from v to z of length $l(qp) = l(q) + l(p)$.

Remark 13 Walks and paths can be treated at once if we define a walk from v to w in a graph Γ as a path from v to w in any digraph $\vec{\Gamma}$ such that $\Gamma = |\vec{\Gamma}|$. We shall also use this variant if it is more appropriate to the concrete situation.

Lemma 101 *If Γ is a graph, the binary relation $v \sim w$ iff there is a walk from v to w is an equivalence relation. The subgraph induced on an equivalence class is called a* connected component *of Γ.*

If Γ is a directed graph, its connected components are the directed subgraphs induced on the equivalence classes of vertexes defined by the associated graph $|\Gamma|$. Since there are no paths or walks between any two distinct connected components, two (di)graphs Γ and Δ are isomorphic iff there is an enumeration $\Gamma_i, \Delta_i, i = 1, \ldots k$ of their connected components such that component Γ_i is isomorphic to component Δ_i.

Proof The relation $\sim$ is evidently reflexive, just take the lazy walk. It is symmetric since, if $p : |n| \to \Gamma$ is a walk from v to w, then the "reverse" walk $r : |n| \to |n|$ with $r(i) = n - i$ turns p into the walk $p \to r$ from w to v. The rest is clear. $\square$

Remark 14 If Γ is a tree with vertex set V, and if v is a vertex, then the graph induced on $V - \{v\}$ is a disjoint union of connected components Γ_i which are also trees. Therefore, the tree Γ is determined by these subtrees Γ_i, together with the vertexes being joined by an edge from v to determined vertexes v_i in Γ_i (check this: in fact, if there were more than one such vertex, we would have cycles, and Γ would not be a tree). This allows us to define the tree concept recursively by the total vertex set V, the selected vertex v, the edges $v \text{———} v_i$, and the subtrees Γ_i. This is the definition of a tree given by Donald Knuth in his famous book, "The Art of Computer Programming" [31]. It has however the defect that it distinguishes a vertex v, which is not the intrinsic property of a tree. For computer science it has the advantage that it is constructive and recursive.

10.3 Cycles

In this section, we want to state a number of elementary facts concerning the existence or absence of cycles in directed and undirected graphs.

Definition 75 *An* Euler cycle e *in a digraph/graph Γ is a cycle such that every vertex and every arrow/line of Γ lies on the cycle (i.e., e is surjective on the vertexes and on the arrows/lines), but every arrow/line appears only once (i.e., e is bijective on the arrows/lines).*

A Hamilton cycle h in a digraph/graph Γ is a cycle which contains every vertex (i.e., h is surjective on the vertexes), but only hits the start and end vertex twice (i.e., is a bijection on the vertexes $0, 1, 2, \ldots l(h) - 1$).

The condition for the existence of Euler cycles in digraphs is described using the degree of a vertex:

Definition 76 *If v is a vertex of a digraph $\Gamma : A \to V^2$, the head degree of v is the number*

$$deg^-(v) = card(\{h \mid h \in A, head(h) = v\}),$$

the tail degree of v is the number

$$deg^+(v) = card(\{h \mid h \in A, tail(h) = v\}),$$

and the degree of v is the number

$$deg(v) = deg^-(v) + deg^+(v).$$

For a graph Γ, the degree $deg(v)$ of a vertex v is defined as the degree of v in any of the digraphs $\vec{\Gamma}$ such that $|\vec{\Gamma}| = \Gamma$; observe that this number is independent of the choice of the digraph $\vec{\Gamma}$.

Here are two classical results by Leonhard Euler:

Proposition 102 *Let Γ be a digraph without loops. Then it has an Euler cycle iff it is connected and for every vertex v, we have $deg^-(v) = deg^+(v)$.*

Proof If Γ has an Euler cycle $c : [n] \to \Gamma$, any two vertexes are connected by a subwalk associated to this cycle, so Γ is connected. If v is a vertex of Γ, then if $w \xrightarrow{a} v, w \neq v$ is an arrow of Γ, it appears right before a subsequent arrow $v \xrightarrow{a'} w$ in the cycle c. But these two arrows appear exactly one each in c, and therefore $deg(v)$ is even.

Conversely, if Γ is connected and for every vertex v, $deg^-(v) = deg^+(v)$, then we first construct a covering of Γ by cycles c_i. Here we need only the hypothesis that the degrees are all even. Let a be an arrow of Γ, then there must exist an arrow a' starting from $head(a)$. The head of a' must also have an arrow a'' the tail of which is the head of a'. After a finite number of such steps, we end up at a vertex which we had already encountered before, and this defines a first cycle. Now omit all these cycle's arrows, and the hypothesis about the even number of degrees still holds, but for the smaller digraph obtained after elimination of the cycle's arrows. By induction we have a covering of this digraph by cycles, and, together with our first cycle obtain the desired covering. Clearly, a union of cycles with disjoint arrows for a connected digraph is again a cycle. To see this observe that (1) if a cycle c that starts and ends at a vertex v, then these vertexes are also transgressed by a cycle c' which starts and ends at any of the vertexes of c; (2) if two cycles c and d in Γ have the vertex v in common, but have disjoint arrow sets, then there is also a cycle containing the union of these arrow sets; (3) if Γ is connected, any given cycle in Γ can be extended by adding to it another cycle of the given covering, having disjoint arrows. □

Proposition 103 *Let Γ be a graph without loops. Then it has an Euler cycle iff it is connected and for every vertex v, $deg(v)$ is an even number (a multiple of 2).*

Proof The case of a graph is clear except for the existence of an Euler cycle for an even number $deg(v)$ in each vertex v. Here, we may couple lines at v in pairs of lines and associate with each pair one incoming and one outgoing arrow. This procedure yields a digraph over the given graph, and we may apply proposition 102. □

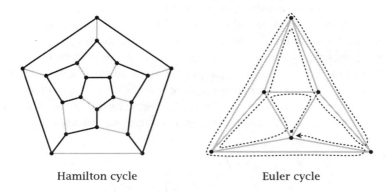

Hamilton cycle   Euler cycle

Fig. 10.12. A Hamilton cycle (heavy lines) on the flattened dodecahedron and a Euler cycle (dashed line) on the flattened octahedron. The dodecahedron has no Euler cycle, since there are vertexes with odd degree.

Exercise 47 A political group wants to make a demonstration in Zurich and asks for official permission. The defined street portions of the demonstration path are the following lines, connecting these places in Zurich: Z = Zweierplatz, S = Sihlporte, HB = Hauptbahnhof, C = Central, HW = Hauptwache, Pr = Predigerplatz, M = Marktgasse, Pa = Paradeplatz, B = Bellevue.

$$Z \text{ ——— } S, Z \text{ ——— } Pa, S \text{ ——— } HB, S \text{ ——— } Pa, HB \text{ ——— } C,$$
$$HB \text{ ——— } HW, Pa \text{ ——— } HW, Pa \text{ ——— } M, Pa \xrightarrow{1} B, Pa \xrightarrow{2} B,$$
$$HW \text{ ——— } Pr, Pr \text{ ——— } M, M \text{ ——— } B.$$

The permission is given if a walk can be defined such that each connection is passed not more than once. Will the permission be given?

A *spanning sub(di)graph* of a (di)graph is a sub(di)graph which contains all vertexes.

Proposition 104 *Every connected graph has a spanning tree, i.e., a spanning subgraph which is a tree. Every connected digraph has a spanning directed tree.*

Proof It suffices to prove the proposition for graphs, then the corresponding digraphs will also yield a spanning directed tree. We prove that case by induction on the number of lines. If there is only one vertex, we are done. Else, there are at least two different vertexes v and w which are connected by a line a. Omit that line. Then, if the remaining graph is still connected, we are done. Else, every vertex must be connectible to either v or w. In fact, before eliminating a, every vertex x was connectible to v. So if x is still connectible to v it pertains to the connected component of v, else, there any walk from x to v must traverse w via a. But then, x is in the connected component of w. So the graph without a has exactly the two connected components C_v of v and C_w of w. By induction, each component C_v, C_w has a spanning tree T_v, T_w, respectively. Adding the line a to the union of the disjoint T_v and T_w still yields a tree, and we are done.   □

Fig. 10.13. Spanning tree of an undirected graph.

Exercise 48 Show that the statement about spanning directed trees follows from the statement about spanning trees in undirected graphs.

CHAPTER 11

Construction of Graphs

We already know about some constructions of new (di)graphs from given ones: The dual digraph is such a construction. The spanning tree is a second one. Here are some more of these constructions.

We have seen in set theory that there are universal constructions of new sets from given ones, such as the Cartesian product $a \times b$, the coproduct $a \sqcup b$, and the powerset a^b. In graph theory, one also has such constructions which we discuss now. They are of very practical use, especially in computer science where the systematic construction of objects is a core business of software object-oriented engineering.

Given two digraphs $\Gamma : A \to V^2$ and $\Delta : B \to W^2$, we have the Cartesian product $\Gamma \times \Delta : A \times B \to (V \times W)^2$, defined by the canonical[1] isomorphism of sets $t : V^2 \times W^2 \to (V \times W)^2 : ((e_1, e_2), (f_1, f_2)) \mapsto ((e_1, f_1), (e_2, f_2))$. In other words, we have

$$head(a, b) = (head(a), head(b))$$
$$tail(a, b) = (tail(a), tail(b)).$$

Example 41 Figure 11.1 shows the Cartesian product of two digraphs Γ and Δ.

---

[1] In mathematics, a construction is called "canonical" if no particular trick is necessary for its elaboration, it is realized by the given "surface structures". Attention: the attribute "natural", which we would also like to use instead, is reserved for a technical term in category theory.

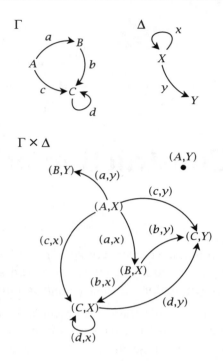

Fig. 11.1. The Cartesian product of two digraphs Γ and Δ.

Exercise 49 Show that the Cartesian product shares formal properties which are completely analogous to those stated for sets. Use the commutative diagram introduced in that context.

We also have a coproduct $\Gamma \sqcup \Delta : A \sqcup B \to (V \sqcup W)^2$, defined by the coproduct $A \sqcup B \to V^2 \sqcup W^2$ of the given maps, followed by the obvious injection $V^2 \sqcup W^2 \to (V \sqcup W)^2$.

Exercise 50 Show that the coproduct shares formal properties which are completely analogous to those stated for sets. Use the commutative diagram introduced in that context.

For undirected graphs, the coproduct construction works exactly as for digraphs, except that we have to replace the right exponent X^2 by the left one ^2X and do the analog mappings. The coproduct of (di)graphs is also called the *disjoint union of (di)graphs*.

Exercise 51 However, for the Cartesian product of graphs Γ and Δ, the analogy does not work. In this case, using exercise 46, we can perform a construction, which yields the best approximation to a product: Take two digraphs $\vec{\Gamma}$ and $\vec{\Delta}$, then their Cartesian product, together with the two projections $\vec{\Gamma} \times \vec{\Delta} \to \vec{\Gamma}$ and $\vec{\Gamma} \times \vec{\Delta} \to \vec{\Delta}$, and then the associated undirected configuration $|\vec{\Gamma} \times \vec{\Delta}| \to \Gamma$ and $|\vec{\Gamma} \times \vec{\Delta}| \to \Delta$. Denote the product $|\vec{\Gamma} \times \vec{\Delta}|$ by $\Gamma\vec{\times}\Delta$ to stress that this object is *not* well defined. Calculate this object for $\Gamma = \Delta = K_3$. What exactly does not work for this object in order to obtain a Cartesian product (i.e., an object sharing those properties of the Cartesian product of digraphs or of sets)?

Given two digraphs Γ and Δ, their *join BipDi*(Γ, Δ) is the digraph obtained from $\Gamma \sqcup \Delta$ by adding two arrows $v \to w$ and $w \to v$ between any pair of vertexes v of Γ and w of Δ (both directions). This generalizes the *BipDi*-construction from example 25. A similar construction works for two graphs Γ and Δ, their *join Bip*(Γ, Δ) adds one edge $v \relbar\joinrel\relbar w$ between any two vertexes v in Γ and w in Δ.

Graphs may also be constructed from non-graphical data, one of which is defined by a covering of a set:

Definition 77 *A covering of a non-empty finite set X is a subset V of 2^X, consisting of non-empty sets, such that $\bigcup V = X$.*

A covering gives rise to a graph as follows:

Definition 78 *Let V be a covering. Then the* line skeleton *$LSK(V)$ of V is the graph the vertex set of which is V, while the edge set is the set of two-element sets $\{v, w\} \subset V$ such that $v \cap w \neq \varnothing$.*

A large class of graphs is indeed derived from coverings:

Proposition 105 *Every graph $\Gamma : A \to {}^2V$ without loops and multiple edges (i.e., has no values of form $\{v\}$, and Γ is injective) is isomorphic to the line skeleton of a covering.*

Proof In fact, let $\Gamma : A \to {}^2V$ be such a graph. Then for each vertex x, we set $A_x = \{a \mid a \in A, x \in \Gamma(a)\}$ for the set of lines joining x to another vertex. Then the subsets $A_x \subset A$ define a covering C of A, and for two different vertexes x and y, $A_x \cap A_y$ is a singleton set containing exactly the line joining x to y. This means $LSK(C) \overset{\sim}{\to} \Gamma$. $\qquad\square$

Example 42 Figure 11.2 shows a covering $V = \{A, B, C, D, E, F, G\}$ of the set of letters from a to q, and, on the right, its line skeleton $LSK(V)$.

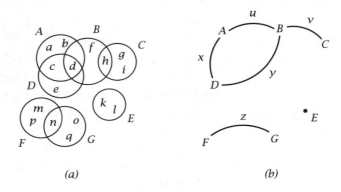

(a)   (b)

Fig. 11.2. A covering (a), and its $LSK(b)$.

Another example is taken from music theory. Consider the C major scale $C = \{c, d, e, f, g, h\}$ and denote the seven triadic degrees by $I = \{c, e, g\}$, $II = \{d, f, a\}$, $III = \{e, g, h\}$, $IV = \{f, a, c\}$, $V = \{g, h, d\}$, $VI = \{a, c, e\}$ and $VII = \{h, d, f\}$. These triads obviously form a covering of C. The line skeleton $LSK(\{I, II, III, IV, V, VI, VII\})$ is illustrated in figure 11.3.

We have chosen a 3-dimensional representation, in order to emphasize the geometric structure induced by the triangles formed by three adjacent triads. These triangles have as their vertexes three triads sharing exactly one element. Such a geometric structure is called a *Möbius strip*.

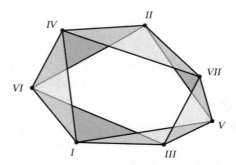

Fig. 11.3. The line skeleton of the covering of a major scale by triads.

CHAPTER 12

Some Special Graphs

We shall discuss two types of special graphs: n-ary trees and Moore graphs.

12.1 n-ary Trees

Binary trees and, more generally, n-ary trees are very frequent in computer algorithms. Intuitively, they formalize a decision hierarchy, where at each step, there is a limited number of possibilities to make a decision. To formalize the decision alternatives, we first need the digraph of n-ary alternatives for natural $n \geq 2$. This is the *loop digraph Loop(n)* : $n \to 1$, consisting of n loops $0, 1, \ldots n - 1$ and one vertex 0. More generally, the loop digraph of a set L is the unique digraph $Loop(L) : L \to 1$.

Definition 79 *For a natural number $n \geq 2$, an n-ary tree is a morphism of digraphs $N : \Gamma \to Loop(n)$, such that*

   (i) *$|\Gamma|$ is an undirected tree,*

  (ii) *Γ has a root,*

 (iii) *each vertex v of Γ has $deg^+(v) \leq n$,*

 (iv) *for any two arrows a and b with common tail v, $N(a) \neq N(b)$.*

For an arrow $v \xrightarrow{a} w$ in an n-ary tree, w is called a child *of v, whereas the necessarily unique v for a given w is called the* parent *of w.*

For $n = 2$, i.e., for binary trees, the labeling has values 0 and 1, but one often calls the value 0 the *left* and 1 the *right alternative*.

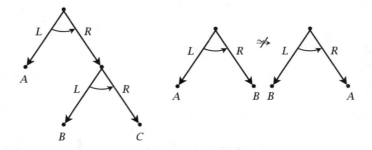

Fig. 12.1. A binary tree where the left alternative is labeled L, and the right alternative R. The right side of the figure illustrates, that the order, indicated by the arrows in the angles, plays a crucial role.

Often, the value $N(a)$ of an arrow in an n-ary tree is denoted as a label of a, but it is more than just a labeling convention. In fact, the comparison of n-ary trees is defined by special morphisms:

Definition 80 *Given two n-ary trees* $N : \Gamma \to Loop(n)$ *and* $R : \Delta \to Loop(n)$, *a morphism of n-ary trees* $f : N \to R$ *is a morphism* $f : \Gamma \to \Delta$ *of digraphs such* $N = R \circ f$.

In particular, for binary trees, this means that a left/right arrow of Γ must be mapped to a left/right arrow of Δ. So it may happen that Γ and Δ are isomorphic digraphs without being isomorphic with the added n-labels of alternatives.

12.2 Moore Graphs

Moore graphs are a special type of process digraphs as already presented in a generic shape in example 24. They arise in the theory of sequential machines and automata, which we introduce here since their concepts are ideally related to the digraph theory thus far developed.

To begin with, we denote by $Path(\Gamma)$ the set of paths in a digraph Γ, including the composition qp of paths p and q if possible (see exercise 45 for this construction). Also denote by $Path_v(\Gamma)$ the set of paths starting at vertex v. For a loop digraph $Loop(A)$, the composition of any two paths is possible and associative, moreover, the lazy path e at vertex 0 is a "neutral element" for this composition, i.e., $e \cdot p = p \cdot e = p$ for any path $p \in Path(Loop(A))$. This structure is also denoted by $Word(A)$ and called the *word monoid over A* (a justification of this terminology is given in chapter 15). Any path p in $Word(A)$ is defined by the (possibly empty) sequence $a_1, \ldots a_k$ of its arrows (the letters of the word) from A as the vertex 0 is uniquely determined. Further, If $p = a_1 \ldots a_k$ and $q = b_1 \ldots b_l$, the composition is exactly the juxtaposition of the letters of the two words: $q \cdot p = b_1 \ldots b_l a_1 \ldots a_k$.

In automata theory, if n is a positive natural number, one considers as "input set" the *n-cube* $Q^n = 2^n$, the elements of which are the "*n*-bit words" such as $w = (0, 0, 1, 0, 1, 1, \ldots, 1)$ (see also figure 13.3 on page 128). For $n = 1$ the two words (0) and (1) are called *bits*, so elements of Q^n are also defined as sequences of n bits. An 8-bit word is called a *byte*. Evidently $card(Q^n) = 2^n$. The word monoid $Word(Q^n)$ is called the *input monoid*, its elements are called *tapes*, so tapes in $Word(Q^n)$ are just words consisting of n-bit words.

Definition 81 *An automaton of n variables is a set function*

$$A : Word(Q^n) \to Q.$$

All automata can be constructed by a standard procedure which is defined by sequential machines:

Definition 82 *A sequential machine of n variables is a map $M : S \times Q^n \to S$, where S is called the state space of the machine M. If M is clear from the context, we also write $s \cdot q$ instead of $M(s, q)$.*

The Moore graph of a sequential machine M is the digraph $Moore(M) : S \times Q^n \to S^2$ defined by $Moore(M)(s, q) = (s, M(s, q))$.

Here is the description of paths in the Moore graph of a sequential machine:

Proposition 106 *For a sequential machine $M : S \times Q^n \to S$, a canonical bijection*

$$PW : Path(Moore(M)) \to S \times Word(Q^n)$$

is given as follows. If

$$p = s_1 \xrightarrow{(s_1, q_1)} s_2 \xrightarrow{(s_2, q_2)} s_3 \ldots \xrightarrow{(s_{m-1}, q_{m-1})} s_m,$$

then $PW(p) = (s_1, q_1 q_2 \ldots q_{m-1})$.

Under this bijection, for a given state $s \in S$, the set $Path_s(Moore(M))$ corresponds to the set $\{s\} \times Word(Q^n)$.

Proof The map PW is injective since from the word $q_1 q_2 \ldots q_{m-1}$ and the state s_1 we can read all the letters $q_1, q_2, \ldots q_{m-1}$, and then the other states by $s_2 = s_1 q_1, s_3 = s_2 q_2, \ldots s_m = s_{m-1} q_{m-1}$. It is surjective, since for any pair $(s_1, q_1 q_2 \ldots q_{m-1})$ the above reconstruction yields a preimage. The last statement is immediate from the definition of PW. $\qquad\square$

We are now ready to define automata associated with sequential machines. To this end, we fix a sequential machine M in n variables over the state space S, an "initial" state s, and a so-called *output function* $O : S \to Q$. The automaton $Automaton(M, s, O) : Word(Q^n) \to Q$ is defined as follows. Denote by $head : Path_s(Moore(M)) \to S : p \mapsto head(p)$ the map associating with each path the head of its last arrow. Also denote $(s, ?) : Word(Q^n) \to \{s\} \times Word(Q^n) : w \mapsto (s, w)$. Then we define

$$Automaton(M, s, O) = head \circ PW^{-1} \circ (s, ?) : Word(Q^n) \to Q.$$

leads to the following result:

Proposition 107 *For every automaton A, there is a sequential machine M, an initial state s, and an output function O such that*

$$A = Automaton(M, s, O).$$

Proof In fact, given the automaton $A : Word(Q^n) \to Q$, take the state space $S = Word(Q^n)$, the output function $O = A$ and the sequential machine $M : S \times Q^n \to Q$ defined by $M(s, q) = sq$. With the initial state e (the empty word) we have the automaton $Automaton(M, e, A)$, and this is the given automaton O. $\qquad\square$

CHAPTER 13

Planarity

Planarity deals with the problem of how graphs can be drawn on a particular surface, such as the plane $\mathbb{R}^2$, the sphere $S^2 = \{(x, y, z) \mid x^2 + y^2 + z^2 = 1\} \subset \mathbb{R}^3$, or the torus, which intuitively looks like the surface of a car tube. We shall use here several concepts which will only be explained rigorously in the chapter on topology in the second volume of this book. However, the elementary character of the results and the important problem of drawing graphs suggests a preliminary treatment of the subject in this first part of the course.

13.1 Euler's Formula for Polyhedra

To begin with, this chapter only deals with *undirected graphs which have no loops and no multiple edges.* In fact, drawing such graphs immediately implies drawing of any more general graphs. In view of proposition 105, we shall call such graphs *skeletal graphs.* So skeletal graphs are characterized by their set V of vertexes, together with a subset $A \subset {}^2V$ comprising only two-element edge sets.

The ad hoc definition we use now (and explain in a more general way in the chapter on limits) is that of continuity:

Definition 83 *A continuous curve in $\mathbb{R}^2$ (or on the sphere S^2) is an injective map $c : [0, 1] \to \mathbb{R}^2$ (or $[0, 1] \to S^2$) defined on the unit interval $[0, 1] = \{x \mid 0 \le x \le 1\} \subset \mathbb{R}$ such that for any $\varepsilon > 0$ and any $s \in [0, 1]$, there is a $\delta > 0$ such that for any $t \in [0, 1]$, if $|s - t| < \delta$, then $|c(t) - c(s)| < \varepsilon$.*

Intuitively, continuity of such a curve means that you can draw the curve with a pencil without lifting from the beginning (curve parameter 0) to the end (curve parameter 1). We shall only consider continuous curves here and omit this adjective in the following discussion. Denote by $]0,1[$ the unit interval without the two endpoints 0 and 1, the so-called *interior of the unit interval*. Here is the definition of a drawing of a skeletal graph:

Definition 84 *A* drawing *D of a skeletal graph* $\Gamma : A \to {}^2V$ *in* $\mathbb{R}^2$ *(or in any more general space X, where continuity is reasonably defined, such as* $X = S^2$*) is given by* $D = (r, c = (c_a)_{a \in A})$:

(i) *an injection* $r : V \to \mathbb{R}^2$ *(or* $V \to X$*)*,

(ii) *for each edge* $v \overset{a}{\longrightarrow} w$ *in A, a curve* $c_a : [0,1] \to \mathbb{R}^2$ *(or in X) such that*

(iii) $\{c_a(0), c_a(1)\} = r(\Gamma(a))$,

(iv) *for any two different lines a and b, the image* $c_a(]0,1[)$ *of the interior of the first curve is disjoint from the image* $c_b([0,1])$ *of the entire second curve.*

A skeletal graph is planar *iff there exists a drawing in* $\mathbb{R}^2$ *of this graph.*

The image $\bigcup_{a \in A} Im(c_a)$ *is denoted by* $D(\Gamma)$ *and is called the* drawn graph.

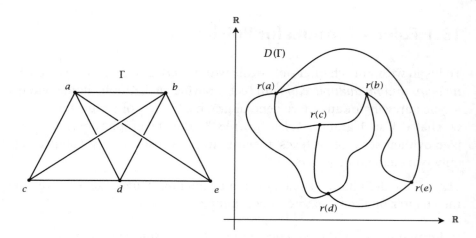

Fig. 13.1. A graph Γ and its drawing $D(\Gamma)$. Intuitively speaking, in a drawing, no two lines intersect except at their endpoints.

Remark 15 It is easy to prove this proposition: *A graph has a drawing in* $\mathbb{R}^2$ *iff it has a drawing in* S^2. The proof uses the stereographic projection, known from geography, in fact a bijection $S^2 - NorthPole \rightarrow \mathbb{R}^2$, which will be introduced in the chapter on topology. This implies that planarity is equivalent to the existence of a drawing on the sphere S^2.

A drawing of a connected skeletal graph on S^2 is also called a *polyhedron*. This is justified by the fact that if you are positioned within a polyhedron (such as a cube or a tetrahedron), you see the edges of the polyhedron on your visual sphere as if they were edges of a graph drawing on S^2.

If D is a drawing of a skeletal graph Γ, then the complement $\mathbb{R}^2 - D(\Gamma)$ of the drawn graph is the disjoint union of a number of regions which are separated by the drawn graph. They are defined as follows: On any subset $X \subset \mathbb{R}^2$, we consider the relation $x \sim y$ iff there is any curve c (attention: such a curve has nothing to do with curves used in drawings of graphs!) such that $c(0) = x$ and $c(1) = y$.

Exercise 52 Show that $\sim$ is an equivalence relation.

The equivalence classes of $\sim$ are called the *connected components of X*. We are now able to write down the famous Euler formula for polyhedra:

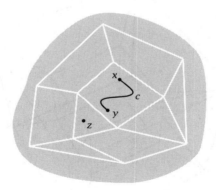

Fig. 13.2. The gray regions are the connected components of a graph. Here x and y are in the same equivalence class, whereas x and z are not.

Proposition 108 *For a skeletal graph* $\Gamma : A \to {}^2V$, *let* $D(\Gamma)$ *be a polyhedron, and* C *the set of connected components of the drawing's complement* $\mathbb{R}^2 - D(\Gamma)$, *and set*

1. $\varepsilon = card(V)$,

2. $\varphi = card(A)$,

3. $\sigma = card(C)$.

Then we have

$$\varepsilon - \varphi + \sigma = 2.$$

Proof Postponed to the chapter on topology in volume II.   □

Example 43 The n-cube Q^n introduced in section 12.2 gives rise to the n-cube graph, also denoted by Q^n. It is the skeletal graph, the vertex set of which is Q^n, whereas the edges are the pairs $\{x, y\}$, where x and y differ exactly by one bit. Find a drawing D of Q^3. Show that in this case, the numbers in the Euler formula are $\varepsilon = 8, \varphi = 12, \sigma = 6$. Can you find a drawing of Q^4?

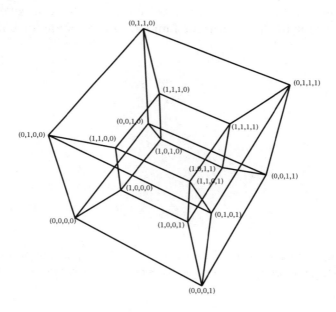

Fig. 13.3. Hypercube Q^4.

Example 44 In figure 13.4 a dodecahedron has been flattened to show the number of vertexes (circled numbers), edges (small-sized numbers) and faces (large-sized numbers). Euler's formula can be verified using $\varepsilon = 20$, $\varphi = 30$ and $\sigma = 12$, i.e., $20 - 30 + 12 = 2$. This shows that the flattened dodecahedron can be drawn with no intersecting edges.

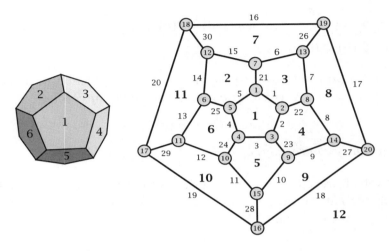

Fig. 13.4. The dodecahedron and its flattened and stretched representation as a graph.

Observe that in Euler's formula there is a rather deep insight into drawings: In fact, the numbers ε and φ are defined by the "abstract" graph, whereas the number σ is a function of the specific drawing. Therefore, all drawings must show the same number of connected components of the drawing's complement!

13.2 Kuratowski's Planarity Theorem

The previous formula is applicable only if we are sure that the graph is planar. Evidently, planarity of a graph Γ is a property which remains conserved if we pass to an isomorphic graph. In other words, there must exist an abstract criterion which tells us when a skeletal graph is planar. We shall present the criterion proved by Kazimierz Kuratowski (1896–1980).

To this end we first need the following construction:

Definition 85 *If $\Gamma \rightarrow {}^2V$ is a skeletal graph, and if $\Gamma(a) = \{x, y\} \in V$ is any two-element set defined by an edge a, we denote by $\Gamma_a : A' \rightarrow {}^2V'$ the skeletal graph with $A' = A - \{a\}$ and $V' = (V - \{x, y\}) \sqcup \{a\}$, and we have*

1. *if $\Gamma(b) \cap \{x, y\} = \varnothing$, then $\Gamma_a(b) = \Gamma(b)$,*
2. *else $\Gamma_a(b) = (\Gamma(b) - \{x, y\}) \cup \{a\}$.*

Γ_a is called an elementary contraction of Γ. A contraction of a skeletal graph Γ *is a graph Δ which is isomorphic to a finite succession*

$$(\ldots((\Gamma_{a_1})_{a_2})\ldots)_{a_m}$$

of elementary contractions of Γ.

Intuitively, Γ_a results from the removal of the edge a in Γ, the insertion of a new point, and the connection of all open-ended edges from the removal to the new point.

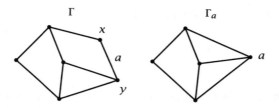

Fig. 13.5. A graph Γ and its elementary contraction Γ_a.

Proposition 109 *If Δ is a contraction of Γ, then, if Γ is planar, so is Δ.*

Exercise 53 Give a proof of proposition 109. Draw this for an elementary contraction.

This is the main theorem:

Proposition 110 *A skeletal graph is planar iff it contains no subgraph that has a contraction which is isomorphic to one of the graphs $K_{3,3}$ or K_5.*

Proof Postponed to the chapter on topology in volume II. $\square$

CHAPTER 14

First Advanced Topic

14.1 Floating Point Arithmetic

In this section, we give an overview of computer-oriented arithmetic based on the adic representation of real numbers as described in proposition 92. Evidently, representation of numbers and arithmetic calculations on a computer cannot be based upon an infinity of information. Since real numbers are based upon infinite information, computer arithmetic must be limited to finite truncations of real arithmetic. Floating point arithmetic is one approach to this problem. It is based on the well-known *approximative* representation of real numbers in scientific contexts, such as Avogadro's number $N \approx 6.02214 \times 10^{23}$, Planck's constant $h \approx 6.6261 \times 10^{-27}$ erg sec or the circle circumference number $\pi \approx 3.14159$. Here we consent that any number can be written as a small positive or negative number f (in fact $1 \le |f| < 10$ or $f = 0$ in the decimal system) times a power 10^e of the basis (here 10) for a positive or negative integer exponent e. The term "floating point" stems from the difference to the "fixed point" representation given in proposition 92 of chapter 9, where the dot is relative to the 0-th power of the base, whereas here, the dot is a relative information related to the variable power of the base.

Remark 16 Observe that the limitation to a finite number of digits in the adic representation of real numbers implies that we are limited to special rational numbers (not even $1/3 = 0.333\ldots$ is permitted in the decimal system!). So why not just take the fractional representation which is a precise one? The point is that evidently, numerator and denominator of a fraction also would grow to unlimited size when doing arithmetic. So the problem is not the fractional representation (though it would make arithmetic precise in some cases), but the general program for calculating with numbers within an absolutely limited memory space.

The IEEE (Institute for Electrical and Electronics Engineers) #754 standard defines a specific way of floating point representation of real numbers. It is specified relating to the binary base and therein by a triple of (s, e, f), where $s \in \{0, 1\}$) is the bit for the sign, i.e., $s = 0$ iff the number is not negative, $s = 1$ iff it is negative. The number e is the exponent and takes integer values $e \in [-127, 128]$, corresponding to a range of 8-bits ($2^8 = 256$). The exponent is encoded by a bias integer $bias = 127$. This means that we have to consider $e + bias$ instead of e in the computerized representation. The extra value -127 is set to 0. This exponent is reserved to represent number zero. We will make this more precise in a moment. And the maximal value $e = 128$, shifted to 255 by the bias, is used to represent different variants of infinity and "not-a-number" symbols. We therefore may vary the exponents in the integer interval $[-126, 127]$. In the binary representation by eight bits, and with regard to the bias, this amounts to considering the zero number exponent symbol 00000000 whereas the infinities and not-a-number symbols are represented by the maximum 11111111 ($= 2^8 - 1 = 255$). So the common number exponent takes shifted values in the binary interval $[00000001, 11111110]$, corresponding to the unshifted interval $[-126, 127]$.

The third number f is called the *mantissa* (an official, but mathematically an ill-chosen terminology) or fraction (non-official, but correct terminology, also adapted by Knuth), we stick to the latter. The fraction f is encoded in binary representation $f = f_1 f_2 f_3 \ldots f_{23} \in [0, 2^{23} - 1]$, i.e., in binary representation

00000000000000000000000 $\leq f_1 f_2 \ldots f_{23} \leq$ 11111111111111111111111

and means the fraction $1.f_1 f_2 \ldots f_{23}$ in binary representation.

Given these three ingredients, the number $\langle s, e, f \rangle$ denoted by the triple (s, e, f) is the following 32-bit word:

$$\langle s, e, f \rangle = (-1)^s \times 2^{e-bias} \times 1.f$$

This is the *normalized floating point representation* of the IEEE standard #754. It is also called *single precision* representation since a finer representation using two 32-bit words exists, but we shall not deal with this refinement here. The values $s = e = f = 0$ represent zero:

$$0 = (0, 00000000, 0000000000000000000000000).$$

Non-zero numbers are given by these 32-bit words:

- $s \in \{0, 1\}$, one bit,
- $e \in [00000001, 11111110]$, eight bits,
- $f \in [00000000000000000000000, 11111111111111111111111]$, 23 bits.

The largest positive number that can be stored is

$$1.11111111111111111111111 \times 2^{127} = 3.402823 \ldots \times 10^{38},$$

whereas the smallest positive number is

$$1.00000000000000000000000 \times 2^{-126} = 1.175494 \ldots \times 10^{-38}.$$

The special infinities and not-a-number (NaN) values are these 32-bit words:

- $(0, 11111111, 0000000000000000000000000)$ = Inf, positive infinity,
- $(1, 11111111, 0000000000000000000000000)$ = −Inf, negative infinity,
- $(0, 11111111, 0111111111111111111111111)$ = NaNS, NaN generating a "trap" for the compiler, while
- $(0, 11111111, 1000000000000000000000000)$ = NaNQ, where Q means "quiet", calculation proceeds without generating a trap.

Given this standardized normal representation, we shall now discuss the arithmetical routines. The general procedure is a very simple two-step algorithm: first, one of the operations of addition, subtraction, multiplication, or division is performed, and then, in a second step, the result is recast to the normalized representation shown above.

Regarding the first step, we shall only discuss the addition here, since the other operations run in a completely similar way. To make the algorithm more transparent, we use another (also Knuth's, but not IEEE-standardized) normalization by replacing the representation $1.f \times 2^e$ by

$0.1f \times 2^{e+1}$. Denote this representation by $\langle\langle s, e, g \rangle\rangle = (-1)^s \times 2^e \times g$ with $0 \leq g < 1$, i.e., we have $\langle s, e, f \rangle = \langle\langle s, e+1, 0.1f \rangle\rangle$. Observe that we now need 24 bits after the dot in order to represent this second normalized representation. Observe also that the third coordinate g of the second representation is the real number, including zero 0.0, and not just the 23-bit word to the right of the dot as with the IEEE standard. Given two normalized representations $\langle\langle s_u, e_u, g_u \rangle\rangle, \langle\langle s_v, e_v, g_v \rangle\rangle$, the computer calculation does not yield the exact sum, but an approximation $\langle\langle s_w, e_w, g_w \rangle\rangle$. In general, it is not normalized anymore. This will be settled by the subsequent normalization algorithm. However, this latter will perform a further rounding approximation of the intermediate approximation!

Here is the algorithm **A** for **addition**:

A.1: We check whether $e_u \geq e_v$. If not, we exchange the summands and proceed. As addition should be commutative, this is a reasonable step, and it will also be the reason why floating point addition is in fact commutative.

A.2: Set $e_w = e_u$.

A.3: If $e_u - e_v \geq 24 + 2 = 26$, we have a large difference of exponents, i.e., the smaller summand has its highest digit below the 24 digits admitted in the normalized representation of $\langle\langle s_u, e_u, g_u \rangle\rangle$. We therefore set $g_w = g_u$. In this case, go to **A.6**. Actually, in this case, the result is already normalized, and we could terminate the entire addition here.

A.4: Divide g_v by $2^{e_u - e_v}$, i.e., shift the binary representation by up to 9 places to the right. Attention! This procedure requires the computer memory to hold up to $24 + 9 = 33$ places temporarily!

A.5: Set $g_w = g_u + g_v$. This last step gives us the correct sum $\langle\langle s_w, e_w, g_w \rangle\rangle$, but we have an g_w which might not be normalized, i.e., we could have $g_w \geq 1$.

A.6: Normalize, i.e., apply the normalization algorithm **N** to $\langle\langle s_w, e_w, g_w \rangle\rangle$.

The **normalization** algorithm **N** runs as follows, it converts a "raw" exponent e and a "raw" fraction $0 \leq g$ into a normalized representation.

N.1: If $g \geq 1$ (fractional overflow), then go to step **N.4**. If $g = 0$, set e to the lowest possible exponent in the normalized representation (in fact $e = -127$ or 00000000 in the biased binary IEEE representation and $e = -126$ in the second normalized representation).

N.2: (Normalization of g, i.e., $g < 1$ but large enough) If $g \geq 1/2$, go to step **N.5**.

N.3: (f is too small, but does not vanish) Multiply g by 2 and return to step **N.2**.

N.4: Divide g by 2 and increase e by 1 and return to step **N.1**.

N.5: (Round g to 24 places) This means that we want to change g to the nearest multiple of 2^{-24}. One looks at $2^{24} \times g = \ldots g_{-24}.h$ and checks the part h after the dot. According to whether h is less than $1/2$ or not, this part is omitted and $2^{24} \times g$ is replaced by $\ldots g_{-24}.0 + 1$ or $\ldots g_{-24}.0$, see the following remark 17 for a comment. If the rounded g is 1, return to step **N.1**.

N.6: (check e) If e is too large (more than 127), an *exponent overflow* condition is sensed. If e is too small (less than -126 in the IEEE standard, -125 in the second normalized form $\langle\langle\rangle\rangle$ used in this algorithm), then an *exponent underflow* condition is sensed.

This terminates the normalization algorithm (except for the actions to be taken for the over- and underflow situations).

Remark 17 There is an axiomatic version of this theory, where the rounding $round(x)$ of a number x is given axiomatically with conditions $round(-x) = -round(x)$ and $x \leq y$ implies $round(x) \leq round(y)$. We then define rounded arithmetic operations by $u \oplus v = round(u + v), u \otimes v = round(u \times v), u \ominus v = round(u - v), u \oslash v = round(u/v)$.

Attention! The floating point operations lose virtually all of the nice properties of the original operations. In particular, associativity and distributivity are lost. However, all floating point operations remain commutative if the originals were so, in particular addition and multiplication with floating point numbers are commutative. See [32].

14.2 Example for an Addition

Let $u = 235.5$ and $v = 22.6$ as an example for the addition algorithm. We expect the result to be $w = 258.1$.

Binary representation

$$u = 1.1101011100000000000000_2 \times 2^7$$

and

$$v = 1.011010011001100110011\ldots_2 \times 2^4$$

Note that v cannot be exactly represented as a binary number with finitely many non-vanishing digits: it is an infinite fraction with periodically recurrent digits.

IEEE notation

$$u = \langle 0, 0000\ 0111, 1101\ 0111\ 0000\ 0000\ 0000\ 000 \rangle$$

and

$$v = \langle 0, 0000\ 0100, 0110\ 1001\ 1001\ 1001\ 1001\ 101 \rangle$$

Note that in this representation the last digit of v has been rounded up.

Knuth's notation

$$u = \langle \langle 0, 0000\ 1000, 0.1110\ 1011\ 1000\ 0000\ 0000\ 0000 \rangle \rangle$$

and

$$v = \langle \langle 0, 0000\ 0101, 0.1011\ 0100\ 1100\ 1100\ 1100\ 1101 \rangle \rangle$$

In particular

$$g_u = 0.11101011100000000000000$$

and

$$g_v = 0.101101001100110011001101$$

Adding u and v

A.1: nothing to do: u is already greater than v

A.2: $e_w = e_u = 0000\ 1000$

A.3: nothing to do: the exponents differ only by 3

A.4: the division results in $g'_v = 0.000101101001100110011001101$, we now need 27 places to work with this number

A.5: performing the addition $g_w = g_u + g'_v$ we have

$$
\begin{array}{r}
0.111010111000000000000000 \\
+\ 0.000101101001100110011001101 \\
\hline
g_w = 1.000000100001100110011001101
\end{array}
$$

A.6: normalization is necessary, because $g_w > 1$.

Normalization

N.1: $g_w \geq 1$, so we continue at **N.4**.

N.4: $g'_w = g_w/2 = 0.100000010000110011001101$, and
$e'_w = e_w + 1 = 0000\ 1001$
do **N.1** again.

N.1: Now, $g'_w < 1$.

N.2: Since $g'_w \geq 1/2$, we continue at **N.5**.

N.5: $2^{24} \cdot g''_w = 100000010000110011001100.1101$.
Since the first place after the decimal point is 1, so we need to increase this product by 1, divide it by 2^{24} and truncate it after 24 digits:
$$g'''_w = 0.100000010000110011001101$$

N.6: Nothing to do, since the exponent is small.

Knuth's notation

Assembling the various parts $(s_w,\ e'_w,\ g'''_w)$ we get:

$$w = \langle\langle 0,\ 0000\ 1001,\ 0.1000\ 0001\ 0000\ 1100\ 1100\ 1101\rangle\rangle$$

IEEE notation

$$w = \langle 0, 0000\ 1000, 0000\ 0010\ 0001\ 1001\ 1001\ 101 \rangle$$

Binary representation

$$w = 1.00000010000110011001101_2 \times 2^8$$

which translates to

$$1.0082031488418579101\,5625 \times 256 = 258.100006103515625$$

Remark 18 The result of the addition algorithm is quite close, but not exactly equal, to our expected result 258.1. This fact shows that calculations with floating point numbers are always only approximations.

PART II

Algebra, Formal Logic, and Linear Geometry

CHAPTER 15

Monoids, Groups, Rings, and Fields

This chapter introduces the indispensable minimal algebraic structures needed for any further discussion, be it for formal logic and data structures, linear equation solving, geometry or differential calculus. The students are asked to study this omnipresent material with particular care.

We shall also see that many of the following structures are indeed already implicitly present in the previous theory. We have encountered recurrent "laws" such as associativity, or identity properties. The following theory therefore is also an action of abstraction, a commonly recognized essential activity in computer science!

15.1 Monoids

Definition 86 *A monoid is a couple* $(M, *)$ *where M is a set, and where* $* : M \times M \to M$ *is a "composition" map sharing these properties:*

  (i) *(Associativity) For any triple $k, l, m \in M$, we have $(k * l) * m = k * (l * m)$, which we also write as $k * l * m$.*

  (ii) *(Neutral element) There is an element $e \in M$ such that $m * e = e * m = m$ for all $m \in M$. Evidently, this element is uniquely determined by this property. It is called the* neutral element *of the monoid.*

A monoid $(M, *)$ *is called* commutative *iff* $m * n = n * m$ *for all* $m, n \in M$.

Usually, we identify a monoid $(M, *)$ with its set M if the composition $*$ is clear. The notation of the product $m * n$ may vary according to the concrete situation, sometimes even without any notation, such as mn for $m * n$. For commutative monoids, one often uses the symbol $+$ instead of $*$ and says that the monoid is *additive*.

Example 45 Denoting by $*$ the multiplication on $\mathbb{N}, \mathbb{Z}, \mathbb{Q}, \mathbb{R}, \mathbb{C}$, these sets define monoids, all of them commutative, with 1 being the neutral element in each of them. They are usually called the *multiplicative monoids* of these number domains.

The subset $U = S(\mathbb{C}) = \{z \mid |z| = 1\}$ of $\mathbb{C}$ is a monoid under ordinary multiplication since $|1| = 1$ and $|z \cdot w| = |z| \cdot |w|$, whence $z \cdot w \in U$ if $z, w \in U$. The monoid U is also called the *unit circle*.

Example 46 Given a set X, the set $End(X)$ of set maps $f : X \rightarrow X$, together with the usual composition of maps, is a monoid with neutral element $e = Id_X$. If we restrict the situation to the bijective maps in $End(X)$, we obtain a monoid $Sym(X)$ (same neutral element). This latter monoid will play a crucial role in the sequel, we shall discuss it extensively in the following section about groups.

Example 47 If $\Gamma : A \rightarrow V^2$ is a directed graph, the set $End(\Gamma)$ of endomorphisms of Γ, i.e., of morphisms $f = (u, v) : \Gamma \rightarrow \Gamma$, together with the composition $\circ$ of digraph morphisms, is a monoid. In general this is not a commutative monoid. We have a number of important special cases of this construction.

To begin with, if the arrow set A of Γ is empty, the endomorphism monoid identifies with the monoid $End(V)$.

In section 12.2 on Moore graphs, we introduced the word monoid $Word(A)$ of a set A. A path $p \in Word(A)$ is just a sequence of elements $p = (a_1, a_2, \ldots a_k)$, and also the composition $p = a_1 \cdot a_2 \cdot \ldots a_k$ of the paths a_i of length one. This is why this monoid is also called the *word monoid over the alphabet A*. A word is then a synonym for "path", and the letters of the word are the elements of A. The lazy path is called the *empty word*.

As with sets, where we introduced functions or set maps, we need the formalism for comparing different instances of the monoid concept. Here is the evident definition:

Definition 87 *Given two monoids* $(M, *_M), (N, *_N)$, *a* homomorphism $f :$ $(M, *_M) \to (N, *_N)$ *is a set map* $f : M \to N$ *such that*

(i) $f(m *_M n) = f(m) *_N f(n)$ *for all* $m, n \in M$,

(ii) $f(e_M) = e_N$ *for the neutral elements* $e_M \in M$ *and* $e_N \in N$.

Again, if the multiplications are clear, then we just write $f : M \to N$ *to denote a monoid homomorphism. The set of monoid homomorphisms* $f :$ $M \to N$ *is denoted by* $Monoid(M, N)$.

If we are given three monoids M, N, O *and two monoid homomorphisms* $f : M \to N$ *and* $g : N \to O$, *their* composition $g \circ f : M \to O$ *is defined by the underlying set map composition, and this is also a monoid homomorphism.*

Exercise 54 Show that the composition of three monoid homomorphisms, if defined, is associative, and that the identity map $Id_M : M \to M$ of any monoid M is a monoid homomorphism, the *identity homomorphism of* M.

Show that for a monoid homomorphism $f : M \to N$ the following statements are equivalent:

(i) There is a homomorphism $g : N \to M$ such that $g \circ f = 1_M$ and $f \circ g = 1_N$,

(ii) f is a bijection of sets.

A homomorphism satisfying these properties is called an *isomorphism* of monoids. One says that two monoids M and N are isomorphic iff there is an isomorphism $f : M \to N$. The homomorphism g is uniquely determined and is called the inverse of f, it is denoted by $g = f^{-1}$. If $M = N$ (same monoids, not only sets!), a homomorphism is called an *endomorphism*. The set of monoid endomorphisms, together with the composition of monoid homomorphisms and the identity on M, is a monoid, it is denoted by $End(M)$. An isomorphism which is also an endomorphism is called *automorphism*. The subset of automorphisms in $End(M)$ is also a monoid, it is denoted by $Aut(M)$.

Here is the "universal property" of the word monoid construction:

Proposition 111 (Universal Property of the Word Monoid) *Let A be a set, and N a monoid. Then the following map of sets is a bijection:*

$$r : Monoid(Word(A), N) \to Set(A, N) : f \mapsto f|_A$$

Proof The map r is injective since, if $f(a)$ is given for all letters $a \in A$, then $f(a_1 \cdot a_2 \cdot \ldots a_k) = f(a_1) \cdot f(a_2) \cdot \ldots f(a_k)$, for any word $a_1 \cdot a_2 \cdot \ldots a_k \in Word(A)$, the empty path v must be mapped to the neutral element e_N, and we know f from its action on letters. Conversely, if $g : A \to N$ is a set map, define $f(a_1 \cdot a_2 \cdot \ldots a_k) = g(a_1) \cdot g(a_2) \cdot \ldots g(a_k)$, which is well defined since the letters of the word $a_1 \cdot a_2 \cdot \ldots a_k$ are uniquely determined. Also, set $f(v) = e_N$. Then this yields a monoid homomorphism, since the multiplication of words is its juxtaposition. So r is surjective. $\qquad\qquad\qquad\qquad\qquad\square$

The word monoid is therefore a kind of "free" object in the sense that any "wild" map on its letters extends uniquely to a monoid homomorphism!

A special type of homomorphisms are the so-called submonoids:

Definition 88 *If $(M, *)$ is a monoid with neutral element e, a submonoid $(M', *')$ of M is a subset $M' \subset M$ such that for all $m, n \in M', m * n \in M'$ and $e \in M'$, while the multiplication $*'$ is the restriction of $*$ to M'. A submonoid therefore gives rise to the evident embedding homomorphism $i_M : M' \subset M$.*

Exercise 55 Given a monoid $(M, *)$ and a (possibly empty) subset $S \subset M$, there is a unique minimal submonoid M' of M such that $S \subset M'$. It is denoted by $\langle S \rangle$ and is called the *submonoid generated by S*. Show that $\langle S \rangle$ consists of all finite products $s_1 * s_2 * \ldots s_n$, $s_i \in S$, and of the neutral element e.

Example 48 We have plenty of submonoids among our previous examples (with the ordinary multiplications): $\mathbb{N} \subset \mathbb{Z} \subset \mathbb{Q} \subset \mathbb{R} \subset \mathbb{C}$, $U \subset \mathbb{C}$. If Γ is a digraph, we have $Aut(\Gamma) \subset End(\Gamma)$.

Exercise 56 The additive monoids, i.e., where $*$ is $+$, also define a chain of submonoids $\mathbb{N} \subset \mathbb{Z} \subset \mathbb{Q} \subset \mathbb{R} \subset \mathbb{C}$. If A is any set, the map $l : Word(A) \to (\mathbb{N}, +)$ defined by the length of a word is a monoid homomorphism, i.e., $l(pq) = l(p) + l(q)$ and $l(e) = 0$.

15.2 Groups

Definition 89 *A monoid $(G, *)$ is called a* group *if every $g \in G$ is invertible, i.e.,*

$$\text{there is } h \in G \text{ such that } g * h = h * g = e.$$

The element h is uniquely determined by g and is called the inverse *of g and denoted by g^{-1}. A* commutative *or* abelian *group is a group which is a commutative monoid.*

If G and H are two groups, a monoid homomorphism $f : G \to H$ is called a group homomorphism. *The set of group homomorphisms $f : G \to H$ is denoted by Group(G, H). A* group isomorphism *is a monoid isomorphism among groups. Accordingly, the set of monoid endomorphisms of a group G is denoted by End(G), while the monoid of automorphisms is denoted by Aut(G). A* subgroup *$G \subset H$ is a submonoid $G \subset H$ where the involved monoids G and H are groups.*

Example 49 The symmetries on the square form a non-commutative group. The eight elements of the group are: the identity i, the three clockwise rotations r_1, r_2 and r_3 by angles 90°, 180° and 270°, respectively; further, the four reflections about the horizontal axis h, the vertical axis v, the first diagonal d_1 and the second diagonal d_2. Figure 15.1 shows a graphical representation of those eight operations. The product of two

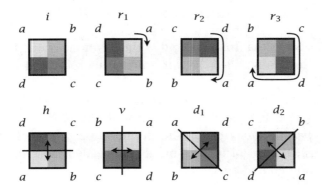

Fig. 15.1. The symmetries on the square.

elements x and y, i.e., $x \cdot y$, is defined as applying the operation y first, then x. The multiplication table for this group is shown in figure 15.2. Although this group is not commutative, it has commutative subgroups, for example $\{i, r_1, r_2, r_3\}$.

Example 50 The inclusion chain of additive monoids $\mathbb{Z} \subset \mathbb{Q} \subset \mathbb{R} \subset \mathbb{C}$ are all commutative groups. The set $\mathbb{Z}^* = \{1, -1\}$ is a multiplicative group

·	i	r_1	r_2	r_3	h	v	d_1	d_2
i	i	r_1	r_2	r_3	h	v	d_1	d_2
r_1	r_1	r_2	r_3	i	d_1	d_2	v	h
r_2	r_2	r_3	i	r_1	v	h	d_2	d_1
r_3	r_3	i	r_1	r_2	d_2	d_1	h	v
h	h	d_2	v	d_1	i	r_2	r_3	r_1
v	v	d_1	h	d_2	r_2	i	r_1	r_3
d_1	d_1	h	d_2	v	r_1	r_3	i	r_2
d_2	d_2	v	d_1	h	r_3	r_1	r_2	i

Fig. 15.2. The multiplication table of the group of symmetries on the square. The entry in row x and column y is the product $x \cdot y$

in the multiplicative monoid $\mathbb{Z}$. More generally, if $(M, *)$ is a monoid, the subset M^* of invertible elements in the sense of definition 89 is a subgroup of M, it is called the *group of invertible elements of M*. Then we have this inclusion chain of groups of invertible elements within the multiplicative monoids of numbers: $\mathbb{Z}^* \subset \mathbb{Q}^* \subset \mathbb{R}^* \subset \mathbb{C}^*$. Observe that $\mathbb{Q}^* = \mathbb{Q} - \{0\}, \mathbb{R}^* = \mathbb{R} - \{0\}, \mathbb{C}^* = \mathbb{C} - \{0\}$.

Exercise 57 Show that for a monoid M the submonoid $Aut(M) \subset End(M)$ is a group. $Aut(M)$ is called the *automorphism group of M*.

The submonoid $Sym(X) \subset End(X)$ of bijections of a set X is a very important group, it is called the *symmetric group of X*, its elements are called the *permutations of X*. The permutation group of the interval $[1, n] = \{1, 2, 3, \ldots n\}$ of natural numbers is called the *symmetric group of rank n* and denoted by S_n.

Exercise 58 If $card(X) = n$ is finite, then there is an isomorphism of groups $Sym(X) \xrightarrow{\sim} S_n$.

A permutation $p \in S_n$ is best described by cycles. A *cycle* for p is a sequence $C = (c_1, c_2, \ldots c_k)$ of pairwise different elements of $[1, n]$ such that $p(c_1) = c_2, p(c_2) = c_3, \ldots p(c_{k-1}) = c_k, p(c_k) = c_1$, where k is called the length of C and denoted by $l(C)$. The underlying set $\{c_1, c_2, \ldots c_k\}$ is denoted by $|C|$, i.e., $l(C) = card(|C|)$. Conversely, each sequence $C = (c_1, c_2, \ldots c_k)$ of pairwise different elements of $[1, n]$ denotes by definition a permutation p such that $p(c_1) = c_2, p(c_2) = c_3, \ldots p(c_{k-1}) =$

$c_k, p(c_k) = c_1$, while the other elements of $[1, n]$ are left fixed under p. We also denote this p by C. Given such cycles $C_1, C_2, \ldots C_r$, one denotes by $C_1 C_2 \ldots C_r$ the permutation $C_1 \circ C_2 \circ \ldots C_r$. If a cycle C has length 2, i.e., $C = (x, y)$, then it is called a *transposition*, it just exchanges the two elements x and y, and $(x, y)(x, y) = Id$.

Exercise 59 Let $n = 4$, take $C_1 = (124)$ and $C_2 = (23)$. Calculate the permutation $C_1 C_2$.

Proposition 112 *Let $p \in S_n$. Then there is a sequence $C_1, C_2, \ldots C_r$ of cycles of S_n such that*

(i) *the underlying sets $|C_i|$ are mutually disjoint,*

(ii) $[1, n] = \bigcup_i |C_i|$,

(iii) $p = C_1 C_2 \ldots C_r$.

The sets $|C_i|$ are uniquely determined by p, and for any two such representations $C_1, C_2, \ldots C_r$ and $C'_1, C'_2, \ldots C'_r$ of p, if $|C_i| = |C'_j|$, then there is an index $1 \le t \le l = l(C_i) = l(C'_j)$ such that $C_i = (c_1, c_2, \ldots c_l)$ and $C'_j = (c_t, c_{t+1}, \ldots c_l, c_1, c_2, \ldots c_{t-1})$.

This representation of p is called the cycle representation.

Proof If such a cycle representation of p exists, then if $i \in C_j$, then $C_j = \{p^k(i) \mid k = 0, 1, 2 \ldots\}$, and conversely, every such so-called *orbit* set $\{p^k(r) \mid k = 0, 1, 2 \ldots\}$ defines the cycle containing r. So the cycles are identified by these orbits which are defined uniquely by p. Such sets define a partition of $[1, n]$. In fact, if $\{p^k(i) \mid k = 0, 1, 2 \ldots\} \cap \{p^k(i') \mid k = 0, 1, 2 \ldots\} \ne \varnothing$, then there are k and k' such that $p^k(i) = p^{k'}(i')$. Suppose $k \ge k'$. Then multiplying by $(p^{-1})^{k'}$, we obtain $p^{k-k'}(i) = i'$, i.e., the orbit of i' is contained in that of i. But since S_n is finite, not all powers $p, p^2, p^3, \ldots$ can be different. If $p^l = p^{l+t}, t \ge 1$, then by multiplication with $(p^{-1})^l$, we have $p^t = Id$, i.e., $p^{-1} = p^{t-1}$. Therefore $p^{k'-k}(i') = (p^{-1})^{k-k'} = (p^{t-1})^{k-k'} = i$, and both orbits coincide. So the cycle representation is the uniquely determined orbit representation, which is a partition of $[1, n]$, thus (i) and (ii) follow. Part (iii) follows since the cycles act like p on their elements. The last statement is clear. □

Exercise 60 Beside the cycle representation of permutations, a tabular representation can be used, where an element on the top line is mapped to the element below it. Find the cycle representation for the following permutations:

$$\frac{1\ 2\ 3\ 4\ 5\ 6\ 7\ 8}{4\ 6\ 5\ 8\ 7\ 2\ 3\ 1} \in S_8 \qquad \frac{1\ 2\ 3\ 4\ 5\ 6}{6\ 1\ 4\ 2\ 5\ 3} \in S_6$$

Definition 90 *The cardinality of a group G is called the* order *of G, it is denoted by ord(G). A group with finite order is called a* finite *group. A group G with ord(G) = 1 is called* trivial.

Exercise 61 Show that any two trivial groups are isomorphic.

Definition 91 *If $n \in \mathbb{N}$, we define $n! = 1$ if $n = 0$, and $n! = 1 \cdot 2 \cdot 3 \cdot 4 \cdot \ldots n$ else. The number $n!$ is called* n-factorial.

Exercise 62 This definition is an inductive definition. Give a rigorous version thereof.

Proposition 113 *If $n \in \mathbb{N}$, then*

$$ord(S_n) = n!$$

Proof The image $i = p(n)$ of n under a permutation $p \in S_n$ can be any of the n elements in $[1, n]$. For each element, the other elements in $[1, n-1]$ are sent to the complement $[1, n] - \{i\}$ having $n - 1$ elements. By induction, this gives $(n-1)!$ possibilities, and we have $n \cdot ((n-1)!) = n!$, as required, and since the induction start $n = 1$ is evident, we are done.   □

The following result is a fundamental tool for constructing subgroups of a given group:

Proposition 114 *Let $X \subset G$ be any subset of a group G. Then there is a unique minimal subgroup H of G such that $X \subset H$. The group H consists of the neutral element e of G and of all finite products $x_1 * x_2 * \ldots x_k$, where either $x_i \in X$ or $x_i = y_i^{-1}, y_i \in X$. The subgroup H is denoted by $\langle X \rangle$ and is called the* subgroup generated by (the elements of) *X.*

If there is a finite set $X \subset G$ such that G is generated by X, $G = \langle X \rangle$, then G is called a finitely generated *group. In particular, if $X = \{x\}$, one writes $G = \langle x \rangle$ and calls G a* cyclic *group.*

Proof The uniqueness of the minimal subgroup H results from the fact that for any non-empty family (G_i) of subgroups of G, their intersection $\bigcap_i G_i$ is a subgroup of G. The family of all subgroups G_i containing X is not empty (G is in the family), so the intersection is this unique minimal subgroup H. Clearly, H contains all finite products of the described type. Moreover, this set is a subgroup, the inverse of a product $x_1 * x_2 * \cdots * x_k$ being $x_k^{-1} * x_{k-1}^{-1} * \ldots x_1^{-1}$, which is of the required type.   □

Definition 92 *For an element x of a group G, the order of $\langle x \rangle$ is called the* order *of x.*

Proposition 115 *Let $n \geq 2$ be a natural number, then S_n is generated by the set $\{(1, k) \mid k = 2, 3, \ldots n\}$ of transpositions. Since a transposition has order 2, any $x \in S_n$ can be written as a product $x = (1, k_1)(1, k_2) \ldots (1, k_r)$.*

The subset A_n of the elements $x \in S_n$ which can be written as a product of an even number (a multiple of 2) of transpositions is a subgroup of of S_n of order $n!/2$, it is called the alternating group of rank n.

Proof For $n = 2$, S_n is obviously generated by $(1, 2)$. For the general case, let $p(n) = i$ for a given permutation p. Then $(i, n)p$ fixes n and, by induction, is a product of transpositions $(1, r), r < n$. But $(i, n) = (1, i)(1, n)(1, i)$, so $p = (1, i)(1, n)(1, i)((i, n)p)$, and we are done.

Clearly, A_n is a subgroup. If $(1, i)$ is any transposition, then $A_n \to S_n : p \mapsto (1, i)p$ is an injective map from A_n onto the set B_n of permutations which are products of an odd number of transpositions. If we can show that $A_n \cap B_n = \varnothing$, then S_n is the disjoint union of A_n and B_n, and therefore $S_n = A_n \sqcup B_n$, whence $card(A_n) = n!/2$. To show this, given a permutation p, denote by $i(p) = (i_1, i_2, \ldots i_n)$ the sequence with $j = p(i_j)$. Let $s(p)$ be 1 if the number of pairs $u < v$ such that $i_u > i_v$ in $i(p)$ is even, and -1 if it is odd. Clearly, $s(Id) = 1$. If $(i, i + 1)$ is a transposition of neighboring numbers, then evidently $s((i, i + 1)) \cdot p = -s(p)$ Moreover, a general transposition (i, j) is the product of an odd number of transpositions $(k, k + 1)$ of neighboring numbers. In fact, if $j = i + 1$, we are done, and for $j > i + 1$, we have $(i, j) = (i, k)(k, j)(i, k)$ for $i < k < j$, so our claim follows by induction. Therefore $s((i, j)) = -1$. If p is a product of r transpositions, we have $s(p) = -1^r$, and r cannot be even and odd at the same time. So $A_n \cap B_n = \varnothing$. □

The alternating group A_n is a typical example of a group construction which has far-reaching consequences for all mathematical theories which involve groups, and this is virtually the entire mathematical science. The observation is that A_n is related to a particular group homomorphism $sig : S_n \to \mathbb{Z}^*$ defined by $sig(x) = 1$ if $x \in A_n$, and $sig(x) = -1$ else (verify that sig is indeed a group homomorphism). The point is that, by definition, $A_n = \{x \in S_n \mid sig(x) = 1\}$, i.e., A_n is the subgroup of elements being sent to the neutral element 1 of the codomain group. The systematic context is this:

Definition 93 *The* kernel *of a group homomorphism* $f : G \to H$ *is the subgroup* $Ker(f) = \{x \in G \mid f(x) = e_H\}$. *The* image *of* f *is the set-theoretic image* $Im(f)$, *a subgroup of* H.

Proposition 116 *A group homomorphism* f *is an injective map iff its kernel is trivial.*

Proof If f is injective, then $Ker(f) = f^{-1}(e)$ must be a singleton, so it is trivial. Conversely, if $Ker(f) = e$, then $f(x) = f(y)$ implies $f(xy^{-1}) = e$, whence $xy^{-1} = e$. $\qquad\square$

Therefore we have $A_n = Ker(sig)$. But there is more: We have two equipollent complementary subsets A_n and $S_n - A_n$ of the full group S_n, and the image group $\mathbb{Z}^*$ is also of order 2. This suggests that we may try to reconstruct the codomain group from the domain group and the kernel group. This can effectively be done, but we need a small preliminary discussion.

Definition 94 *Given a subgroup* $H \subset G$ *and an element* $g \in G$, *the* left (right) H-*coset of* g *is the set* $gH = \{gh \mid h \in H\}$ $(Hg = \{hg \mid h \in H\})$. *The set of left (right) H-cosets of G is denoted by G/H $(H\backslash G)$.*

Sorite 117 *Given a subgroup $H \subset G$, we have these facts:*

(i) *The relation $x \sim y$ iff $x \in yH$ is an equivalence relation, the equivalence classes are the left cosets, i.e., two cosets are either disjoint or equal, and $G/H = G/\sim$. Each left coset gH is equipollent with H by means of $h \mapsto gh$. This means that we have a set bijection $G/H \times H \overset{\sim}{\to} G$.*

(ii) *The relation $x \sim y$ iff $x \in Hy$ is an equivalence relation, the equivalence classes are the right cosets, i.e., two cosets are either disjoint or equal, and $H\backslash G = G/\sim$. Each right coset Hg is equipollent with H by means of $h \mapsto hg$. We have a set bijection $H\backslash G \overset{\sim}{\to} G/H$, and therefore also $H\backslash G \times H \overset{\sim}{\to} G$.*

The common cardinality of G/H or $H\backslash G$ is denoted by $(G : H)$ and is called the index *of H in G.*

(iii) *Let G be a finite group, then we have the* Lagrange equation

$$card(G) = card(H) \cdot (G : H).$$

In particular, the order of any subgroup H of a finite group G divides the order of G. More specifically, if $x \in G$ is any element and if G is finite, then the order of x divides the order of G.

Proof (i) The relation $x \sim y$ means $x = yh, h \in H$. Taking $h = e$, we obtain $x \sim x$, and from $x = yh$ one deduces $y = xh^{-1}$, i.e., $y \sim x$. Finally, $x = yh$ and $y = zk$, $k \in H$, implies $x = zkh$, whence $x \sim z$. The surjection $g : H \to gH : h \mapsto gh$ is an injection because $h = g^{-1}(gh)$ gives us h back. So $G/H \times H \xrightarrow{\sim} G$. The proof of (ii) works similarly with "left" and "right" being exchanged. The Lagrange equation (iii) now follows from (ii) since $card(G) = card(G/H \times H) = card(G/H) \cdot card(H) = (G : H) \cdot card(H)$. $\square$

The kernel of a homomorphism is more than a subgroup, it is exactly such that the coset set G/H can be made into a group in a canonical way.

Proposition 118 *Let H be a subgroup of a group G. Then the following properties are equivalent:*

   (i) *There is a group homomorphism $f : G \to K$ such that $H = Ker(f)$.*

  (ii) *Left and right cosets coincide, i.e., for all $x \in G$, $xH = Hx$, and the composition $xH \cdot yH = xyH$ defines a group structure, the quotient group, on G/H. The group H is the kernel of the group homomorphism $G \to G/H : x \mapsto xH$.*

Proof (i) implies (ii): If $H = Ker(f)$, then $f(xHx^{-1}) = e$ for all x, so $xHx^{-1} \subset H$, but also $x^{-1}Hx \subset H$, whence $xHx^{-1} = H$, i.e., $xH = Hx$, all x. (ii) implies (i): The composition $xH \cdot yH = xyH$ is well defined since if $xH = x'H$, then $xyH = xHy = x'Hy = x'yH$, and if $yH = y'H$, then $xyH = xy'H$. It is a group composition having H as neutral element and $x^{-1}H$ as the inverse of xH. The map $f : G \mapsto G/H : g \mapsto gH$ is a surjective homomorphism, and $gH = H$ iff $g \in H$. So $H = Ker(f)$. $\square$

Definition 95 *A subgroup $H \subset G$ with the equivalent properties of proposition 118 is called a* normal *subgroup of G; the group G/H is called the* quotient group *of G modulo H.*

Therefore, the alternating group A_n is normal in S_n.

Exercise 63 Show that every subgroup of a commutative group is normal. For example, given $n \in \mathbb{N}$, we consider the additive subgroup $\langle n \rangle \subset \mathbb{Z}$. Show that this is the group consisting of all multiples $z \cdot n, z \in \mathbb{Z}$ of n; therefore, we also write $\mathbb{Z} \cdot n$ for $\langle n \rangle$. The quotient group $\mathbb{Z}/\mathbb{Z} \cdot n$ is denoted by $\mathbb{Z}_n$ and is called the cyclic group of order n. Its order is effectively n. In fact, by the Euclidean division theorem, every integer w is written uniquely in the form $w = t \cdot n + r, 0 \le r < n$. Therefore every coset is of the form $r + \mathbb{Z} \cdot n, 0 \le r < n$, and if $r + \mathbb{Z} \cdot n = r' + \mathbb{Z} \cdot n$, then $r = r' + t \cdot n$, therefore $t = 0$, and $r' = r$.

If we deal with elements of $\mathbb{Z}_n$, they are represented by cosets $r + \mathbb{Z}$. Often, one makes calculations on $\mathbb{Z}_n$ but works with such representatives. In order to tell that two representatives $r, s \in \mathbb{Z}$ represent the same coset, one writes $r \equiv s \bmod n$ or $r \equiv s \pmod{n}$ or else $r \equiv s\ (n)$ or even $r = s(n)$.

We are now ready for the reconstruction of the image $Im(f)$ of a group homomorphism f by means of the domain group and the kernel $Ker(f)$:

Proposition 119 *If $f : G \to H$ is a group homomorphism, there is a canonical isomorphism of groups*

$$G/Ker(f) \xrightarrow{\sim} Im(f).$$

It is given by the map $gKer(f) \mapsto f(g)$ and satisfies the following commutative diagram:

$$
\begin{array}{ccc}
G & \xrightarrow{\ f\ } & H \\
\downarrow & & \uparrow \\
G/Ker(f) & \xrightarrow{\ \sim\ } & Im(f)
\end{array}
$$

Proof The map $p : G/Ker(f) \to Im(f)$ is well defined since $gKer(f) = hKer(f)$ iff $g^{-1}h \in Ker(f)$, so $e = f(g^{-1}h) = f(g^{-1})f(h)$, i.e., $f(g) = f(h)$. It is surjective by construction and also a group homomorphism by the definition of the quotient group. Its kernel is the set of those cosets $gKer(f)$ such that $f(g) = e$, i.e., $g \in Ker(f)$, i.e., $gKer(f) = Ker(f)$, the neutral element of the quotient group $G/Ker(f)$. So by proposition 116 it is injective. □

Example 51 Reconsidering the homomorphism $sig : S_n \to \mathbb{Z}^*$, we know that $Ker(sig) = A_n$, and $Im(sig) = \mathbb{Z}^*$. Therefore, by proposition 119, $S_n/A_n \xrightarrow{\sim} \mathbb{Z}^*$.

15.3 Rings

Although groups are fundamental to mathematics, their structure is somewhat too poor for realistic applications. We have seen in chapter 9 that the important number domains $\mathbb{Z}, \mathbb{Q}, \mathbb{R}, \mathbb{C}$ share a simultaneous presence of two kinds of operations: addition and multiplication. We have even learned that the very construction of the real numbers needs the auxiliary space of Cauchy sequences, which also shares addition and multiplication with the other spaces. Here is the precise statement about the common properties of these combined operations:

Definition 96 *A* ring *is a triple* $(R, +, *)$ *such that* $(R, +)$ *is an additive abelian group with neutral element* 0_R *(or 0 if the context is clear), and where* $(R, *)$ *is a multiplicative monoid with neutral element* 1_R *(or 1 if the context is clear). These two structures are coupled by the* distributivity *law:*

$$\text{for all } x, y, z \in R, \quad x * (y + z) = x * y + x * z$$
$$\text{and} \quad (x + y) * z = x * z + y * z.$$

One refers to $(R, +)$ *when referring to the additive group of a ring and to* $(R, *)$ *when referring to the multiplicative monoid of the ring. Usually, one simply writes R for the ring* $(R, +, *)$ *if addition and multiplication are clear from the context. The group of multiplicatively invertible elements of a ring is denoted by* R^*. *Also, multiplication* $*$ *is often denoted by ab or* $a \cdot b$ *or a.b instead of* $a * b$, *if the context is clear.*

A ring is commutative *iff its multiplicative monoid is so.*

A subring *of a ring is a ring which is simultaneously an additive subgroup and a multiplicative submonoid.*

Example 52 As already announced, we have a number of prominent rings from the previous theory. If we denote by + and $*$, respectively, the ordinary addition and multiplication, respectively, in the number domains $\mathbb{Z}, \mathbb{Q}, \mathbb{R}, \mathbb{C}$, then each of $(\mathbb{Z}, +, *), (\mathbb{Q}, +, *), (\mathbb{R}, +, *), (\mathbb{C}, +, *)$ is a commutative ring. Moreover, the set of Cauchy sequences C, together with the defined sum and product of Cauchy sequences, is a commutative ring. Check the corresponding sections!

The additive cyclic groups $(\mathbb{Z}_n, +)$ can also be turned into commutative rings by defining the multiplication $(r + n \cdot \mathbb{Z}) * (s + n \cdot \mathbb{Z}) = rs + n \cdot \mathbb{Z}$. Evidently, this multiplication is well defined, since two representations $r \equiv r' \pmod{n}$ have their difference in $n \cdot \mathbb{Z}$.

Exercise 64 Verify that multiplication in $\mathbb{Z}_n$ is well defined. What is the multiplicative neutral element $1_{\mathbb{Z}_n}$?

Rings are also related to each other by ring homomorphisms:

Definition 97 *A set map* $f : R \to S$ *of rings* R *and* S *is a* ring homomorphism *if* f *is a group homomorphism of the additive groups of* R *and* S *and if* f *is a monoid homomorphism of the multiplicative monoids of* R *and* S. *The set of ring homomorphisms from* R *to* S *is denoted by* $\text{Ring}(R,S)$.

The composition *of two ring homomorphisms* $f : R \to S$ *and* $g : S \to T$ *is the set-theoretic composition* $g \circ f$. *By the facts from group and monoid homomorphisms, the composition is also a ring homomorphism.*

An isomorphism f *of rings is a homomorphism of rings which is an isomorphism of additive groups and of multiplicative monoids (which is equivalent to the condition that* f *is a bijective set map). An* endomorphism or rings *is a homomorphisms of one and the same ring. A* ring automorphism *is an endomorphism which is an isomorphism of rings.*

Example 53 The inclusions $\mathbb{Z} \subset \mathbb{Q} \subset \mathbb{R} \subset \mathbb{C}$ are ring homomorphisms. The conjugation $\overline{?} : \mathbb{C} \xrightarrow{\sim} \mathbb{C}$ is an automorphism of $\mathbb{C}$. The canonical maps $\mathbb{Z} \to \mathbb{Z}_n$ are ring homomorphisms.

Example 54 A crucial construction of rings which are not commutative in general is the so-called "monoid algebra". A very important special case is the omnipresent polynomial ring, so the following construction is far from an academic exercise!

To do the construction, we need a commutative ring $(R, +, \cdot)$ and a monoid $(M, *)$. The monoid algebra of R and M is the following ring: Its set is the subset $R\langle M \rangle$ of the powerset R^M consisting of all functions $f : M \to R$ such that $f(m) = 0_R$ for all but a finite number of arguments $m \in M$. Given a couple $(r, m) \in R \times M$, the special function $f : M \to M$ with $f(m) = r$ and $f(n) = 0$ for all $n \neq m$ is denoted by $r \cdot m$. The addition on $R\langle M \rangle$ is defined by $(f + g)(m) = f(m) + g(m)$. Clearly this function evaluates to 0_R for all but a finite number of arguments if f, g do so. The product is defined by the sum $(f \cdot g)(m) = \sum_{n,t} g(n) \cdot f(t)$, where the sum is taken over the finite number of couples (n, t) such that $n * t = m$ and either $g(n)$ or $f(t)$ differs from 0_R. If there is no such non-zero value, the product is defined to be the zero function $0_R \cdot e_M$.

Exercise 65 Given a ring R and a monoid M as above, show that the monoid algebra $R\langle M \rangle$ defined in example 54 is a ring with additive neutral element $0_R \cdot e_M$ and multiplicative neutral element $1_R \cdot e_M$. It is commutative iff M is so. Every element $f \in R\langle M \rangle$ can be written as a finite sum $f = \sum_{i=1,\ldots k} f_i \cdot m_i = f_1 \cdot m_1 + f_2 \cdot m_2 + \ldots f_k \cdot m_k$, with $f_i = f(m_i)$.

If $f \neq 0$, the the summands $f_i \cdot m_i$ are uniquely determined up to permutations if one only adds up those with $f_i \neq 0$. One therefore also uses the notation $f = \sum_m f_m \cdot m$ where it is tacitly supposed that only a finite number of summands is considered, of course comprising only those m where $f(m) \neq 0$.

There is an injective ring homomorphism $R \to R\langle M \rangle : r \mapsto r \cdot e_M$, and an injective homomorphism $M \to R\langle M \rangle : m \mapsto 1_R \cdot m$ of multiplicative monoids. One therefore also identifies elements of R and M, respectively with their images in the monoid algebra. We observe that under this identification, any two elements $r \in R$ and $f \in R\langle M \rangle$ commute, i.e., $r \cdot f = f \cdot r$. This is the theoretical reason why the word "R-algebra" comes into this construction.

Example 55 If in particular $M = Word(A)$ is the word monoid of an alphabet A, then the monoid algebra is called the *R-algebra of noncommutative polynomials in the indeterminates from A* So every element is a sum of so-called *monomials* $r \cdot X_1 X_2 \ldots X_k$ in the indeterminates $X_i \in A$. In particular, if $A = \{X_1, X_2, \ldots X_n\}$ is a finite set of so-called "indeterminates" X_i (the term is slightly misleading, the elements are completely normal sets), then the monoid algebra is denoted by $R\langle X_1, X_2, \ldots X_n \rangle$.

The most prominent such algebra is the case for $A = \{X\}$. We then get the words $X^k = XXX \ldots X$, the k-fold juxtaposition of the unique letter X. A polynomial in the indeterminate X is then written in the form

$$f = r_k \cdot X^k + r_{k-1} \cdot X^{k-1} + r_{k-2} \cdot X^{k-2} + \ldots r_2 \cdot X^2 + r_1 \cdot X + r_0,$$

or else $f(X)$ in order to put the indeterminate X in evidence. The monoid algebra is then commonly denoted by $R[X]$ instead of $R\langle X \rangle$. Compare addition and multiplication of such polynomials in X to what you know from high school mathematics!

In particular, if the maximal coefficient is r_3 the polynomial $f(X)$ is called *cubic*, if r_2 is the maximal non-zero coefficient, $f(X)$ is called *quadratic*, if r_1 is the maximal non-zero coefficient, $f(X)$ is called *linear*, and if $f(X) = r_0 \in R$, then $f(X)$ is called a *constant* polynomial. For a non-zero polynomial in X, the maximal power X^k of X such that the coefficient r_k is different from zero is called the degree of f, in symbols $deg(f)$.

Remark 19 Consider the monoid $ComWord(A)$ whose elements are equivalence classes of words over A in the sense that $w \sim w'$ iff the words' letters are just permutations of each other. Then the product of words from $Word(A)$ is well defined on equivalence classes, which are also called commutative words. Taking the monoid $ComWord(A)$, the monoid algebra is commutative and is denoted by $R[A] = R\langle ComWord(A)\rangle$. It is called the *R-algebra of commutative polynomials in the indeterminates from A*.

The power of the monoid algebra construction is shown in this proposition:

Proposition 120 (Universal Property of Monoid Algebras)   *Let R be a commutative ring, M a monoid, and S a (not necessarily commutative) ring. Suppose that $f : R \to S$ is a ring homomorphism such that for all $r \in R$ and $s \in S$, $f(r) \cdot s = s \cdot f(r)$. Then for any monoid homomorphism $\mu : M \to S$ into the multiplicative monoid of S, there is exactly one ring homomorphism $f\langle \mu \rangle : R\langle M \rangle \to S$ which extends f and μ, i.e., $f\langle \mu \rangle|_R = f$ and $f\langle \mu \rangle|_M = \mu$.*

Proof On an argument $\sum_m g_m \cdot m$, any ring homomorphism $f\langle \mu \rangle$ which extends f and μ as required, must be defined by $f\langle \mu \rangle(\sum_m g_m \cdot m) = \sum_m f\langle \mu \rangle(g_m) \cdot f\langle \mu \rangle(m) = \sum_m f(g_m) \cdot \mu(m)$. But this is a well defined map since the sum representation of the argument is unique. It is now straightforward to check that this map is indeed a ring homomorphism.   □

The most important consequence of this proposition is this corollary:

Corollary 121 *If R is a commutative ring, and if $\{X_1, X_2, \ldots X_n\}$ is a finite set, then every set map $v : X_i \mapsto x_i \in R$ extends to a ring homomorphism $R\langle v \rangle : R\langle X_1, X_2, \ldots X_n \rangle \to R$, whose value on a monomial $r \cdot X_{i_1} X_{i_2} \ldots X_{i_k}$ is $r \cdot x_{i_1} x_{i_2} \ldots x_{i_k}$. This homomorphism is called the* evaluation *of polynomials with respect to v.*

Proof This follows at once from proposition 111 and proposition 120, applying the universal property of the word monoid over the symbols X_i to obtain μ, and the universal property of a monoid algebra for $f = Id_R$ and μ.   □

Example 56 If $f = f(X) = r_k \cdot X^k + r_{k-1} \cdot X^{k-1} + r_{k-2} \cdot X^{k-2} + \ldots r_2 \cdot X^2 + r_1 \cdot X + r_0$ is a polynomial in $\mathbb{C}[X]$, then the map $X \mapsto x \in \mathbb{C}$ defines the *evaluation* $f(x) = r_k \cdot x^k + r_{k-1} \cdot x^{k-1} + r_{k-2} \cdot x^{k-2} + \ldots r_2 \cdot x^2 + r_1 \cdot x + r_0$ *of* $f(X)$ *at x*. This means that a polynomial $f(X)$ defines a function $f(?) : \mathbb{C} \to \mathbb{C}$, this is what is called a *polynomial function*. Generally speaking, a polynomial function is simply a function defined by a polynomial and an evaluation of its indeterminates, as guaranteed by the above proposition 120.

One might think of having similar structures as normal subgroups, which are the kernels of group homomorphisms, for rings. This is indeed the case. But we shall only need the theory for commutative rings. Here is the characterization:

Proposition 122 *Let J be an additive subgroup of a commutative ring R. Then the following properties are equivalent:*

(i) *There is a homomorphism of commutative rings $f : R \to S$ such that $J = Ker(f)$ for the underlying additive groups.*

(ii) *J is a subgroup of the additive group of R, and for every $r \in R$, if $x \in J$, then $r \cdot x \in J$. The multiplication $(x + J) \cdot (y + J) = xy + J$ defines the multiplication of a ring structure on the quotient group R/J, the quotient ring. The group J is the kernel of the ring homomorphism $R \to R/J : x \mapsto x + J$.*

Proof The proof is completely analogous to that regarding normal subgroups and kernels of group homomorphisms. We leave it as an exercise to the reader. □

Definition 98 *A subgroup $J \subset R$ with the equivalent properties of proposition 122 is called an* ideal *of R.*

Example 57 We are now in state of better understanding the construction of the ring of real numbers $\mathbb{R}$ from Cauchy sequences. The set C of Cauchy sequences as defined in definition 50 is a commutative ring, this is statement (i) in proposition 79, together with the constant sequence $(1)_i$ as multiplicative neutral element, while statements (i) and (ii) in proposition 79 tell us that the zero sequences $\mathcal{O}$ define an ideal. Finally, by lemma 81, the equivalence relation in lemma 80 is precisely the relation defined by the ideal $\mathcal{O}$, and the quotient structure $C/\mathcal{O}$ is the same as the structure defined in definition 54.

Example 58 Reconsidering the construction of the cyclic groups $\mathbb{Z}_n$, we recognize that the subgroup $\mathbb{Z} \cdot n$ of $\mathbb{Z}$ is in fact an ideal. The construction of the ring multiplication on the quotient group $\mathbb{Z}/\mathbb{Z} \cdot n$ is exactly the one we just defined for the quotient ring.

The type of ideal in the previous example is of major importance for the theory of prime numbers and follows from the Euclidean algorithm:

Definition 99 *An ideal J of a commutative ring R is called a* principal ideal *if there is an element $x \in J$ such that $J = R \cdot x = \{r \cdot x \mid r \in R\}$.*

Such an ideal is also denoted by $J = (x)$. A ring the ideals of which are all principal is called a principal ideal ring.

Proposition 123 *The ring $\mathbb{Z}$ is principal. If J is an ideal of $\mathbb{Z}$, then either $J = (0)$ or $J = (n)$, where $n \in \mathbb{N}$ is the smallest positive integer in J.*

Proof If J is an non-zero ideal in $\mathbb{Z}$, then there is a positive element n in J, take the smallest such element. For any $j \in J$, we have $j = an + b$, $0 \le b < n$, by elementary natural arithmetic. But then $b = j - an \in J$, whence $b = 0$, therefore $J = (n)$.   □

We also have the analogous ring-theoretic proposition to proposition 119 for groups:

Proposition 124 *If $f : R \to S$ is a homomorphism of commutative rings, there is a canonical isomorphism of rings*

$$R/Ker(f) \xrightarrow{\sim} Im(f).$$

It is given by the map $g + Ker(f) \mapsto f(g)$ and satisfies the commutative diagram

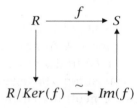

Proof The proof of this proposition follows the same line as that of proposition 119, we leave it as an exercise.   □

We shall see in the following section that this proposition entails the remarkable isomorphism of rings $\mathbb{C} \xrightarrow{\sim} \mathbb{R}[X]/(X^2 + 1)$. To this end we need some specific extra properties of rings such as $\mathbb{R}$.

15.4 Fields

Definition 100 *A ring $K \ne 0$ is called a* skew field *if every non-zero element is invertible. A commutative skew field is also called a* field.

Example 59 The rings $\mathbb{Q}$, $\mathbb{R}$, $\mathbb{C}$ are fields. This follows right from the properties of these rings which we have exhibited in chapter 9.

The main fact about polynomial rings $K[X]$ over fields K is that they are principal. To see this, we need the following lemma:

Lemma 125 *If K is a field, and if f and g are non-zero polynomials in $K[X]$, then $deg(f \cdot g) = deg(f) + deg(g)$.*

Proof If $f = a_n X^n + a_{n-1} X^{n-1} + \ldots a_0$ and $g = b_m X^m + b_{n-1} X^{n-1} + \ldots b_0$ with $a_n, b_m \neq 0$, then $f \cdot g = a_n b_m X^{n+m} + \ldots a_0 b_0$, the highest coefficient $a_n b_m \neq 0$ because we are in a field, so $deg(f \cdot g) = deg(f) + deg(g)$. $\square$

This entails Euclid's division theorem for polynomial rings over fields:

Proposition 126 (Division Theorem) *If f and g are non-zero polynomials in the polynomial algebra $K[X]$ over a field K, then there is a polynomial h such that either $f = h \cdot g$ or $f = h \cdot g + r$, where $deg(r) < deg(g)$. The polynomials h and r are uniquely determined.*

Proof Let $deg(f) < deg(g)$, then $f = 0 \cdot g + f$ is a solution. If $deg(f) \geq deg(g)$, then for $f = a_n X^n + a_{n-1} X^{n-1} + \ldots a_0$ and $g = b_m X^m + b_{n-1} X^{n-1} + \ldots b_0$, we consider $f' = f - \frac{a_n}{b_m} X^{n-m} \cdot g$. The polynomial f' has smaller degree than f, and we may proceed by induction to find a solution $f' = h' \cdot g + r'$. Then $f = (\frac{a_n}{b_m} X^{n-m} + h') \cdot g + r'$ solves the problem. As to uniqueness, if we have two decompositions $f = h_1 \cdot g + r_1 = h_2 \cdot g + r_2$, then $(h_1 - h_2) \cdot g = r_2 - r_1$, which, for degree reasons, only works if $h_1 - h_2 = r_2 - r_1 = 0$. $\square$

This result implies the announced result:

Proposition 127 *If K is a field, then the polynomial ring $K[X]$ is principal. If J is an ideal of $K[X]$, then either $J = (0)$ or else, $J = (g)$, where g is a polynomial in J of minimal degree.*

Proof In fact, the proof works like for integers. Take a minimal polynomial f of positive degree in an ideal J, then by proposition 126, $J = (f)$. $\square$

We may now demonstrate the announced isomorphism between $\mathbb{C}$ and $\mathbb{R}[X]/(X^2 + 1)$. Consider the ring homomorphism $v : \mathbb{R}[X] \to \mathbb{C}$ defined by the natural embedding $\mathbb{R} \subset \mathbb{C}$ and the evaluation $X \mapsto i = \sqrt{-1}$. Clearly, v is surjective, i.e., $Im(v) = \mathbb{C}$. The kernel of v is principal, $Ker(v) = (t)$. But no non-zero linear polynomial $a \cdot X + b$ is in this kernel, since $a \cdot i + b = 0$ iff $a = b = 0$. On the other hand, $X^2 + 1 \in Ker(v)$, therefore by proposition 127, $Ker(v) = (X^2 + 1)$. By proposition 124, this implies $\mathbb{R}[X]/(X^2 + 1) \xrightarrow{\sim} \mathbb{C}$.

Exercise 66 Using the isomorphism $\mathbb{R}[X]/(X^2 + 1) \xrightarrow{\sim} \mathbb{C}$, try to describe the conjugation on $\mathbb{C}$ by means of the quotient ring $\mathbb{R}[X]/(X^2 + 1)$.

For both constructions, that of $\mathbb{R}$ from Cauchy sequences, and that of $\mathbb{C}$ from polynomials over $\mathbb{R}$, we are left with two secrets in the understanding of quotient rings: Why do these quotient rings yield fields? Here is the answer:

Definition 101 *An ideal J of a commutative ring R is called* maximal *if there is no ideal I such that $J \subsetneq I \subsetneq R$.*

Proposition 128 *An ideal J is maximal in R iff the quotient R/J is a field.*

Proof If J is not maximal, than a strictly larger ideal $J \subsetneq I \subsetneq R$ has an image I' in R/J which is a proper ideal $0 \subsetneq I' \subsetneq R/J$, so R/J cannot be a field. If J is maximal, then any $x \notin J$ must generate the improper ideal R together with J, in particular there are $u \in R$ and $v \in J$ such that $1 = ux + v$. Then the image $1 + J = ux + J = (u + J) \cdot (x + J)$, i.e., every non-zero element $x + J$ in R/J is invertible, and we have a field. $\square$

Now, the missing links are easily inserted:

Example 60 The ideal $(X^2 + 1)$ in $\mathbb{R}[X]$ is maximal. In fact, a strictly larger ideal must be of shape $(X + a)$, but then we must have the factorization $X^2 + 1 = (X + a) \cdot (X + b)$. Evaluating $(X^2 + 1)$ at $x = -a$ yields $a^2 + 1 = 0$, an impossibility in $\mathbb{R}$. Turning to Cauchy sequences, any non-zero sequence $(a_i)_i$ has a sequence $(b_i)_i$ such that $(a_i)_i \cdot (b_i)_i$ converges to 1, this was shown in statement (viii) of sorite 84. So every ideal which is strictly larger than $\mathcal{O}$ must be the principal ideal (1), i.e., the entire ring, and we are done.

CHAPTER 16

Primes

This chapter is devoted to prime numbers and prime polynomials, a type of objects which play an important role in cryptography. Every computer scientist should therefore have a minimal knowledge about the structure of prime factorization for the ring of integers $\mathbb{Z}$ and the polynomial algebras $K[X]$ over one indeterminate X with coefficients in a field K.

16.1 Prime Factorization

We first need a small preliminary discussion about the construction of ideals by generators, a generalization of the concept of a principal ideal introduced in the last chapter.

Definition 102 *If $G \subset R$ is a non-empty subset of a commutative ring R, the ideal generated by G is defined as the set of all sums $r_1 \cdot g_1 + r_2 \cdot g_2 + \ldots r_k \cdot g_k$ where $g_i \in G$ and $r_i \in R$. It is denoted by (G) or $(h_1, h_2, \ldots h_t)$ if $G = \{h_1, h_2, \ldots h_t\}$ is finite.*

Exercise 67 Check that the set (G) is effectively an ideal.

Here is the definition of a prime element in a commutative ring:

Definition 103 *An element $p \in R$ in a commutative ring $R \neq 0$ is* prime *if it is not invertible and if $p = q \cdot r$ implies that either q or r is invertible. For polynomial algebras, prime polynomials are also called* irreducible.

To begin with, we make sure that in the most interesting rings, every non-invertible element has a factorization into prime elements:

Lemma 129 *If $R = \mathbb{Z}$ or $R = K[X]$, K a field, then every non-invertible element $x \neq 0$ has a factorization $x = p_1 \cdot p_2 \cdot \ldots p_k$ as a product of primes p_i.*

Proof For a non-invertible, non-zero $p \in \mathbb{Z}$, we proceed by induction on $|p| \geq 2$. A factorization $p = q \cdot r$ implies $|p| = |q| \cdot |r|$. So if $|p| = 2$, then $p = \pm 2$ is already prime, since one of its factors in $2 = q \cdot r$ must be 1. In the general case, a factorization $p = q \cdot r$ with non-invertible factors implies $|q|, |r| < |p|$, so by induction on these factors, we are done. For a non-invertible, non-zero $p \in K[X]$, we proceed by induction on $deg(p) \geq 1$. A factorization $p = q \cdot r$ implies $deg(p) = deg(q) + deg(r)$. If $deg(p) = 1$, then either q or r is a constant (degree zero), and therefore invertible. If $deg(p) > 1$, then if all factorizations have either q or r of degree zero, then these factors are invertible and p is prime. Else, degrees decrease and induction applies to the factors. $\square$

It could however happen that only a finite number of such primes occur in the prime factorization. But we have the very classical result about prime numbers in $\mathbb{Z}$, Euclid's theorem:

Proposition 130 *The set of primes in $\mathbb{Z}$ is infinite.*

Proof Suppose that $p_1 = 2, p_2 = 3, \ldots p_k$ is the set of all finitely many positive primes in $\mathbb{Z}$. Then the prime factorization of $q = 1 + \prod_{i=1,2,\ldots k} p_i$ must contain one of these primes, say p_t. Then we have $q = p_t \cdot u = 1 + \prod_{i=1,2,\ldots k} p_i$, and therefore[1] $1 = p_t \cdot (u - \prod_{i=1,2,\ldots \hat{t},\ldots k} p_i)$, which is impossible since p_t is not invertible by hypothesis. So there are infinitely many primes in $\mathbb{Z}$. $\square$

Example 61 In particular 0_R is not prime since $0_R = 0_R \cdot 0_R$, and $1_R, -1_R$ are not prime since they are invertible. The numbers ± 2, ± 3, ± 5, ± 7, ± 11, ± 13, ± 17, ± 19 in $\mathbb{Z}$ are primes, while $12, -24, 15$ are not. In $\mathbb{R}[X]$ all linear polynomials $a \cdot X + b, a \neq 0$, and all quadratic polynomials $a \cdot X^2 + b$ with $a, b > 0$ are prime. (Use the argumentation from example 60 used to see that $X^2 + 1 = (u \cdot X + v)(r \cdot X + s)$ is impossible in $\mathbb{R}[X]$.)

Clearly, prime factorization is not unique because of the commutativity of the ring and the existence of invertibles, e.g., $12 = 2 \cdot 2 \cdot 3 = 3 \cdot$

---

[1] The convention $\hat{x}_i$ means that in an indexed sequence, the object x_i with index i is omitted, and only the earlier and later terms are considered. For example, in $x_1, x_2, x_3, \ldots \hat{x}_9, \ldots x_{20}$, means that we take the sequence $x_1, x_2, x_3, \ldots x_8, x_{10}, \ldots x_{20}$.

$(-2) \cdot (-2)$. However, this is the only reason of ambiguities in prime factorization for our prominent rings.

Definition 104 *If x and y are two elements in a commutative ring such that there is an element z with $y = x \cdot z$, then we say that x divides y or that x is a divisor of y and write $x|y$. If x does not divide y, we write $x \nmid y$. Clearly, $x|y$ is equivalent to the ideal inclusion $(y) \subset (x)$.*

If $0 = x \cdot y$ for $x, y \neq 0$, then x is called a zero divisor. *A commutative ring without zero divisors is called an* integral domain.

Lemma 131 *In an integral domain R, the generator of a principal ideal a is unique up to invertible elements, i.e., $(a) = (b)$ iff there is $e \in R^*$ such that $b = e \cdot a$.*

Proof Clearly the existence of such an $e \in R^*$ is sufficient. Conversely, $(a) = (b)$ implies $a = xb, b = ya$, whence $a = xya$, i.e., $xy = 1$ if $a \neq 0$. But if $a = 0$, then also $b = 0$ and we have discussed all cases. □

Notation 9 *If R is a principal integral domain, the ideal generated by two elements a and b is principal, $(a, b) = (d)$. This is equivalent to the two facts that (1) $d|a$ and $d|b$, (2) there are two elements u and v such that $d = u \cdot a + v \cdot b$. Such a d is called the* greatest common divisor *of a and b, in symbols $d = gcd(a, b)$. The ideal $(a) \cap (b)$ is also principal, $(a) \cap (b) = (l)$. This means that (1) $a|l$ and $b|l$, (2) whenever $a|x$ and $b|x$, then $l|x$. Such an l is called the* least common multiple *of a and b, in symbols $l = lcm(a, b)$. Observe that gcd and lcm are only determined up to invertible elements.*

Proposition 132 (Euclidean Algorithm) *For two integers a and b the $gcd(a, b)$ is calculated by this algorithm: If $b = 0$ we are done, i.e., $gcd(a, b) = a$. Else, apply the division theorem $a = r \cdot b + s$ and proceed with the couple (b, s) instead of (a, b).*

Proof Suppose that we have this chain of successive divisions: $a = r \cdot b + s, b = r_1 \cdot s + s_1, s = r_2 \cdot s_1 + s_2, s_1 = r_3 \cdot s_2 + s_4, \ldots s_k = r_{k+2} \cdot s_{k+1}$. Then the claim is that $s_{k+1} = gcd(a, b)$. In fact, clearly $s_{k+1}|a$ and $s_{k+1}|b$, by successive replacement in these formulas from the end. Moreover, each remainder s, s_1, etc. is a combination $u \cdot a + v \cdot b$, and so is in particular the last, i.e., s_{k+1}, whence the proposition. □

Example 62 A field and any of its subrings is an integral domain. In particular, all the rings $\mathbb{Z}, \mathbb{Q}, \mathbb{R}, \mathbb{C}$ are integral domains, and so are the polynomial algebras $K[X]$ over a field K, by the degree formula $deg(f \cdot g) = deg(f) + deg(g)$ for non-zero polynomials.

Remark 20 The converse is also true: Any integral domain is a subring of a field.

Lemma 133 *If the ring R is a principal integral domain, then an ideal I is maximal iff there is a prime p with $I = (p)$. If q is another prime with $I = (q)$, then there is an invertible element e such that $q = e \cdot p$.*

Proof If $I = (p)$ is maximal, then $p = q \cdot r$ implies $(p) \subset (q)$, so either $(q) = R$, and $q \in R^*$, or else $(p) = (q)$, whence $r \in R^*$. If I is not maximal, there is q such that $I \subsetneq (q) \subsetneq R$, i.e., $p = q \cdot r$, and $q, r \notin R^*$.   □

Proposition 134 *For a prime number $p \in \mathbb{Z}$, the finite ring $\mathbb{Z}_p$ is a field. In particular, the multiplicative group $\mathbb{Z}_p^*$ has $p - 1$ elements, and therefore for every integer x, we have $x^p \equiv x$ mod (p), and $x^{p-1} \equiv 1$ mod (p) if $p \nmid x$. The latter statement is called* Fermat's little theorem.

Proof The proposition is immediate from lemma 133.   □

The next result leads us towards the uniqueness property of prime factorization:

Proposition 135 *Let R be a principal integral domain and $p \in R$ prime. Then if $p | a_1 \cdot a_2 \cdot \ldots a_n$, then there is an index i such that $p | a_i$.*

Proof In fact, by lemma 133, $R/(p)$ is a field, and therefore, the division $p | a_1 \cdot a_2 \cdot \ldots a_n$ reduces to $0 = \bar{a}_1 \cdot \bar{a}_2 \cdot \ldots \bar{a}_n$ in the field $R/(p)$, but this implies that one of these factors vanishes, i.e., p divides one of the factors a_i.   □

Proposition 136 *Let R be a principal integral domain. Then, if an element a has two prime factorizations $a = p_1 \cdot p_2 \cdot \ldots p_k = q_1 \cdot q_2 \cdot \ldots q_l$, then $k = l$, and there is a permutation $\pi \in S_k$ and a sequence $e_1, e_2, \ldots e_k$ of invertible elements such that $q_i = e_i \cdot p_{\pi(i)}, i = 1, 2, \ldots k$.*

Proof This is clear from the preceding proposition 135 if we take the factorization $q_1 \cdot q_2 \cdot \ldots q_l$ and a divisor p_j of this product. This means $p_j = e_i \cdot q_i$ for an invertible e_i. Then, dividing everything by p_j reduces to a shorter product, and we are done by induction.   □

Proposition 137 *In $\mathbb{Z}$, every non-invertible element $a \neq 0$ has a unique factorization by positive primes $p_1 < p_2 < \ldots p_r$ and positive powers $t_1, t_2, \ldots t_r$ of the form*

$$a = \pm p_1^{t_1} \cdot p_2^{t_2} \cdot \ldots p_r^{t_r}.$$

Proof This follows directly from the above proposition 136, adding the ordering by size of the intervening prime numbers. □

Corollary 138 *In the polynomial algebra $K[X]$ over a field K, every polynomial of positive degree is a product of irreducible polynomials, and this factorization is unique in the sense of proposition 136. In particular, every linear polynomial is irreducible.*

Proof This follows from the existence of a prime factorization following lemma 129 and the uniqueness statement in proposition 136. □

Remark 21 It can be shown that in $\mathbb{C}[X]$, the irreducible polynomials are exactly the linear polynomials. This means that every polynomial $f(X)$ of positive degree is a product $f(X) = a(X - b_1)(X - b_2) \ldots (X - b_k), a \neq 0$. This is a quite profound theorem, the so-called *fundamental theorem of algebra*.

Exercise 68 Show that $\sqrt{2}$ is not a rational number. Use the prime factorization of numerator a and denominator b in a fictitious representation $\sqrt{2} = \frac{a}{b}$.

Exercise 69 Use exercise 68 to show that the set $\mathbb{Q}(\sqrt{2})$ consisting of the real numbers of form $z = a\sqrt{2} + b, a, b \in \mathbb{Q}$, is a subfield of $\mathbb{R}$. Show that $\mathbb{Q}(\sqrt{2}) \overset{\sim}{\to} \mathbb{Q}[X]/(X^2 - 2)$.

16.2 Roots of Polynomials and Interpolation

In this section, let K be a field. Let $x \in K$. Then we know from example 56 that for a polynomial $f(X) \in K[X]$, there is an evaluation $f(x) \in K$. We now want to discuss the relation between the polynomial $f(X)$ and the polynomial function $f(?) : K \to K : x \mapsto f(x)$.

Definition 105 *If $f(X) \in K[X]$, then an element $x \in K$ is a root of $f(X)$ if $f(x) = 0$.*

Lemma 139 *If $x \in K$ is a root of a polynomial $f(X) \in K[X]$, then $(X - x) | f(X)$.*

Proof We have the division with remainder $f(X) = Q(X) \cdot (X - x) + c, c \in K$. And evaluation f at x yields $c = 0$, whence the claim. □

Proposition 140 *If $x_1, x_2, \ldots x_r$ are r different roots of $f(X) \in K[X]$, then*

$$(X - x_1)(X - x_2) \ldots (X - x_r) | f(X).$$

In particular, a non-zero polynomial $f(X)$ has at most $\deg(f)$ different roots.

Proof The proof uses induction on r. By the above lemma 139, the claim is true for $r = 1$. But since each $X - x_r$ is prime and differs from all $X - x_j, j \neq i$, we have a factorization $f(X) = g(X) \cdot (X - x_r)$. So all $x_i, i < r$ are roots of $g(X)$, and by the induction hypothesis, we are done. □

Corollary 141 *If two polynomials $f, g \in K[X]$ of degrees $\deg(f) < n$ and $\deg(g) < n$ have n different arguments $x_i, i = 1, 2, \ldots n$ with common values $f(x_i) = g(x_i)$, then $f = g$.*

Proof This is an easy exercise using proposition 140. □

Corollary 142 *If for $K = \mathbb{Q}, \mathbb{R}, \mathbb{C}$, two polynomial functions $f(?), g(?)$ coincide, then the polynomials $f(X), g(X) \in K[X]$ are equal.*

Proof This follows from the fact that we have infinitely many arguments, where the functions coincide. □

This allows us to identify polynomials and their associated functions, but this is only a special situation, which does not generalize to arbitrary polynomial algebras.

However, we do not know if there is always a polynomial $f(X)$ of degree strictly less than n such that its values $f(x_i) = y_i$ can be prescribed for n different arguments $x_1, x_2, \ldots x_n$. This is guaranteed by different so-called *interpolation formulas*, the best known being those by Lagrange and Newton. Since the result must be unique by corollary 141, we may pick one such formula.

Proposition 143 (Newton Interpolation Formula) *Suppose that we are given a sequence $(x_1, y_1), (x_2, y_2), \ldots (x_n, y_n)$ of couples $(x_i, y_i) \in K^2$*

for a field K, where $x_i \neq x_j$ for $i \neq j$. Then there is a (necessarily unique) polynomial $f(X) \in K[X]$ of degree $\deg(f) < n$ such that $f(x_i) = y_i$, $i = 1, 2, \ldots n$. It is given by the Newton interpolation formula

$$f(X) = a_0 + a_1(X - x_1) + a_2(X - x_1)(X - x_2) + \ldots$$
$$a_{n-1}(X - x_1)(X - x_2) \ldots (X - x_{n-1}).$$

Exercise 70 Give a proof of proposition 143. Start with the evaluation at x_1 and calculate the coefficient a_0. Then proceed successively with the calculation of all coefficients $a_1, a_2, \ldots a_{n-1}$.

Why are such formulas called "interpolation formulas"? The point is that we are often given a series of "values" y_i for arguments x_i, but we do not know which function $f : K \to K$ takes these values, $y_i = f(x_i)$. In most cases there is a large number of solutions for this problem. Any solution, such as the polynomial solution given in proposition 143, will also give us values for all other $x \in K$. For example, if $K = \mathbb{R}$, we get all the evaluations $f(x)$ for the intervals $x \in [x_i, x_{i+1}]$. This means that f can also be evaluated on values 'between' the given arguments, which is the very meaning of the word 'interpolation'.

Example 63 Given are four points $p_1 = (-2, 3)$, $p_2 = (-\frac{1}{2}, -\frac{1}{2})$, $p_3 = (1, \frac{1}{2})$ and $p_4 = (2, -1)$ in $\mathbb{R}^2$. The goal is to construct the interpolation polynomial $f(X) \in \mathbb{R}[X]$ through these points. Proposition 143 ensures that $f(X)$ is of the form:

$$f(X) = a_0 + a_1(X - x_1) + a_2(X - x_1)(X - x_2) +$$
$$a_3(X - x_1)(X - x_2)(X - x_3)$$

Now, setting $X = x_i$ and $f(X) = y_i$ for $i = 0, 1, 2, 3, 4$, the a_i, $i = 0, 1, 2, 3$ are calculated as follows:

For p_1, every term but the first, a_0, vanishes, thus

$$a_0 = 3$$

For p_2:

$$-\tfrac{1}{2} = a_0 + a_1(-\tfrac{1}{2} - (-2))$$

thus, after substituting the known value for a_0, and solving for a_1:

$$a_1 = -\tfrac{7}{3}$$

For p_3:
$$\tfrac{1}{2} = a_0 + a_1(1 - (-2)) + a_2(1 - (-2))(1 - (-\tfrac{1}{2}))$$
which yields, using the previously calculated values for a_0 and a_1:
$$a_2 = 1$$
And finally, for p_4:
$$-1 = a_0 + a_1(2 - (-2)) + a_2(2 - (-2))(2 - (-\tfrac{1}{2})) +$$
$$a_3(2 - (-2))(2 - (-\tfrac{1}{2}))(2 - 1)$$
produces
$$a_3 = -\tfrac{7}{15}$$
Putting everything together, and expanding the polynomial:
$$f(X) = 3 - \tfrac{7}{3}(X + 2) + (X + 2)(X + \tfrac{1}{2}) - \tfrac{7}{15}(X + 2)(X + \tfrac{1}{2})(X - 1)$$
$$= -\tfrac{7}{15}X^3 + \tfrac{3}{10}X^2 + \tfrac{13}{15}X - \tfrac{1}{5}$$

The polynomial $f(X)$ is drawn in figure 16.1.

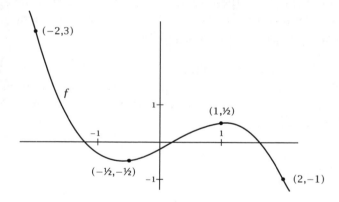

Fig. 16.1. The polynomial $f(X) = -\tfrac{7}{15}X^3 + \tfrac{3}{10}X^2 + \tfrac{13}{15}X - \tfrac{1}{5}$.

Observe that the interpolation polynomial may not necessarily satisfy specific conditions on the "smoothness" of the curve. Indeed, if n points are specified, the resulting polynomial will have degree $n - 1$, its shape becoming increasingly "bumpy" as the degree rises, with interpolated values straying widely off the mark.

Therefore, for practical purposes, more flexible interpolation techniques, like splines, are used (see [6] for examples).

CHAPTER 17

Formal Propositional Logic

Until now, we have been using logic in its meaning as a guiding principle of human thought—as such it was described under the title of "propositional logic" in the introductory section 1.1. We are not going to replace this fundamental human activity, but we are interested in the problem of mimicking logical reasoning on a purely formal level. What does this mean? The idea is that one would like to incorporate logic in the mathematical theory, which means that we want to simulate the process of logical evaluation of propositions by a mathematical setup. Essentially, this encompasses three tasks:

1. Give a mathematical construction which describes the way in which logically reasonable propositions should be built. For example, as we have learned from section 1.1, if $\mathcal{A}, \mathcal{B}$ are propositions, one may want to build "NOT $\mathcal{A}$", "$\mathcal{A}$ IMPLIES $\mathcal{B}$" and so forth. But the construction should not deal with contents, it should just describe the combinatorial way a new expression is constructed, in other words, this is a problem of syntax (relating to how expressions of a sign system are combined independent of their specific meaning). So the original logical work is split into different subtasks, the first of which is the management of reasonable propositional expressions. *The first task is: How can such a syntactical building scheme be described?* This subtask is evidently required if we want to delegate some logical work to machines such as computers, where content is not feasible without a very precise and mechanical control of form.

2. Suppose that the first subtask of formalization has been achieved. One then needs to rebuild the truth values which propositions should express. This refers to the *meaning* or *semantics* of such propositions. This is a radical point of view: The meaning of a proposition is not its specific contents, but one of the two truth values: "true" or "false", or, equivalently, "is the case" or "is not the case". For example, the meaning of the proposition "if it is raining I am tired" has nothing to do with rain or fatigue, it is only the truth value of the implication. It is this semantical issue which is mimicked by the formal theory of propositional logic. So one needs a domain of truth values, which is addressed by a formal propositional expression in order to evaluate its 'meaning'. Such domains will be called *logical algebras. Therefore, the second task is to give a rigorous mathematical description of what a logical algebra is which manages the semantic issue of propositional logic.*

3. The third subtask deals with the connection of the first two subtasks, i.e., given a syntactical description of how to build logically correct propositions, as well as a logical algebra which manages the semantics of truth values, we need to specify how these two levels are related with each other. *This third task is that of "signification", i.e., the (mathematically rigorous definition of an) association of meaning with a given propositional expression.*

We should, however, pay attention to the power of such a formalization procedure, since often a naive understanding thereof claims a replacement of our human reasoning and its contents by purely formal devices. We shall see on different occasions that and how this is wrong. There is however a first and fundamental error, which is easily explained: A formal, i.e., mathematical theory of propositional logic as sketched above, must rely on given mathematical structures and results, the most important one being set theory, the theory of natural numbers, and—above all— the recursion theorem which is a fundamental tool for the construction of formal expressions and for the proof of properties which are shared by formal expressions. In order to keep the level of human thought and the formal logical level separate, we never use symbols of formal logic in formulas which are meant as a human thought, although this abuse of symbols is a widespread mixup of scientific levels! In a famous 1931 paper *On Formally Undecidable Propositions of Principia Mathematica and Related Systems*, the Czech-born mathematician Kurt Gödel demonstrated that within any given branch of mathematics, there would always be some propositions that could not be proved either true or false using the rules and axioms of that mathematical branch itself. Gödel himself alluded to machine supported logical reasoning. But subsequently, many philosophers, theologists and other would-be scientists concluded that human thought, as well, is tainted by these incompleteness properties, actually derived for formal systems only. We hope that the reader may understand in the following that these formalizations of human thought are just low-level simulations of its mechanical flattening, and not of proper thought. In fact, it is easy to show that there are propositions which are formally undecidable, but by true thought can be decided with ease. We come back to this issue in an exercise.

17.1 Syntactics: The Language of Formal Propositional Logic

In the language of formal propositional logic, the scope is to create the variety of possible propositions, but there are no proper parts of propositions which are not propositions: this is an elementary concept. The only possibility to build new propositions is to start with given ones and to combine them by logical operations. Here is the formal setup of the repertory of expressions in which we shall work in order to build logically reasonable propositions:

Definition 106 *A propositional alphabet is a triple $A = (V, B, C)$ of mutually disjoint sets, where*

(i) *the set V is supposed to be denumerable, i.e., equipollent to $\mathbb{N}$, and the elements $v_0, v_1, \ldots \in V$ are called propositional variables;*

(ii) *the set $B = \{(,)\}$ of $($, the left bracket and $)$, the right bracket;*

(iii) *the set $C = \{!, \&, |, ->\}$ consists of these four logical connective symbols, the negation symbol !, the conjunction symbol &, the disjunction symbol |, and the implication symbol ->.*

The monoid $EX = EX(A) = Word(V \sqcup B \sqcup C)$ is called the monoid of expressions over A. Within EX, the subset $S(EX)$ of sentences over A is the smallest subset $S(EX) \subset EX$ with the following properties:

(i) $V \subset S(EX)$;

(ii) *if $\alpha \in S(EX)$, then $(!\alpha) \in S(EX)$;*

(iii) *if $\alpha, \beta \in S(EX)$, then $(\alpha \,\&\, \beta) \in S(EX)$;*

(iv) *if $\alpha, \beta \in S(EX)$, then $(\alpha \mid \beta) \in S(EX)$;*

(v) *if $\alpha, \beta \in S(EX)$, then $(\alpha \rightarrow \beta) \in S(EX)$.*

The set $S(EX)$ is also called the propositional language defined by A.

Observe that the symbols of A have been chosen for the sake of the logical context, but strictly speaking, they are just sets with no special properties whatsoever.

As it is defined, the language $S(EX(A))$ is not effectively described, we can only tell that, for example, expressions such as $v_0, (!v_3), (v_2 \,\&\, v_6)$ are sentences. But there is no general control of what is possible and what is not.

Exercise 71 Show that the expression $!(!v_1)$ is not a sentence. Is $(v_3 \rightarrow (!v_2))$ a sentence?

We now give a recursive construction of the subset $S(EX) \subset EX$ for a given alphabet A. To this end, we use the fact that we have the disjoint union $S(EX) = \bigsqcup_{n \geq 0} S_n(EX)$, where $S_n(EX) = S(EX) \cap EX_n$, with $EX_n = \{w \in EX \mid l(w) = n\}$ the set of expressions of word length n. Here is the structure of $S(EX)$ in these terms:

Definition 107 *Given a propositional alphabet A, let S_n be the following subsets of EX_n, which we define by recursion on n:*

(i) *For $n = 0, 2, 3$, we set $S_n = \varnothing$;*

(ii) *we set $S_1 = V$;*

(iii) *we set $S_4 = \{(!v_i) \mid v_i \in V\}$;*

(iv) *we set S_5 to the set of words of one of these three types: $(v_i \,\&\, v_j)$, $(v_i \mid v_j), (v_i \rightarrow v_j)$, where $v_i, v_j \in V$ are any two propositional variables;*

(v) *for $n > 5$, we set $S_n = S_n^! \cup S_n^{\&} \cup S_n^{\mid} \cup S_n^{\rightarrow}$, where $S_n^* = \{(\alpha * \beta) \mid \alpha, \beta \in \bigsqcup_{i < n} S_i, l(\alpha) + l(\beta) = n - 3\}$ for the symbols $* \in \{\&, \mid, \rightarrow\}$, and where $S_n^! = \{(!\alpha) \mid \alpha \in S_{n-3}\}$.*

We then set $S = \bigsqcup_{n \geq 0} S_n$.

Proposition 144 *Given a propositional alphabet A, we have*

$$S(EX) = S,$$

where S is the set defined in definition 107.

Proof Clearly, S is contained in $S(EX)$. Let us check that it fulfills the five axioms (i) through (v). By construction, $V = S_1$, whence (i). If $\alpha \in S$, then for $l(\alpha) \leq 3$, we only have $\alpha \in V$, and S_4 covers that case, else $l(\alpha) > 3$, and this case is covered by construction v of the definition of S, whence (ii). The cases (iii)-(v) are similar: for $l(\alpha) + l(\beta) \leq 3$, we must have $\alpha, \beta \in V$, this is covered by (iv) in the definition of S. For $l(\alpha) + l(\beta) > 3$, construction (v) in the definition does the job. □

We now know how the construction of the sentences over A works, but we still do not know how many sentences α and β could give rise to one and the same sentence, say $w = (\alpha \& \beta)$.

Lemma 145 *If left(w) and right(w) denote the numbers of left and right brackets, respectively, in a sentence $w \in S(EX)$, then we have left(w) = right(w).*

Proof The description of $S(EX)$ by S guaranteed by proposition 144 yields a straightforward inductive proof by the length of a sentence. □

Exercise 72 Give a proof of lemma 145 by recursion on the length of w.

Lemma 146 *Let $w \in S(EX)$, and suppose that we have a left bracket (in w, i.e., $w = u(x, l(u) > 0$. Then if the number of left brackets in (x is l and the number of right brackets in (x is r, we have $l < r$.*

Proof Induction on $l(w)$: For $l(w) \leq 5$, it is clear by the explicit words from rules (i)-(iv) in the definition of S. For general w, if $w = u(x = (!\alpha)$, then $u = (!$ or $u = (!u'$, where $u'(v' = \alpha$. By induction, the number of left brackets in (v' is smaller than the number of right brackets. So the same is true for ($x = (v')$. A similar argument is used for the other connectives. □

Proposition 147 *Let $w \in S(EX)$. Then exactly one of the following decompositions is the case: $w \in V$, or $w = (!\alpha)$, or $w = (\alpha \& \beta)$, or $w = (\alpha \mid \beta)$, or $w = (\alpha \to \beta)$, where α and β are uniquely determined sentences. They are called the* components *of w.*

Proof First, suppose $(!\alpha) = (!\alpha')$, then clearly the inner words α and α' must be equal. Second, suppose $(!\alpha) = (\beta * \gamma)$, then the letter ! must be the first letter of β, which is impossible for any sentence. Then suppose $(\alpha * \beta) = (\gamma * \delta)$. If

$l(\alpha) = l(\gamma)$, then $\alpha = \gamma$, therefore also $\beta = \delta$. So suppose wlog (without loss of generality) $l(\alpha) < l(\gamma)$. Then $\beta = (\gamma' * \delta$, where $\gamma = x(\gamma'$. So by lemma 146, $(\gamma'$ has fewer left than right brackets. But this contradicts the fact that β and δ have the same number of left and right brackets.   □

This proposition has deep implications. In fact, it allows the definition of functions on $S(EX)$ by recursion on the length of sentences and and in function of the unique logical connectives defining compound sentences. But let us first discuss the announced "logical algebras".

17.2 Semantics: Logical Algebras

On the syntactic level of a formal propositional language $S(EX)$ over a propositional alphabet A, the sentences look like meaningful expressions, but they are not, their shape only looks like being logically reasonable. In order to load such expressions with logical meaning, we need to provide the system with logical values. In order to have a first orientation of what a logical algebra should be, we refer to the "Boolean algebra" of subsets $L = 2^a$ of a set a as discussed in chapter 3, and especially in proposition 7. More specifically, in the special case of $a = 1 = \{0\}$, we have two subsets, $\bot = \varnothing = 0$ and $\top = 1$. Following a classical idea of the great mathematician and philosopher Gottfried Wilhelm Leibniz (1646-1716) and its elaboration by the mathematician George Boole (1815-1864), one can mimic truth values on the Boolean algebra $L = 2 = 2^1$ as described in proposition 7. The value "true" is mimicked by $\top$, whereas the value "false" is mimicked by $\bot$. The truth table of the conjunction $\mathcal{A}$ AND $\mathcal{B}$ is mimicked by the Boolean operation of intersection of the truth values assigned to the components, i.e., if $\mathcal{A}$ is true and $\mathcal{B}$ is false, the truth value of the conjunction is $\top \cap \bot = \bot$, and so on with all other combinations. In other words, we are combining the values $value(\mathcal{A}), value(\mathcal{B}) \in 2$ under Boolean operations. We see immediately that the Boolean operation "$\cap$" mimics conjunction, that "$\cup$" mimics disjunction, that the complementation "$-$" mimics negation, whereas the truth value of $\mathcal{A}$ IMPLIES $\mathcal{B}$ is mimicked by the value of (NOT $\mathcal{A}$) OR $\mathcal{B}$, i.e., by $(-value(\mathcal{A})) \cup value(\mathcal{B})$. However, it is not always reasonable to deduce logical implication from negation and disjunction, we rather would like to leave this operation as an autonomous operation. This is very important in order to cope with non-classical logical approaches, such as fuzzy logic. It was formalized by the mathematician Arend Heyting (1898-1980). The formalism needed for this approach is this:

Definition 108 *A* Heyting algebra (HA) *is a partially ordered set* $(L, \leq)$, *together with*

- *three binary operations: the* join $a \vee b$, *the* meet $a \wedge b$, *and the* implication $a \Rightarrow b$ *for $a, b \in L$, and*

- *two distinguished elements* $\perp$ *and* $\top$, *called "False" and "True", respectively.*

These data are subjected to the following properties:

(i) *$\perp$ and $\top$ are the minimal and maximal element with respect to the given partial ordering, i.e., $\perp \leq x \leq \top$ for all $x \in L$.*

(ii) *The operations join and meet are commutative.*

(iii) *Join and meet are mutually* distributive, *i.e., $x \wedge (y \vee z) = (x \wedge y) \vee (x \wedge z)$ and $x \vee (y \wedge z) = (x \vee y) \wedge (x \vee z)$, for all $x, y, z \in L$.*

(iv) *The join $x \vee y$ is a* least upper bound (l.u.b) *of x and y, i.e., $x, y \leq x \vee y$ and for any z with $x, y \leq z$, we have $x \vee y \leq z$.*

(v) *The meet $x \wedge y$ is a* greatest lower bound (g.l.b.) *of x and y, i.e., $x \wedge y \leq x, y$ and for any z with $z \leq x, y$, we have $z \leq x \wedge y$.*

(vi) (Adjunction) *For any $z \in L$, $z \leq (x \Rightarrow y)$ iff $z \wedge x \leq y$.*

For an element x of a Heyting algebra L, we define the negation *of x by $\neg x = (x \Rightarrow \perp)$.*

Exercise 73 Show that in a Heyting algebra, we always have $y \leq \neg x$ iff $y \wedge x = \perp$. Deduce from this that always

$$\neg x \wedge x = \perp.$$

Use the adjunction characterization (vi) of definition 108 to prove that

$$((x \vee y) \Rightarrow z) = ((x \Rightarrow z) \wedge (y \Rightarrow z)).$$

More specifically, prove *De Morgan's first law*:

$$\neg(x \vee y) = \neg x \wedge \neg y.$$

Definition 109 *In a Heyting algebra L, a* complement *of an element x is an element a such that $x \wedge a = \perp$ and $x \vee a = \top$.*

Lemma 148 *The complement of an element x in a Heyting algebra L, if it exists, is uniquely determined.*

Proof If b is another complement of x, we have $b = b \wedge \top = b \wedge (x \vee a) = (b \wedge x) \vee (b \wedge a) = (x \wedge a) \vee (b \wedge a) = (x \vee b) \wedge a = a$. $\qquad\square$

Exercise 74 Show that $y \leq \neg x \Leftrightarrow y \wedge x = \bot$.

An important and classical type of special Heyting algebras is defined by these properties:

Lemma 149 *For a Heyting algebra L, the following properties are equivalent: For all $x \in L$,*

(i) $x \vee \neg x = \top$,

(ii) *we have $\neg\neg x = x$.*

Proof Observe that always $\neg\top = \bot$ and $\neg\bot = \top$. (i) implies (ii): If $x \vee \neg x = \top$, then by De Morgan's first law (exercise 73), $\neg x \wedge \neg\neg x = \bot$. So $\neg\neg x$ is a complement of $\neg x$, but x is also a complement of $\neg x$, therefore, by uniqueness of complements (lemma 148), $\neg\neg x = x$. Conversely, if $\neg\neg x = x$, then by De Morgan's first law, $\neg x \vee \neg\neg x = \top$, whence $\neg x \vee x = \top$. $\qquad\square$

Definition 110 *A Heyting algebra which has the equivalent properties of lemma 149 is called a* Boolean algebra (BA).

Example 64 The classical example is the "Boolean algebra" of subsets $L = 2^a$ of a given set a, as discussed in chapter 3. Here, the partial ordering is the set inclusion, $x \leq y$ iff $x \subset y$, we have $\bot = \varnothing, \top = a$, $\vee = \cup, \wedge = \cap$, and $x \Rightarrow y = -x \cup y$. Evidently, $\neg x = (-x) \vee \varnothing = -x$, whence $x \Rightarrow y = (\neg x) \vee y$, i.e., implication is deduced from the other operations. It is really a BA since the double complementation is the identity: $-(-(x)) = x$. This is the a posteriori justification for the name "Boolean algebra", which we attributed to the powerset algebra in chapter 3. But observe that for a general set a, many truth values are possible besides the classical $\bot$ and $\top$.

Example 65 A less classic example is the set $L = Fuzzy(0, 1)$ of all intervals $I_x = [0, x[\subset [0, 1[$ of the "half open" real unit interval $[0, 1[= \{x \mid 0 \leq x < 1\} \subset \mathbb{R}$. Its elements are so-called *fuzzy* truth values, meaning that something may be true to $x \cdot 100\%$, not entirely true, and not entirely false. The percentage is given by the upper number x of the interval I_x.

The partial ordering is again that of subset inclusion, and the extremal values are $\bot = I_0 = \varnothing$ and $\top = I_1$. We also take $\vee = \cup$ and $\wedge = \cap$. This means $I_x \vee I_y = I_{max(x,y)}$ and $I_x \wedge I_y = I_{min(x,y)}$.

The implication is a little more tricky. We must have $I_z \leq I_x \Rightarrow I_y$ iff $I_z \cap I_x \subset I_y$. The solution therefore must be $I_x \Rightarrow I_y = \bigcup_{z, I_z \cap I_x \subset I_y} I_z$. Therefore $I_x \Rightarrow I_y = I_w, w = sup\{z \mid I_z \cap I_x \subset I_y\}$. This gives the following implications: $I_x \Rightarrow I_y = \top$ if $x \leq y$, and $I_x \Rightarrow I_y = I_y$ if $x > y$. In particular, $\neg I_x = \bot$ if $x \neq 0$, $\neg \bot = \top$. And finally, $\neg \neg I_x = \top$ if $x \neq 0$, $\neg \neg \bot = \bot$. This also shows that this Heyting algebra $Fuzzy(0, 1)$ is not Boolean. For the general theory of fuzzy systems, see [34] or [48].

We shall henceforth use the term *logical algebra* for a Heyting algebra or, more specifically, for a Boolean algebra. Intuitively, a logical algebra is a structure which provides us with those operations "negation", "conjunction", "disjunction", and "implication" which are preconized by the formal logical setup and should simulate the "reality of logical contents".

Here are the general properties of logical algebras:

Sorite 150 *Let L be a Heyting algebra and $x, y \in L$. Then*

(i) $x \leq \neg \neg x$,

(ii) $x \leq y$ *implies* $\neg y \leq \neg x$,

(iii) $\neg x = \neg \neg \neg x$,

(iv) (De Morgan's first law) $\neg(x \vee y) = \neg x \wedge \neg y$.

(v) $\neg \neg(x \wedge y) = \neg \neg x \wedge \neg \neg y$,

(vi) $x \wedge y = \bot$ *iff* $y \leq \neg x$,

(vii) $x \wedge \neg x = \bot$,

(viii) $(x \Rightarrow x) = \top$,

(ix) $x \wedge (x \Rightarrow y) = x \wedge y$,

(x) $y \wedge (x \Rightarrow y) = y$,

(xi) $x \Rightarrow (y \wedge z) = (x \Rightarrow y) \wedge (x \Rightarrow z)$.

  In particular, if L is a Boolean algebra, then

(xii) $x = \neg \neg x$,

(xiii) $x \vee \neg x = \top$,

(xiv) $x \Rightarrow y = \neg x \vee y$,

(xv) $x \leq y$ *iff* $\neg y \leq \neg x$,

(xvi) (De Morgan's second law) $\neg(x \wedge y) = \neg x \vee \neg y$.

Proof By exercise 73, $x \leq \neg\neg x$ iff $x \wedge \neg x = \bot$, but the latter is always true. If $x \leq y$, then also $x \wedge \neg y \leq y \wedge \neg y = \bot$, so $\neg y \wedge x = \bot$, therefore $\neg y \leq \neg x$. Statement (iii) follows immediately from (i) and (ii). (iv) is De Morgan's first law (see exercise 73). The proof of (v) is quite technical and is omitted, see [36]. Statements (vi) and (vii) follow from exercise 73. Statements (viii) to (xi) immediately follow from the characterization of the g.l.b. and adjointness for implication. (xii) and (xiii) is the characterization from lemma 149. For (xiv), we show $z \leq (\neg x \vee y)$ iff $z \wedge x \leq y$. If $z \wedge x \leq y$, then $z = z \wedge 1 = z \wedge (\neg x \vee x) = (z \wedge \neg x) \vee (z \wedge x) \leq \neg x \vee y$. Conversely, if $z \leq (\neg x \vee y)$, then $z \wedge y \leq (\neg x \vee y) \wedge x \leq y \wedge x \leq y$. (xv) follows from (ii) and (xii).   □

17.3 Signification: Valuations

We may now turn to the third component of a semiotic (sign-theoretic) system: the signification process from the expressive surface down to the semantic depth. Our modeling of such a system by use of a propositional alphabet A and a logical algebra L must provide us with a function $value : S(EX) \to L$ such that each sentence $w \in S(EX)$ is assigned a truth value $value(w) \in L$. Such a valuation should however keep track with the requirement that logical evaluation is related to the logical composition of these sentences. Here is the precise setup:

Definition 111 *Given a propositional alphabet A and a logical algebra L, a* valuation *is a map value : $S(EX) \to L$ such that for all sentences $\alpha, \beta \in S(EX)$, we have*

   (i) *$value((!\alpha)) = \neg(value(\alpha))$,*

   (ii) *$value((\alpha \mid \beta)) = value(\alpha) \vee value(\beta)$,*

   (iii) *$value((\alpha \& \beta)) = value(\alpha) \wedge value(\beta)$,*

   (iv) *$value((\alpha \mathrel{-\!>} \beta)) = value(\alpha) \Rightarrow value(\beta)$.*

The set of valuations value : $S(EX) \to L$ over the propositional alphabet A with values in the logical algebra L is denoted by $V(A, L)$.

And here is the existence theorem for valuations:

Proposition 151 *Given a propositional alphabet A and a logical algebra L, the functional restriction map*

$$var : V(A, L) \to Set(V, L) : value \mapsto value|_V$$

is a bijection, i.e., for each set map $v : V \to L$ *defined on the propositional variables, there is exists exactly one valuation* $value : S(EX) \to L$ *such that* $value|_V = v$.

Proof Injectivity is proved by induction on the length of a sentence w. For length 1, we have values from V, and this is what we want. For $l(w) > 1$, we have one of the forms $w = (!\alpha), (\alpha \mid \beta), (\alpha \& \beta), (\alpha \to \beta)$, and the induction hypothesis on α, β, together with the axiomatic properties (i)-(iv) of valuations solve the problem. Conversely, if a map $v_V : V \to L$ is given, its extension to any sentence $w \in S(EX)$ may be defined by recursion following the axiomatic properties (i)-(iv) of valuations. This is indeed well defined, because by proposition 147, the form and the components of a compound sentence are uniquely determined by w. □

Exercise 75 Given a propositional alphabet A, let $a = 3 = \{0, 1, 2\}$ and consider the powerset Boolean algebra $L = 2^3$. Let a value map $value \in V(A, L)$ be defined by $value(v_0) = \bot, value(v_1) = \{0, 2\}, value(v_2) = \{1\}, \ldots$. Calculate the value

$$value((((!(v_2 \& (!v_0)))) \mid ((!v_1) \to v_2))).$$

Notation 10 *For long expressions, such as the still rather short example in the previous exercise 75, the number of brackets becomes cumbersome to keep under control. To avoid this effect, one uses the same trick as for ordinary arithmetic: strength of binding, i.e., the formal rules are replaced by some implicit bracket constructions which enable us to omit brackets in the notation. The rules of binding strength are these:* ! *binds stronger than* & *and* |, *and these bind stronger than* $\to$. *Under this system of rules, the above example*

$$((!(v_2 \& (!v_0)))) \mid ((!v_1) \to v_2))$$

would be shortened to

$$!(v_2 \& !v_0) \mid (!v_1 \to v_2)$$

The main problem of formal propositional logic is to understand which kind of sentences $w \in S(EX)$ have the value $value(w) = \top$ under certain valuations. The most prominent of these questions is the problem of so-called tautologies, i.e., sentences, which have the value $\top$ under all possible valuations of a given class. Here is the formalism:

Definition 112 *Given a propositional alphabet A, a logical algebra L, a sentence $s \in S(EX)$, and a valuation $v \in V(A, L)$, one says that s is v-valid if $v(s) = \top$, in signs: $v \vDash s$. If $v \vDash s$ for all $v \in V(A, L)$, one says that s is L-valid, in signs $L \vDash s$. In particular, if $2 \vDash s$, one says that s is classically valid or that s is a tautology and writes $\vDash s$. If s is valid for all Boolean algebras, one says that s is Boolean valid and writes $BA \vDash s$. If s is valid for all Heyting algebras, one says that s is Heyting valid and writes $HA \vDash s$.*

Exercise 76 Show that for any sentence $s \in S(EX)$, we have $HA \vDash s \rightarrow !!s$ and $BA \vDash !!s \rightarrow s$. Give an example of a HA L such that $L \vDash !!s \rightarrow s$ is false.

In order to control what kind of sentences are valid for given valuation classes, one needs a more constructive approach. In fact, the very definition of sentences has already been understood by its recursive construction, and this method is what we now shall also apply to our question.

17.4 Axiomatics

Axiomatics is about the construction of a set of new sentences from given ones by use of a predetermined system of inference rules. This setup is a very particular case of so-called production grammars, which will be discussed in chapter 19. Here, we stay rather intuitive as to the general nature of such an inference rule system and will only concentrate on a very special type, the one defined by the classical inference rule *modus ponens* (the Latin name is a marker for the Medieval tradition of formal logic). It is the rule which we apply incessantly: If $\mathcal{A}$ is the case (i.e., true), and if the implication $\mathcal{A}$ IMPLIES $\mathcal{B}$ is the case, then also $\mathcal{B}$ is the case. In fact, otherwise, by absolute logic, $\mathcal{B}$ would not be the case (false), but then $\mathcal{A}$ IMPLIES $\mathcal{B}$ cannot be the case by very definition of the implication. The formal restatement of this inference rule defines classical axiomatics as follows.

Definition 113 *One is given a set $AX \subset S(EX)$ of sentences, called axioms. A proof sequence (with respect to AX) is a finite sequence $p = (s_i)_{i=1,2...n} \in S(EX)^n$ of positive length $l(p) = n$ such that $s_1 \in AX$, and for $i > 1$, either $s_i \in AX$, or there are two sentences s_k and $s_l = s_k \rightarrow s_i$ where $k, l < i$. A terminal sentence in a proof chain is called a theorem with respect to AX.*

The set of theorems is denoted by $S_{AX}(EX)$. If $S(EX)$ is clear, the fact that $s \in S_{AX}(EX)$ is also denoted by $\underset{AX}{\vdash} s$.

Intuitively, the role of axioms is this: One accepts axioms as being true a priori, i.e., one only looks for classes of valuations which map all axioms to $\top$. One then wants to look for all sentences which are also true if the axioms are so. The only rule is the formalized modus ponens: Given two theorems s and $s \rightarrow t$, then t is also a theorem, and evidently, this production process can be managed by a machine. The point of axiomatics is that a particular set of axioms yields the tautologies. Here it is:

Definition 114 *The axioms of* classical logic (CL) *are the sentences which can be built from any given three sentences $\alpha, \beta, \gamma \in S(EX)$ by one of the following constructions:*

  (i) $(\alpha \rightarrow (\alpha \,\&\, \alpha))$

  (ii) $((\alpha \,\&\, \beta) \rightarrow (\beta \,\&\, \alpha))$

  (iii) $((\alpha \rightarrow \beta) \rightarrow ((\alpha \,\&\, \gamma) \rightarrow (\beta \,\&\, \gamma)))$

  (iv) $(((\alpha \rightarrow \beta) \,\&\, (\beta \rightarrow \gamma)) \rightarrow (\alpha \rightarrow \gamma))$

  (v) $(\beta \rightarrow (\alpha \rightarrow \beta))$

  (vi) $((\alpha \,\&\, (\alpha \rightarrow \beta)) \rightarrow \beta)$

  (vii) $(\alpha \rightarrow (\alpha \mid \beta))$

 (viii) $((\alpha \mid \beta) \rightarrow (\beta \mid \alpha))$

  (ix) $(((\alpha \rightarrow \beta) \,\&\, (\beta \rightarrow \gamma)) \rightarrow ((\alpha \mid \beta) \rightarrow \gamma))$

  (x) $((!\alpha) \rightarrow (\alpha \rightarrow \beta))$

  (xi) $(((\alpha \rightarrow \beta) \,\&\, (\alpha \rightarrow (!\beta))) \rightarrow (!\alpha))$

 (xii) $(\alpha \mid (!\alpha))$

The axioms of intuitionistic logic (IL) *are those sentences in CL built from all constructions but the last, $(\alpha \mid (!\alpha))$.*

The intuitionistic axiom system IL contains those axioms which we need to produce sentences which are Heyting valid! Recall that we have in fact Heyting algebras L, for example $L = Fuzzy(0, 1)$, where $x \vee (\neg x) \neq \top$ in general. The crucial proposition is this:

Proposition 152 *Given a propositional alphabet A and a sentence $s \in S(EX)$, the following statements are equivalent:*

For classical logic:

   (i) *The sentence s is a tautology, i.e., $\vDash s$.*

  (ii) *The sentence s is Boolean valid, i.e., $BA \vDash s$.*

 (iii) *s is a theorem with respect to CL, i.e., $\vdash_{CL} s$*

And for intuitionistic logic:

   (i) *The sentence s is Heyting valid, i.e., $HA \vDash s$.*

  (ii) *s is a theorem with respect to IL, i.e., $\vdash_{IL} s$*

Proof If $\vdash_{CL} s$, then $BA \vDash s$, this is similar to proposition 153, i.e., the tautological situation (the BA is 2), we leave it to the reader, too. If the latter holds, then it holds for 2. If s is a tautology, then it is classically valid, $\vdash_{CL} s$, by a more involved demonstration by Emil Post, see the text after proposition 154.

The equivalence of (i) and (ii) in the intuitionistic case follows these lines: Soundness, i.e., (i) implies (ii) follows as easily as soundness for Boolean algebras. Completeness is proved as follows: From $\vdash_{IL} s$, one constructs a special Heyting algebra, the so-called Lindenbaum algebra H_{IL}. Then one shows that the validity for this algebra implies $\vdash_{IL} s$. So finally, as $HA \vDash s$ implies $H_{IL} \vDash s$, we are done. The details are omitted in the frame of this introductory book, but see [38] for details. □

This theorem deserves some comments. Let us concentrate on the equivalence of statements (i) and (iii) in the Boolean part, to fix the ideas. What should one prove in that case? On one hand, one has to show that by the axiom system CL of classical logic, only tautologies are generated. This is the so-called *soundness theorem:*

Proposition 153 (Soundness) *If $\vdash_{CL} s$, then $\vDash s$.*

This is just an exercise:

Exercise 77 Prove proposition 153 as follows: First show that all axioms are classically valid. Then show by induction on the proof chain length that any theorem is classically valid.

The more involved part is the converse, the so-called *completeness theorem:*

Proposition 154 (Completeness) *If $\vDash s$, then $\vdash_{CL} s$.*

We shall not give a proof of the completeness theorem, which is quite involved. The original proof of such a theorem was given in 1921 by Emil Post (see [45]).

Example 66 The sentences $(\alpha \rightarrow (\beta \mid \alpha))$ for $\alpha, \beta \in S(EX)$ seem to be obvious theorems of CL, especially since $(\alpha \rightarrow (\alpha \mid \beta))$ is an axiom schema. But we have to provide a proper proof sequence in order to establish this fact. On the right of each line of the proof sequence we indicate whether we have used an axiom (ax.) or applied modus ponens (m.p.) to two of the previous lines. Circled numbers ① and ② are used as abbreviations to refer to the formulas in line 1 and line 2, respectively. (To be absolutely accurate, we should state that the following proof sequence is really a schema for generating proof sequences of $(\alpha \rightarrow (\beta \mid \alpha))$ for all $\alpha, \beta \in S(EX)$).

1. $(\alpha \rightarrow (\alpha \mid \beta))$ ax. (vii)
2. $((\alpha \mid \beta) \rightarrow (\beta \mid \alpha))$ ax. (viii)
3. $(① \rightarrow (② \rightarrow ①))$ ax. (v)
4. $(② \rightarrow ①)$ m.p. 1, 3
5. $((② \rightarrow ①) \rightarrow ((② \,\&\, ②) \rightarrow (① \,\&\, ②)))$ ax. (iii)
6. $((② \,\&\, ②) \rightarrow (① \,\&\, ②))$ m.p. 4, 5
7. $(② \rightarrow (② \,\&\, ②))$ ax. (i)
8. $(② \,\&\, ②)$ m.p. 2, 7
9. $(① \,\&\, ②)$ m.p. 6, 8
10. $((((\alpha \rightarrow (\alpha \mid \beta)) \,\&\, ((\alpha \mid \beta) \rightarrow (\beta \mid \alpha))) \rightarrow (\alpha \rightarrow (\beta \mid \alpha)))$ ax. (iv)
11. $(\alpha \rightarrow (\beta \mid \alpha))$ m.p. 9, 10

Hence $\vdash_{CL} (\alpha \rightarrow (\beta \mid \alpha))$ for all $\alpha, \beta \in S(EX)$.

This proof was rather easy and short. However, proof sequences for theorems of even a little more complexity tend to become long and intricate. Therefore, whenever one has established the theoremhood of a sentence, one may use it in subsequent proof sequences just as if it were an axiom. If asked, one could then always recursively expand the theorems to their proof sequences to get a sequence in the originally required form.

Exercise 78 Abbreviate $((\alpha \rightarrow \beta) \,\&\, (\beta \rightarrow \alpha))$ by $(\alpha \longleftrightarrow \beta)$. Use the axioms CL of classical logic to show that the following sentences are tautologies (with the usual omission of brackets due to the binding order):

1. (Associativity)

     $((\alpha \mid \beta) \mid \gamma \longleftrightarrow \alpha \mid (\beta \mid \gamma))$ and $((\alpha \,\&\, \beta) \,\&\, \gamma \longleftrightarrow \alpha \,\&\, (\beta \,\&\, \gamma))$

2. (Commutativity) $(\alpha \mid \beta <\!\!-\!\!> \beta \mid \alpha)$ and $(\alpha \,\&\, \beta <\!\!-\!\!> \beta \,\&\, \alpha)$

3. (De Morgan's Laws)

$$((!(\alpha \mid \beta)) <\!\!-\!\!> (!\alpha \,\&\, !\beta)) \text{ and } ((!(\alpha \,\&\, \beta)) <\!\!-\!\!> (!\alpha \mid !\beta))$$

One also calls sentences s and t *equivalent* iff $(s <\!\!-\!\!> t)$ is a tautology. Notice that from the associativity, we may group conjunctions or disjunctions in any admissible way and obtain sentences which are equivalent to each other. We therefore also omit brackets in multiple conjunctions or disjunctions, respectively.

Exercise 79 A sentence s is in *disjunctive normal form* iff $s = s_1 \mid s_2 \mid \ldots s_k$, where each s_i is of the form $s_i = s_{i1} \,\&\, s_{i2} \,\&\, \ldots s_{ik(i)}$ with s_{ij} being a propositional variable $v \in V$ or its negation $(!v)$.

A sentence s is in *conjunctive normal form* iff $s = s_1 \,\&\, s_2 \,\&\, \ldots s_k$, where each s_i is of the form $s_i = s_{i1} \mid s_{i2} \mid \ldots s_{ik(i)}$ with s_{ij} being a propositional variable $v \in V$ or its negation $(!v)$.

Show that every sentence is equivalent to a sentence in disjunctive normal form and also to a sentence in conjunctive normal form.

Exercise 80 Define the *Sheffer stroke operator* by $(\alpha \mid\mid \beta) = (!(\alpha \,\&\, \beta))$. Show that every sentence is equivalent to a sentence, where only the Sheffer operator occurs. In electrical engineering the stroke operator is also known as NAND.

CHAPTER 18

Formal Predicate Logic

Until now, we have succeeded in formalizing basic logic as it is controlled by truth values produced by the propositional connectives !, &, |, and $\rightarrow$ of negation, conjunction, disjunction, and implication. However, nothing has been done to mimic the 'anatomy of propositions', in fact, we had just offered an 'amorphous' set of propositional variables $v_0, v_1, \ldots$ with no further differentiation. So the truth value of sentences was based on the completely arbitrary valuation of propositional variables.

Now, mathematics needs more refined descriptions of how truth and falsity are generated. For example, the simple set-theoretic definition $a \cap b = \{x \mid x \in a \text{ and } x \in b\}$ uses composed **predicates**: (1) $P(x)$ is true iff $x \in a$, (2) $Q(x)$ is true iff $x \in b$, and the combination thereof $(P \& Q)(x)$ is true iff $P(x)$ and $Q(x)$ are both true. So we first of all need to *formalize the concept of a predicate*.

Next, let us look at the pair axiom: "If a and b are two sets, then there is the pair set $\{a, b\}$." This statement uses the predicates (1) $S(x)$ iff x is a set, (2) $E(x, y)$ iff $S(x), S(y)$, and $x \in y$. We implicitly also need the predicate $I(x, y)$ iff $S(x), S(y)$, and $x = y$. This setup transforms the axiom to the shape "If a and b are such that $S(a)$ and $S(b)$, then there is c with $E(a, c)$ and $E(b, c)$, and if x is such that $E(x, c)$, then $I(x, a)$ or $I(x, b)$." Besides the predicative formalization, we here encounter two more specifications: the first part "If a and b are such that $S(a)$ and $S(b)\ldots$", which means "Whenever we take a, $b\ldots$", in other words, we suppose a given universe of objects from where we may select instances a and b and then ask them to comply with certain predicates, viz., being sets $S(a)$ and $S(b)$. This is restated by the so-called *universal quanti-*

fier: "**For all** a, b...". Further, we also recognize an *existence quantifier*: "...there is c with ...", which is restated by "...**there exists** c with ...".

In order to cope with the common mathematical constructions, one usually is a bit more specific in the formalization of predicates. As modeled from set theory, there are two basic types of predicates: relations and functions. This means that we are considering predicates defined by relations and functions in the following sense: Given n sets $A_1, \ldots A_n$ and an n-ary relation, i.e., a subset $R \subset A_1 \times \ldots A_n$ of their Cartesian product, one defines the associated n-ary predicate by $R(x_1, \ldots x_n)$ iff $(x_1, \ldots x_n) \in R$. Observe that n-ary relations generalize the more restrictive concept of an n-ary relation $R \subset a^n$ introduced in definition 33 by varying each of the n factors of a^n. Similarly, if $f : A_1 \times \ldots A_n \to A$ is a set arrow, one defines the predicate $f(x_1, \ldots x_n, y)$ iff $f(x_1, \ldots x_n) = y$. That includes two special cases where $n = 0$. For 0-ary relations, this means that we consider the 'empty Cartesian product' (check the universal property of Cartesian products to understand the following definition!). This means that we are given a subset of the final set 1, in other words, one of the classical truth values $\perp, \top$ of the Boolean algebra 2. In other words, we also include the extremal truth values as basic predicates. As to functions of 0 variables, this means that the domain is again the empty Cartesian product 1. So a 0-ary function $f : 1 \to A$ is identified with the image $y = f(0)$ of the unique argument $0 \in 1$, in other words, 0-ary functions are just 'constant' elements in A.

A last remark must be made concerning variables. We have constantly used some symbols a, b, x, etc. to feed the predicates. The nature of these variables has not been discussed. Relating to the specification of predicates as being generated by relations or functions, we may interpret variables as being taken from the basic involved domain sets A_i, A. We do however not allow to take variables which refer to relations or functions or even higher order objects, such as relations of relations, etc. Together with this last restriction we have what is called *first order (formal) predicate logic*.

We are now ready to set up the formal framework. The methodology is quite the same as for formal propositional logic: One first sets up the syntactical structures, then the objects of semantics, and last the formalization of the signification process.

18.1 Syntactics: First-order Language

The basis of the formalized language is again a set of alphabetic symbols which we then combine to obtain reasonable words. Let us first formalize the relations and functions, together with the corresponding variables. To this end, if S is a set, define by $Sequ(S)$ the set of finite sequences $s = (A_1, \ldots, A_n), A_i \in S$, where for $n = 0$ we take by definition the unique sequence which is defined on the empty index set 0.

Exercise 81 Show that the set $Sequ(S)$ always exists, and that $Sequ(S) = Sequ(T)$ iff $S = T$.

Definition 115 *A* signature *is a triple*

$$\Sigma = (FunType : Fun \to Sequ(S), RelType : Rel \to Sequ(S), (V_A)_{A \in S})$$

of two set maps and a family of denumerable sets V_A. In the uniquely determined S, finite *by hypothesis, the elements $A \in S$ are called* sorts, *the elements $f \in Fun$ (in the uniquely determined set Fun) are called* function *symbols, and the elements $R \in Rel$ (in the uniquely determined set Rel) are called* relation *symbols. One supposes that all the sets Fun, Rel, and V_A are mutually disjoint. The elements $x \in V_A$ are called* variables of sort A. *The value of FunType(f) and RelType(R), respectively, is called the* type *of f and R, respectively. The length $n \geq 0$ of the type $(A_1, \ldots A_n)$ of a relation symbol is called its* arity; *the number $n \geq 0$ in the type $(A_1, A_2 \ldots A_{n+1})$ of a function symbol is called its* arity; *so by definition, function symbols always have at least one sort in their type. In particular, 0-ary functions are called* constants, *whereas 0-ary relations are called* atomic propositions.

Given a signature Σ, a function symbol f, together with its type $(A_1 \ldots A_n, A)$, is denoted by

$$f : A_1 \ldots A_n \to A,$$

where the last sort is denoted by A since its semantic role to be defined later is that of a codomain, but it is a sort much as the others are. A relation symbol R, together with its type $(A_1 \ldots A_n)$, is denoted by

$$R \rightarrowtail A_1 \ldots A_n.$$

In order to denote the sort A of a variable $x \in V_A$ or the $(n + 1)^{st}$ sort A of a function $f : A_1 \ldots A_n \to A$ one also writes $x : A$, or $f : A$, respectively.

Since we are interested in a vocabulary of general usage, we shall moreover suppose that the following special symbols are part of our signature:

- *For each sort A, the relational equality symbol $\stackrel{A}{=}$ with $RelType(\stackrel{A}{=}) = (A, A)$ is an element of Rel, and we usually use the infix notation ($a \stackrel{A}{=} b$) instead of $\stackrel{A}{=} (a, b)$, and, if the sort A is clear, we just write $a = b$, but be warned: this is by no means equality in the sense of set theory, it is just an abbreviation of $\stackrel{A}{=}$ and has no contents whatsoever on the present syntactical level of the theory.*

- *Among the atomic proposition symbols we have the falsity atom $\perp$ and the truth atom $\top$. We shall not invent new symbols for these entities in order to distinguish them from the synonymous entities in logical algebras since there is no danger of confusion.*

Example 67 As a prototypical example, we shall develop the predicate logic which describes Peano's construction of natural arithmetic, a setup, which we have modeled on the set theory of finite ordinal numbers, and which has been described in terms of Peano's five axioms, see propositions 45 and 47.

For Peano's axioms we need this repertory of symbols and operations:

- A symbol for 0;

- a set of variables $x, y, \ldots$ to designate natural numbers;

- a predicate symbol of equality $x = y$ between natural numbers x, y;

- a function symbol for the sum $x + y$ of two natural numbers x, y;

- a function symbol for the product $x \cdot y$ of two natural numbers x, y;

- a function symbol for the successor x^+ of a natural number x.

This requirement analysis yields the following signature: We have a single sort (the natural numbers) A, so the sort set is $S = \{A\}$. Accordingly, we have one (denumerable) variable set V_A, where we choose the variables $x, y, \ldots$. To fix ideas, take the set of ASCII words $V_A = \{x_0, x_1, x_2, \ldots\}$ with indexes being natural numbers in their decimal representation. The relation symbol set is the singleton $Rel = \{\stackrel{A}{=} \rightarrowtail AA, \perp \rightarrowtail 0, \top \rightarrowtail 1\}$, the function symbol set is $Fun = \{\stackrel{A}{+} : AA \rightarrow A, \stackrel{A}{\cdot} : AA \rightarrow A, {}^{+_A} : A \rightarrow A, \stackrel{A}{0} : 1 \rightarrow A\}$, the superscripts being added to indicate that we are only setting up a symbol set, and not real arithmetic operations. The type maps can also been given by the arrow notation, e.g., $FunType(\overset{A}{+}) = (A, A, A)$, or $FunType(\overset{A}{0}) = (A)$, the latter being the constant symbol for zero.

With these symbols, one now defines the alphabet of a predicate language as follows:

Definition 116 *A predicative alphabet is a triple $P = (\Sigma, B, C)$ of these sets:*

(i) *The set Σ is a signature, with the defined sets S of sorts, the set Fun of function symbols, the set Rel of relation symbols, and the family $(V_A)_S$ of variable sets.*

(ii) *The set $B = \{(,), ,\}$ of left and right brackets, and the comma.*

(iii) *The set $C = \{!, \&, |, ->, \forall, \exists\}$ of connectives, where $\forall$ is called the* universal quantifier, *and $\exists$ is called the* existence quantifier.

One again supposes that the sets V_A, Fun, Rel, B, C are mutually disjoint.

We again have this monoid of predicative expressions:

Definition 117 *Given a predicative alphabet P as explicited in definition 116, the* monoid $EX = EX(P)$ *of expressions over P is the word monoid $Word(V_S \sqcup Fun \sqcup Rel \sqcup B \sqcup C)$, with $V_S = \bigsqcup_{A \in S} V_A$.*

Like with sentences, we want to construct reasonable predicative expressions, which this time we call *formulas* instead of sentences. The construction needs an intermediate step.

Definition 118 *Given a predicative alphabet P, the set $Term(P)$, the elements of which are called* terms *is defined as the (uniquely determined) minimal subset $Term(P) \subset EX(P)$ such that the following two conditions hold. Simultaneously we add the recursive definition of the* sort *of a term.*

(i) *$V_S \subset Term(P)$, and a variable $x \in V_A$ qua term has the same sort $x : A$ as the given variable. Attention: the expressions $x : A, f : A$ are not words of $Term(P)$ or of any other formula, they are normal mathematical formulas!*

(ii) *If for $0 \le n$, $t_1 : A_1, \ldots t_n : A_n$ are terms with the respective sort sequence (type) $A_1, \ldots A_n$, and if $f(A_1 \ldots A_n) \to A$ is a function symbol, then the expression $f(t_1, \ldots t_n)$ is a term, and we define $f(t_1, \ldots t_n) : A$. In particular, the constants $f() : A$ are terms*

$(n = 0)$; *for practical reasons we also include the words f (and the notation $f : A$ for $f() : A$) together with the constants $f()$.*

Example 68 Taking up our prototypical example 67 of Peano arithmetic, we have these terms: the variables x_n with their (unique) sort $x_n : A$, then the function symbols, first the constant symbol $\overset{A}{0} : A$, then the function symbols with terms in their argument places, e.g., $^{+_A}(\overset{A}{0})$, which we presently also abbreviate by $\overset{A}{1}$, etc., $n \overset{A}{+} 1$ for $^{+_A}(\overset{A}{n})$, but attention: this is only our informal convention to save space, not the correct setup. Then we have the function expressions $(\overset{A}{0} \overset{A}{+} \overset{A}{0})$, $(\overset{A}{0} \overset{A}{+} \overset{A}{0})$, $(\overset{A}{0} \overset{A}{+} \overset{A}{1})$, $((\overset{A}{0} \overset{A}{+} \overset{A}{0}) \overset{A}{+} \overset{A}{1})$, and so forth.

Similarly to the recursive construction of sentences in definition 107 and proposition 144, one may describe *Term(P)* recursively—this is a nice exercise.

Exercise 82 Give a recursive definition of *Term(P)* in terms of its intersections $Term(P)_n = Term(P) \cap EX(P)_n$, starting from $Term(P)_0 = \varnothing$, $Term(P)_1 = Fun_0 \sqcup V_S$, where $Fun_0 = \{f \mid f \in Fun, FunType(f) = (A)$, i.e., 0-ary$\}$ is the set of constant symbols.

We now may define general formulas for a predicative language as follows by recursion on the word length:

Definition 119 *Given a predicative alphabet P, the set $F(EX)$ over the predicative alphabet P is the smallest subset $F(EX) \subset EX = EX(P)$ containing all these words which are called its* formulas*:*

   (i) *the* relational *formulas $R(t_1, \dots t_n)$ for $R \in Rel$, and terms $t_i \in Term(P)$ with $t_i : A_i$ for the type $RelType(R) = (A_1, \dots A_n)$, including the 0-ary relation words $R()$, which we again also shorten to R; this includes in particular the equality formulas $(t \overset{A}{=} s)$ for terms $s : A$ and $t : A$;*

  (ii) *the* truth *formula $\top$ and the* falsity *formula $\bot$;*

 (iii) negation: *if ϕ is a formula, then so is $(!\phi)$;*

 (iv) disjunction: *if ϕ and ψ are formulas, then so is $(\phi \mid \psi)$;*

  (v) conjunction: *if ϕ and ψ are formulas, then so is $(\phi \,\&\, \psi)$;*

 (vi) implication: *if ϕ and ψ are formulas, then so is $(\phi \rightarrow \psi)$;*

(vii) universal quantification: *if ϕ is a formula, and if x is any variable, then $(\forall x)\phi$ is a formula;*

(viii) existential quantification: *if ϕ is a formula, and if x is any variable, then $(\exists x)\phi$ is a formula.*

The set $F(EX)$ of formulas over the predicative alphabet P is called the predicative language defined by P.

Example 69 Continuing example 68 of Peano arithmetic, we have these formulas: In addition to the falsity and truth formulas $\perp$ and $\top$, we have the equality relation formulas $t \overset{A}{=} s$, for terms s and t, e.g., $\overset{A}{0} \overset{A}{=} (\overset{A}{0} \overset{A}{+} \overset{A}{0})$. We then have the formulas obtained by logical connectives which are recursively applied to formulas, e.g., $(!\perp), (!\overset{A}{0} \overset{A}{=} (\overset{A}{0} \overset{A}{+} x_7)), (\top \,\&\, \overset{A}{0} \overset{A}{=} \overset{A}{0})$. Finally we have the formulas obtained from the universal quantifier, e.g., $(\forall x_3)\overset{A}{0} \overset{A}{=} (x_3 \overset{A}{+} x_7)$, or from the existence quantifier, e.g., $(\exists x_1)\overset{A}{0} \overset{A}{=} {}^{+_A}(x_1)$.

Exercise 83 Give a recursive definition of $F(EX)$ which is based on the formulas of given word length n, namely, $F(EX)_n = F(EX) \cap EX_n$.

Within this vast vocabulary, many formulas are just meaningless 'forms'. For example, the formula $(f(x) = 3)$ has no semantics if the variable x is not specified, even if we know a priori that the sorts $f : B, 3 : B$ coincide. However, if we prepend the existence quantifier, i.e., $(\exists x)(f(x) = 3)$, then the formula may become meaningful, i.e., loaded with a semantic truth value. Therefore one is interested in defining which variables in a formula are "bound" by a quantifier, and which are not, i.e., "free". Here is the precise definition of the set $Free(\phi) \subset V_S$ of *free variables of ϕ*. It is however convenient to start this definition with the set of free variables of terms. Attention: We use the fact that the components of compound formulas, such as $(\forall x)\phi$ or $(\phi \rightarrow \psi)$, are uniquely determined. We have shown such a uniqueness fact in proposition 147, and it is recommended to meditate over this fact in the present context, too! In particular, the *scope of a variable x* is the uniquely determined formula ϕ following $(\forall x)$ or $(\exists x)$ in a formula $\ldots (\forall x)\phi \ldots$ or $\ldots (\exists x)\phi \ldots$, respectively.

- $Free(\top) = Free(\perp) = \varnothing$;

- for a constant term $t = f()$, we set $Free(t) = \varnothing$;

- for a variable x, we set $Free(x) = \{x\}$;

- for a term $t = f(t_1, \ldots t_n), n > 0$, we set $Free(t) = \bigcup_i Free(t_i)$;

- if $\phi = R(t_1, \ldots t_n)$ is a relational formula, then $Free(\phi) = \bigcup_i Free(t_i)$, i.e., the set of all variables appearing in the terms t_i;

- we set $Free((!\phi)) = Free(\phi)$;

- if σ is one of the formulas $(\phi \mid \psi), (\phi \,\&\, \psi), (\phi \to \psi)$, then $Free(\sigma) = Free(\phi) \cup Free(\psi)$;

- we set $Free((\forall x)\phi) = Free((\exists x)\phi) = Free(\phi) - \{x\}$.

The concept of free variables is quite delicate. For example, the variable x is free in the formula $((\forall x)(f(x) = g(y)) \,\&\, (x = x))$ because it is free in the second formula, whereas it is not free in the first formula. Intuitively, the role of x in these two component formulas is completely different: In the first one, x could be replaced by any other variable of the same sort without changing the formula's meaning, in the second component, we could not do so because the usage of this one changes radically if we embed it in a larger context of formulas and variables.

Definition 120 *A formula ϕ without free variables, i.e., $Free(\phi) = \varnothing$, is called a* (predicative) sentence.

The only way to produce non-trivial sentences without free variables is to apply quantifiers. Suppose that the set V_S of variables is linearly ordered (we know that a finite disjoint union of n denumerable sets can be linearly ordered, for example by interpreting the element $j \geq 0$ of the ith set, $i = 0, 1, 2 \ldots n - 1$, as the natural number $j \cdot n + i$). We write this ordering as $x < y$ for variables x and y.

Definition 121 *Given a formula $\phi \in F(EX)$, let $x_1 < x_2 < \ldots x_r$ be the ordered sequence of the elements of $Free(\phi)$. Then we denote by $(\forall)\phi$ the sentence $(\forall x_1)(\forall x_2) \ldots (\forall x_r)\phi$, and by $(\exists)\phi$ the sentence $(\exists x_1)(\exists x_2) \ldots (\exists x_r)\phi$ and call these sentences the* universal or existential closures, *respectively, of ϕ.*

18.2 Semantics: Σ-Structures

Semantics has to provide us with logical algebras, where the truth values can be calculated from the formal data. Here is the framework for this calculation. Given an alphabet, the invariant data are the sets B of brackets and comma, and the set C of connectives. The set which can vary is the signature Σ. The semantic structure is tied to the signature. We need these objects:

Definition 122 *Given a signature Σ, a (set-theoretic) Σ-structure $\mathfrak{M}$ is the datum of these sets:*

(i) *For each sort $A \in S$, we are given a set $\mathfrak{M}_A$.*

(ii) *For each n-ary relational symbol $R \rightarrowtail A_1 \ldots A_n$, we are given a subset $\mathfrak{M}_R \subset \mathfrak{M}_{A_1} \times \ldots \mathfrak{M}_{A_n}$, called a relation (recall that we had defined a relation by a subset of the second power X^2 of a set X, and a graph as a subset of a Cartesian product $X \times Y$ of sets X and Y, the present one is a generalization of those concepts). In particular, for atomic propositions with $n = 0$, e.g., $\bot, \top$, we are given the subsets of the final set 1, i.e., the truth values $\bot = 0, \top = 1$, elements in the Boolean algebra 2.*

(iii) *For each n-ary function symbol $f : A_1 \ldots A_n \rightarrow A$, we are given a set function $\mathfrak{M}_f : \mathfrak{M}_{A_1} \times \ldots \mathfrak{M}_{A_n} \rightarrow \mathfrak{M}_A$. In particular, for $n = 0$ we are given a "constant", i.e., an element $\mathfrak{M}_f \in \mathfrak{M}_A$.*

(iv) *For each equality symbol $\stackrel{A}{=}$, we are given the diagonal $\mathfrak{M}_{\underline{A}} = \Delta_A \subset \mathfrak{M}_A^2$.*

This being, we shall now be able to define truth values in the Boolean algebras of the sets $\mathfrak{M}_{A_1} \times \ldots \mathfrak{M}_{A_n}$ and to define signification of formulas with respect to these logical algebras.

Of course, if we would take a more general Heyting algebra on the powerset of such a Cartesian product, or even on Cartesian products of digraphs and still more 'exotic' objects, we would obtain a more general predicate logic. This can be done in the so-called topos theory (see for example [21]), but for our modest needs, we stick to the classical situation of Boolean powerset algebras. This is what we mean when talking about "set-theoretic" Σ-structures.

Example 70 Following up our prototypical example 69 of Peano arithmetic, we may define a Σ-structure $\mathfrak{M}$ which everybody would expect: For the sort A, take $\mathfrak{M}_A = \mathbb{N}$. For $\bot$ and $\top$, we have no choice by definition, i.e., $\bot = \varnothing \subset 1$ and $\top = 1 \subset 1$. For equality, we have to take $\mathfrak{M}_{\underline{A}} = \Delta_{\mathbb{N}}$. For the function $\stackrel{A}{+}$ we take the ordinary addition $\mathfrak{M}_A : \mathbb{N} \times \mathbb{N} \rightarrow \mathbb{N} : (x, y) \mapsto x + y$, for $\overset{A}{\cdot}$ we take the ordinary multiplication $\mathfrak{M}_A : \mathbb{N} \times \mathbb{N} \to \mathbb{N} : (x, y) \mapsto x \cdot y$, and for $^{+A}$ we take the ordinary successor $\mathfrak{M}_{+A} : \mathbb{N} \to \mathbb{N} : x \mapsto x^+$. Finally, set $\mathfrak{M}_{A \atop 0} = 0 \in \mathbb{N}$.

But we could also take any other structure $\mathfrak{M}'$, for example exchanging addition and multiplication in the above $\mathfrak{M}$, i.e., $\mathfrak{M}'_{A \atop +} = \mathfrak{M}_{A \atop \cdot}$ and $\mathfrak{M}'_{A \atop \cdot} = \mathfrak{M}_{A \atop +}$, and setting $\mathfrak{M}'_{A \atop 0} = 23 \in \mathbb{N}$.

18.3 Signification: Models

The truth values of formulas are constructed as follows.

Definition 123 *Let $\mathfrak{M}$ be a Σ-structure. If t, ϕ is a term or a formula, respectively, and if $x_1 < \ldots x_n$ are its free variables with corresponding sorts $A_1, \ldots A_n$, we denote by $\mathfrak{M}_t$ or $\mathfrak{M}_\phi$, respectively, the Cartesian product $\mathfrak{M}_{A_1} \times \ldots \mathfrak{M}_{A_n}$, including the special case $n = 0$, where we set $\mathfrak{M}_t = 1$ or $\mathfrak{M}_\phi = 1$, respectively. Call this set the* free range *of t and ϕ, respectively.*

Next, we need to define evaluation of a term for specific values under a given Σ-structure. Let t be a term, and $x \in \mathfrak{M}_t$. Then the evaluation $s[x] \in \mathfrak{M}_s$ at x of a term s with $Free(s) \subset Free(t)$ is recursively defined by (1) the component at position s $s[x] = x_s \in \mathfrak{M}_A$ if $s : A$ is a variable of sort A; (2) the value $s[x] = \mathfrak{M}_f(t_1[x], \ldots t_m[x]) \in \mathfrak{M}_A$ for $s = f(t_1, \ldots t_m)$ and we have $f : A$.

We shall now attribute to each formula $\phi \in F(EX)$ a truth value $\top(\phi)$ in the Boolean algebra $2^{\mathfrak{M}_\phi}$, i.e., a subset $\top(\phi) \subset \mathfrak{M}_\phi$.

Definition 124 *If $\phi \in F(EX)$ is a formula, and if $\mathfrak{M}$ is a Σ-structure, one defines $\top(\phi)$ according to these cases:*

(i) *If $\phi = R$ is an atomic proposition, one sets $\top(\phi) = \mathfrak{M}_R \in 2$, in particular, $\top(\top) = \top$ and $\top(\bot) = \bot$.*

(ii) *If $\phi = R(t_1, \ldots t_m), m > 0$, then*

$$\top(\phi) = \{x \mid x \in \mathfrak{M}_\phi, (t_1[x], \ldots t_m[x]) \in \mathfrak{M}_R\}$$

(iii) *If $\phi = (!\psi)$, then $Free(\phi) = Free(\psi)$, and we set $\top(\phi) = \mathfrak{M}_\phi - \top(\psi)$.*

(iv) *For the three cases $\phi = (\psi * \rho)$ of $* =$ conjunction, disjunction, or implication, one has $Free(\phi) = Free(\psi) \cup Free(\rho)$, and therefore* *canonical projections $p_\psi : \mathfrak{M}_\phi \to \mathfrak{M}_\psi$ and $p_\rho : \mathfrak{M}_\phi \to \mathfrak{M}_\rho$. We then use the Boolean connectives and define*

- $\top((\psi \,\&\, \rho)) = p_\psi^{-1}(\top(\psi)) \cap p_\rho^{-1}(\top(\rho))$,
- $\top((\psi \mid \rho)) = p_\psi^{-1}(\top(\psi)) \cup p_\rho^{-1}(\top(\rho))$,
- $\top((\psi \to \rho)) = \top(((!\psi) \mid \rho))$,

(v) *For $\phi = (\forall x)\psi$ or $\phi = (\exists x)\psi$, one has $Free(\phi) = Free(\psi) - \{x\}$ and therefore the projection $p : \mathfrak{M}_\psi \to \mathfrak{M}_\phi$. Then one sets*

- $\top((\forall x)\psi) = \{y \mid y \in \mathfrak{M}_\phi, p^{-1}(y) \subset \top(\psi)\}$,
- $\top((\exists x)\psi) = \{y \mid y \in \mathfrak{M}_\phi, p^{-1}(y) \cap \top(\psi) \neq \varnothing\}$,

including the special case where $x \notin Free(\psi)$. In this case the projection is the identity, and we have $\top((\forall x)\psi) = \top(\psi)$ and $\top((\exists x)\psi) = \top(\psi)$.

Given these truth evaluations of formulas, one can state validity of formulas similarly to propositional validity. If $Free(\phi) = x_1 < x_2 < \ldots x_m$ defines a subsequence of a sequence $y_1 < \ldots y_n$ of variables with sorts $y_i : B_i$, then if $y \in \mathfrak{M}_{B_1} \times \ldots \mathfrak{M}_{B_n}$, we have the projection y_ϕ of y to the coordinate sequence from $\mathfrak{M}_\phi$. We then define that ϕ *is valid in* y, $\mathfrak{M} \vDash \phi[y]$ iff $y_\phi \in \top(\phi)$. If ϕ is a sentence, we write $\mathfrak{M} \vDash \phi$ for this fact (which is now independent of y), and this means that $\top(\phi) = \top$. One then says that the Σ-*structure* $\mathfrak{M}$ *is a model for the sentence* ϕ. If ϕ is not a sentence, one considers its universal closure sentence $(\forall)\phi$, see definition 121, and defines validity of ϕ by $\mathfrak{M} \vDash \phi$ iff $\mathfrak{M} \vDash (\forall)\phi$, which means that $\mathfrak{M} \vDash \phi[y]$ for all y as above.

Example 71 To conclude example 70 of Peano arithmetic, we want to model the formulas which are given by Peano's five axioms. Here are these formulas, including those defining addition and multiplication, which are in fact all sentences (observe that one could omit the quantifiers in the following sentences and then use the universal closure to model the Peano axioms):

(i) (Zero is not a successor) $(\forall x_1)(!\overset{A}{0} \overset{A}{=} {}^{+_A}(x_1))$,

(ii) (Equal successors have equal predecessors)

$$(\forall x_1)(\forall x_2)({}^{+_A}(x_1) \overset{A}{=} {}^{+_A}(x_2) \to x_1 \overset{A}{=} x_2),$$

(iii) (Zero is additive neutral element) $(\forall x_1)x_1 \overset{A}{+} \overset{A}{0} \overset{A}{=} x_1$,

(iv) (Recursive definition of addition)

$$(\forall x_1)(\forall x_2) x_1 \overset{A}{+} {}^{+_A}(x_2) \overset{A}{=} {}^{+_A}(x_1 \overset{A}{+} x_2),$$

(v) (Zero is multiplicative "neutralizer") $(\forall x_1) x_1 \overset{A}{\cdot} \overset{A}{0} \overset{A}{=} \overset{A}{0},$

(vi) (Recursive definition of multiplication)

$$(\forall x_1)(\forall x_2) x_1 \overset{A}{\cdot} {}^{+_A}(x_2) \overset{A}{=} x_1 \overset{A}{\cdot} x_2 \overset{A}{+} x_1,$$

(vii) (Principle of induction) Denote by $\Phi(x_i)$ a formula, where x_i pertains to its free variables. By $\Phi({}^{+_A}(x_i))$ and $\Phi(\overset{A}{0})$, we denote the formula after replacement of each occurrence of x_i by ${}^{+_A}(x_i)$ and $\overset{A}{0}$, respectively. Then, for each formula $\Phi(x_i)$, we have this formula:

$$(\Phi(\overset{A}{0}) \,\&\, (\forall x_i)(\Phi(x_i) \rightarrow \Phi({}^{+_A}(x_i)))) \rightarrow (\forall x_i)\Phi(x_i).$$

The last item (vii) is not one formula, but one formula for each formula Φ. A *Peano structure* is a structure $\mathfrak{M}$ such that for each of the formulas Ψ described in (i)–(vii), we have $\mathfrak{M} \vDash \Psi$.

Let us now check the validity of formula (i) for our structure $\mathfrak{M}$ described in example 70. Formula (i) has the shape $\phi = (\forall x_1)\psi$, which is evaluated according to the projection $p : \mathfrak{M}_\psi \rightarrow \mathfrak{M}_\phi = 1$, since $Free(\phi) = \varnothing$. So let us check the fiber $\mathfrak{M}_\psi = p^{-1}(0)$ and test whether $\mathfrak{M}_\psi \subset \top(\psi)$, i.e., $\mathfrak{M}_\psi = \top(\psi)$. But $\top(\psi) = \top((!\overset{A}{0} \overset{A}{=} {}^{+_A}(x_1))) = \mathbb{N} - \top(\overset{A}{0} \overset{A}{=} {}^{+_A}(x_1))$, and $\top(\overset{A}{0} \overset{A}{=} {}^{+_A}(x_1)) = \{x \mid x \in \mathfrak{M}_\psi = \mathbb{N}, 0 = \overset{A}{0}[x] = {}^{+_A}(x_1)[x] = x + 1\}$, which is the empty set, we are done, i.e., $\top(\psi) = \mathbb{N}$ and $\mathfrak{M} \vDash \phi$.

Exercise 84 Check whether our structure $\mathfrak{M}$ described in example 70 is a Peano structure. Do the same for the structure $\mathfrak{M}'$.

Many of the possible formulas are equivalent in the sense that they yield the same logical values. More precisely:

Definition 125 *Given a Σ-structure $\mathfrak{M}$, two formulas ϕ and ψ are called equivalent iff we have $\mathfrak{M} \vDash (\phi <-> \psi)$ with the usual biimplication $(\phi <-> \psi)$ as an abbreviation for the conjunction $((\phi -> \psi) \,\&\, (\psi -> \phi))$.*

So we are looking for equivalent formulas which look as simple as possible. One such simplified type is the *prenex normal form* which means this:

Definition 126 *A formula is in* prenex normal form *(or shorter:* is prenex*) iff it is an uninterrupted (possible empty) sequence of universal or existential quantifiers, followed by a formula without quantifiers. It is in* Skolem normal form *if it is in prenex normal form such that all existential quantifiers precede all universal quantifiers.*

Example 72 The formula $(\forall x)(\exists y)(\forall z)((x = y) \mid (z < w))$ is in prenex normal form.

Here is the crucial result with regard to Skolem normalization:

Proposition 155 *Every formula ϕ of a Σ-structure $\mathfrak{M}$ is equivalent to a Skolem formula ψ.*

Proof The proof of this theorem is by induction on the length of the formula. It is not difficult, but uses a number of auxiliary lemmas which we have not the place to deal with here. However, the principal ideas are these: To begin with, if a bound quantifier in $\forall x \phi$ is replaced by any other variable z except the free variables of ϕ different from x, the new formula is equivalent to the old one. One then shows that $!(\forall x)\phi$ is equivalent to $(\exists x)(!\phi)$ and that $!(\exists x)\phi$ is equivalent to $(\forall x)(!\phi)$. If $\phi * (\exists x)\psi$, where $* \in \{\mid, \&\}$, is a formula, we may suppose that x is not one of the variables in *Free(ψ)*, and then $\phi * (\exists x)\psi$ is equivalent to $(\exists x)(\phi * \psi)$, similarly for the existence quantifier. Formulas of shape $(\psi \rightarrow \phi)$ are equivalent to $((!\psi) \mid \phi)$ by the very definition of truth values for implications. This gives us the prenex normal form. To construct the Skolem normal form, one needs auxiliary formulas and free variables, which must be added to the given formula in order to enable the existence quantifiers to precede the universal quantifiers, see [27] for details. $\square$

Exercise 85 The mathematician Paul Finsler has proposed a problem in absolute vs. formal mathematical reasoning which has provoked violent reactions among mathematicians, and which we describe here as an intriguing exercise in rigorous thinking. His proposal regards statements which cannot be proved by formal reasoning, but nevertheless can be proved by non-formal mental reasoning, i.e., by non-formalizable human thought. Here is the setup:

Suppose that we are given a finite (or even denumerable) alphabet which is used to write down formal proofs in the shape of finite chains of words (including the empty space to separate words), as described in the preceding chapters. Clearly, the set of these chains is denumerable. Now, we are only interested in such chains of words which are correct (formal) proofs of a very specific type of statement: We consider dual representations of numbers $d = 0.d_1 d_2 \ldots$, and we only consider those chains $Ch \to d$ of words which are correct proofs of the fact that either a specific dual number d has or has not infinitely many zeros. We may order these proofs lexicographically and also order their dual numbers according to this ordering. We then obtain a sequence $d(1), d(2), \ldots d(n), \ldots$ of all dual numbers which admit any proof chain of the given type. So observe that there are no other formal proofs in this formal framework which decide on the infinity or non-infinity of zeros in dual numbers. Now define a new dual number a by an antidiagonal procedure: $a_n = 1 - d(n)_n, n = 1, 2, \ldots$. For this dual number, there is no formal proof in our repertory, since any proof $Ch \to a$ would place a at a position m, say. And if $a = d(m)$, then we have a contradiction $a_m = 1 - d(m)_m = d(m)_m$. So a has no formal proof.

But we may easily give an informal proof, and show that a has an infinity of zeros.

Proof: In fact, take the dual sequence $d = 0.1111\ldots$ having only 1 as entries. By definition, this d has no zeros. A formal proof, say $Ch_0 \to 0.1111\ldots$, of this is immediate. But then the formal proofs $Ch_0 \& Ch_0 \& Ch_0 \ldots Ch_0$, n times, for $n = 1, 2, 3 \ldots$ all also do the job. So the number $0.1111\ldots$ appears an infinity of times (once for each such formal proof), and the antidiagonal therefore has an infinity of zeros.

The point is that by construction of the set of *all* formal proofs, this proof cannot be contained in the set of all formal proofs. (The formal description uses the above letters which we may easily suppose to be part of our alphabet.) Think of this argumentation. Where did we leave formal reasoning in the above proof? Finsler argues that we *must have left formal reasoning because we cannot, by construction, find this special proof in our proof list!*

CHAPTER 19

Languages, Grammars, and Automata

Until now, all formalizations of logic have been presented on a level which does not directly involve computers in the sense of machines which can execute commands and produce an output relating to formalized logical expressions. So our overall plan to incorporate logical reasoning in computerized operations still lacks the conceptual comprehension of how machines may (or may not) tackle such a task. We have learned in chapters 17 and 18 that formalized logic is built upon word monoids over adequate finite or infinite alphabets. We have also learned that the reasonable expressions of formal logic are words which can be defined by recursion on the word length and a defined set of construction rules. More explicitly, this was formalized in the axiomatic setup of formal logic, where the deduction of theorems is cast in a sequential setup starting from a set of axioms and applying a set of deduction rules.

Now, it turns out that this setup of formal logic and its construction methods is not really bound to the logical context. In fact, computers don't care about what words they are dealing with, the only relevant point of view to them is that they are allowed to build new words from given ones, following certain rules, and specific alphabets. The semantical issue of logic is not a conditio sine qua non for the formal control of reasonable expressions (words). In what follows, we shall therefore develop the more comprising context of general formal languages, their generative description by use of so-called phrase structure grammars, and their machine-oriented restatements by automata. So we have a triple perspective: first the "static" description of what is a (formal) language, second its more "dynamic" description by grammatical production systems, and third, the machine processes of automata, which englobe certain languages. *The yoga of this triple approach is that in fact, with respect to the languages they are generating, certain classes of grammars correspond to prominent classes of automata (such as stack automata or Turing machines), classes which are arranged in a four-fold hierarchy which is due to the linguist Noam Chomsky* [14].

19.1 Languages

This section is not more than a souped-up review of what we have already learned about word monoids in section 15.1. To deal with (formal) languages, we need an alphabet and then investigate the set of words generated by some determined procedure. However, it is worth extending the word concept to infinite words, the *streams*, for the sake of completeness of conceptualization. *In this section, we shall always reserve the letter $\mathcal{A}$ for a set, which plays the role of an alphabet.* Attention: The alphabet may be a finite or infinite set, no restriction on the cardinality is assumed in general.

Definition 127 *Given an alphabet $\mathcal{A}$, a* stream *or* infinite word *(over $\mathcal{A}$) (in contrast to a common word, which is also termed* finite stream*) is an infinite sequence $s = (s_0, s_1, \ldots) \in \mathcal{A}^{\mathbb{N}}$ of $\mathcal{A}$-letters. The length of a stream s is said to be infinity, in symbols: $l(s) = \infty$. The (evidently disjoint) union $Word(\mathcal{A}) \cup \mathcal{A}^{\mathbb{N}}$ is denoted by $Stream(\mathcal{A})$ and is called the* stream monoid *over $\mathcal{A}$. Its monoid product is defined as follows: If $x, y \in Word(\mathcal{A})$, we reuse the given product xy in $Word(\mathcal{A})$. If $x \in \mathcal{A}^{\mathbb{N}}$ is a stream and y is any element of $Stream(\mathcal{A})$, we set $xy = x$ and say that streams are* right absorbing*; if $x = a_1 a_2 \ldots a_n \in Word(\mathcal{A})$ and $y = (y_0, y_1, \ldots)$ is a stream, we define $xy = (a_1, a_2, \ldots a_n, y_0, y_1, \ldots)$, i.e., x is prepended to the stream y. In particular, if $x = \varepsilon$ is the neutral element, we set $xy = y$. In theoretical computer science it is common to call a proper left (right) factor x of a word or stream z, i.e., $z = xy$ ($z = yx$) a* prefix (postfix) *of z; if $z = uxv$ with $u, v \neq \varepsilon$, then x is called* infix *of z.*

Exercise 86 Verify that $Stream(\mathcal{A})$ is indeed a monoid, and that we have the submonoid $Word(\mathcal{A}) \subset Stream(\mathcal{A})$ of finite words.

Definition 128 *Given an alphabet* $\mathcal{A}$*, a* stream language (over $\mathcal{A}$) *is a subset* $L \subset Stream(\mathcal{A})$*. If the stream language L is contained in the submonoid* $Word(\mathcal{A})$*, it is called a* word language, *or simply a* language*. The set of languages over* $\mathcal{A}$ *identifies with the powerset* $2^{Word(\mathcal{A})}$ *and is denoted by* $Lang(\mathcal{A})$*.*

We shall mostly deal with (word) languages, except in rare cases, where the exception is explicitly indicated. So languages are completely unstructured subsets of $Word(\mathcal{A})$. Therefore, the boolean operations on the Boolean algebra $Lang(\mathcal{A})$ generate new languages from given ones, in particular, we have the union $L_1 \cup L_2$ and the intersection $L_1 \cap L_2$ of two languages L_1 and L_2, as well as the complement $-L$ of a language L over $\mathcal{A}$. Moreover, if $L_1, L_2 \in Lang(\mathcal{A})$, we have the product language $L_1 L_2 = \{xy \mid x \in L_1, y \in L_2\}$. In particular, for $n \in \mathbb{N}$, we have the powers $L^n = LL \dots L$ (n times) of L, including $L^0 = \{\varepsilon\}$, the *unit language over* $\mathcal{A}$. We say that L is *closed under concatenation* iff $L^2 \subset L$.

Example 73 Let $\mathcal{A} = \{a, b\}$, and $L_1, L_2, L_3 \in Lang(\mathcal{A})$ be defined as follows: L_1 is the language of non-empty words of the form $abab \dots$ of finite or infinite length (We also write $L_1 = \{(ab)^n \mid n > 0\}$). L_2 is the language of words of length ≤ 4. L_3 is the language of non-empty words of the form $baba \dots$ of finite of infinite length ($L_3 = \{(ba)^n \mid n > 0\}$). Then $L_1 \cap L_2$ is the set $\{ab, abab\}$. L_2^2 is the language of words of length ≤ 8 and L_1 is closed under concatenation. $L_1 \cup L_3$ is the language of all non-empty words with alternating letters a and b, finite or infinite, with first letter a or b. The complement $-L_2$ contains all words with length > 4. Finally, $L_1 \cap L_3$ is empty.

Exercise 87 Show that for a given alphabet $\mathcal{A}$, $Lang(\mathcal{A})$, together with the product of languages is a monoid with neutral element $\{\varepsilon\}$.

Definition 129 *Given an alphabet* $\mathcal{A}$*, the* Kleene operator *is the map*

$$* : Lang(\mathcal{A}) \to Lang(\mathcal{A}) : L \mapsto L^* = \langle L \rangle$$

which associates with every language L the monoid L^* *generated by L.*

Example 74 Let $\mathcal{A} = \{a, b, c\}$ and $L = \{aa, ab, ac, ba, bb, bc, ca, cb, cc\}$. Then L^* is the language of all words of even length.

Exercise 88 Show that for a given alphabet $\mathcal{A}$, a language L is closed under concatenation iff $L^* = L \cup \{\varepsilon\}$. Further show that the Kleene operator is *idempotent*, i.e., $L^{**} = L^*$.

Remark 22 We avoid the common, but ill-chosen notation $\mathcal{A}^*$ for the word monoid $Word(\mathcal{A})$ since it conflicts with the Kleene operator, in fact, $Word(L) \neq L^*$ in general! Check this latter inequality.

We shall now give an important example of languages as related to automata, which will be dealt with more extensively in section 19.3. Recall from section 12.2 about Moore graphs that a sequential machine of n variables was a map $M : S \times Q^n \rightarrow S$ involving a state space S and the n-cube Q^n as an input set. More generally, we may define a *sequential machine over an alphabet $\mathcal{A}$ and state space S* as being a map $M : S \times \mathcal{A} \rightarrow S$, and again we write $s \cdot a$ for $M(s, a)$ if M is clear. The *Moore graph of M* is defined as previously in the special case $\mathcal{A} = Q^N$ by $Moore(M) : S \times \mathcal{A} \rightarrow S^2 : (s, a) \mapsto (s, s \cdot a)$. The proposition 106 of section 12.2 discussed in the case of $\mathcal{A} = Q^n$ is also valid for general alphabets:

Proposition 156 *For a sequential machine $M : S \times \mathcal{A} \rightarrow S$, a canonical bijection*

$$PW : Path(Moore(M)) \rightarrow S \times Word(\mathcal{A})$$

is given as follows. If

$$p = s_1 \xrightarrow{(s_1, a_1)} s_2 \xrightarrow{(s_2, a_2)} s_3 \ldots \xrightarrow{(s_{m-1}, a_{m-1})} s_m,$$

then $PW(p) = (s_1, a_1 a_2 \ldots a_{m-1})$.

Under this bijection, for a given state $s \in S$, the set $Path_s(Moore(M))$ of paths starting at s corresponds to the set $\{s\} \times Word(\mathcal{A})$.

Proof The proof is completely analogous to the proof of proposition 106 in the case of $A = Q^n$, we therefore refer to that text.   □

Under the previous bijection PW we can associate with each couple $(s, w) \in S \times Word(\mathcal{A})$ the state $W(M)(s, w)$ resulting from the successive application of the word's letters by this map:

$$W(M) : S \times Word(\mathcal{A}) \rightarrow S : (s, w) \mapsto head(PW^{-1}(s, w))$$

If M is clear, we shall also write $s \cdot w$ instead of $W(M)(s, w)$.

Example 75 Given a sequential machine M over an alphabet $\mathcal{A}$ and state space S, one is often interested in a set $E \subset S$ of *final* states insofar as they may be reached by M from an *initial* state $i \in S$. This means by definition that we look for words $w \in Word(\mathcal{A})$ such that $i \cdot w \in E$. Denote the language (i.e., the set) of these words by $(i : M : E)$, or $(i : E)$ if M is clear from the context, and call these words/language *the words/language which are/is accepted by the sequential machine M*.

To give a concrete example, take $S = \{1, 2, 3, 4\}, \mathcal{A} = \{a, b, c\}$, whereas the machine M is defined by this table (state i is mapped by letter x to the state on the column below x on the row of state i, e.g., $3 \cdot b = 2$):

letter →	a	b	c
state 1	2	2	1
state 2	1	2	3
state 3	1	2	4
state 4	1	3	1

The graph for this machine is shown in figure 19.1.

Take $i = 2, E = \{4, 2\}$ and calculate the language $(i : E)$.

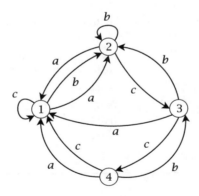

Fig. 19.1. The graph for the sequential machine of example 75.

Before leaving this generic subject, we should make a concluding remark about alphabets as they occur in the real life of computer scientists. Mathematically, the set $\mathcal{A}$ comprising the "letters" $x \in \mathcal{A}$ is quite irrelevant, and this is also generally true for computer science. Therefore standardization committees have agreed to create standard alphabets of natural numbers that represent known sets of letters. The most famous is the *American Standard Code for Information Interchange (ASCII)* character set codification, as made precise by the Standard ANSI X3.4-1986, "US-ASCII. Coded Character Set - 7-Bit American Standard Code for Information Interchange". Figure 19.2 shows a sample of that encoding.[1]

octal	decimal	hexadecimal	Name
⋮	⋮	⋮	⋮
060	48	0x30	0 (zero)
061	49	0x31	1
062	50	0x32	2
063	51	0x33	3
064	52	0x34	4
065	53	0x35	5
066	54	0x36	6
067	55	0x37	7
070	56	0x38	8
071	57	0x39	9
072	58	0x3a	: (colon)
073	59	0x3b	; (semicolon)
074	60	0x3c	< (less than)
075	61	0x3d	= (equals)
076	62	0x3e	> (greater than)
077	63	0x3f	? (question mark)
0100	64	0x40	@ (commercial at)
0101	65	0x41	A
0102	66	0x42	B
0103	67	0x43	C
0104	68	0x44	D
⋮	⋮	⋮	⋮

Fig. 19.2. Excerpt from ASCII encoding.

Here, the octal representation refers to the 8-ary representation of numbers, whereas the hexadecimal refers to basis 16. One has these prefix notations in order to distinguish different adic representations: for bi-

---

[1] See http://www.asciitable.com for the complete table of the $2^7 = 128$ characters encoded by ASCII.

nary, one doesn't put any prefix, for octal one starts numbers with 0, for hexadecimal 0x and for decimal, like binary, also nothing.

Thus ASCII sets up a bijection of the integer interval $[0, 127]$ and a set of relevant characters, predominantly used in the Angloamerican culture. However, as computers have spread over all cultures, more comprehensive character and sign types have been included in the standardization, ultimately leading to *Unicode* standard. This is a 16-bit character set standard, designed and maintained by the non-profit consortium *Unicode Inc.* Parallel to the development of Unicode an ISO/IEC standard was worked on, putting a large emphasis on being compatible with existing character codes such as ASCII. Merging the ISO (International Organization for Standardization) standard effort and Unicode in 1992, the *Basic Multilingual Plane BMP* was created. But presently the BMP is half empty, although it covers all major languages, including Roman, Greek, Cyrillic, Chinese, Hiragana, Katakana, Devanagari, Easter Island "rongo-rongo", and even Elvish (but leaves out Klingon).[2]

19.2 Grammars

We evidently need means to classify languages, since to the date the Babylonian variety of languages is beyond control, just imagine all languages possible with the traditional European alphabet. We are rightly confused by the world's variety of dead or living languages, and by the ever growing variety of computer languages. A natural way to access languages is a rule system which directly produces language items, i.e., a grammatical construction which we also use in natural language to build new sentences from given ones and from phrase building schemes. Observe that this is a totally different approach to languages as compared to the language construction by a sequential machine introduced in the above example 75. We shall however relate these approaches in the following section 19.3 on automata.

Definition 130 *Given an alphabet $\mathcal{A}$, a production grammar over $\mathcal{A}$ is a map*

$$f : Lang(\mathcal{A}) \to Lang(\mathcal{A})$$

---

[2] See http://www.unicode.org for more information about the Unicode standard.

*which commutes with unions, i.e., for any family $(L_i)_{i \in I}$ of languages L_i
over $\mathcal{A}$, we have*

$$f(\bigcup_I L_i) = \bigcup_I f(L_i),$$

*in particular $f(\emptyset) = \emptyset$. If $w \in Word(\mathcal{A})$ is a word, we set $f(w) = f(\{w\})$
and obtain a restricted map $f : Word(\mathcal{A}) \to Lang(\mathcal{A}) : x \mapsto f(x)$, which
we denote by the same symbol. One then has*

$$f(L) = \bigcup_{x \in L} f(x)$$

*for any language $L \in Lang(\mathcal{A})$. Conversely, if we are given any map
$g : Word(\mathcal{A}) \to Lang(\mathcal{A}) : x \mapsto f(x)$, we obtain a production grammar
(again denoted by the same symbol) $f : Lang(\mathcal{A}) \to Lang(\mathcal{A})$ defined by
the above formula $f(L) = \bigcup_{x \in L} f(x)$. In examples, we shall use either
definition according to the concrete situation.*

*If a production grammar is such that $f(x)$ is always a singleton set, i.e.,
$f(x) = \{y\}$ for all $x \in Word(\mathcal{A})$, then one calls f deterministic, otherwise, it is called nondeterministic. For a deterministic f, we also write
$f(x) = y$ instead of $f(x) = \{y\}$.*

Given a production grammar $f : Lang(\mathcal{A}) \to Lang(\mathcal{A})$ and an initial
language $I \in Lang(\mathcal{A})$, *one has the* language $f^{\infty}(I)$ *generated by f starting from I, i.e., $f^{\infty}(I) = \bigcup_{0 \le i} f^i(I)$ with $f^0 = Id$ and $f^i = f \circ f \circ \ldots f$,
i times, for positive i. If we are also given a terminal language T, the
language generated starting from I and terminating in T is defined by
$(I : f : T) = T \cap f^{\infty}(I)$. If the production grammar f is clear, one also
writes $(I : T)$ instead of $(I : f : T)$. For a given alphabet $\mathcal{A}$, two production
grammars f_1 and f_2 with initial and terminal languages I_1 and I_2 and T_1
and T_2, respectively, are called* equivalent *iff $(I_1 : f_1 : T_1) = (I_2 : f_2 : T_2)$.*

Example 76 L-systems are a type of production grammar proposed by
biologist Aristid Lindenmayer in 1968 to give an axiomatic description
of plant growth. We give a simple example of a so-called *turtle graphics*
production grammar t to illustrate the power of L-systems for the production of complex graphical objects associated with the language $t^{\infty}(I)$.
This situation is built upon a small alphabet $\mathcal{A} = \{F, G, +, -\}$, whereas
the grammar $t = t_{w_1, \ldots w_n}$ is defined by a finite set $\{w_1, \ldots w_n\}$ of words
and the function $t = t_{w_1, \ldots w_n}(x) = \{x | w_1, \ldots x | w_n\}$ on words x, where
$x | w$ denotes the word deduced from x by replacing each appearance of

the letter F by w. For example, if $x = F - FG + F, w = FG - F$, then $x|w = FG - F - FG - FG + FG - F$.

Let us consider the deterministic case $t_w, w = F - FG + F$ with one initial word $x_0 = FG$, i.e., $I = \{x_0 = FG\}$. Then the language $t_w^\infty(x_0)$ is the infinite set

$$\{FG, F - FG + FG, F - FG + F - F - FG + FG + F - FG + FG \ldots\}.$$

The turtle language is given by the graphical interpretation of letters and words as follows: Read a word as a sequence of commands from left to right, so $F + FG$ means: First do F, then do $+$, then do F, then do G. The command associated with F is this: You are a turtle moving on a white paper surface. Whenever you move, you leave a trace (of ink, say) on the surface. Now, doing F in a word where we have k appearances of the letter F means that the turtle has a given direction and moves on a straight line of defined length $1/k$. Doing G means that the turtle draws a circle of diameter $1/4k$ around its center, but then recovers its position after drawing the circle. Doing $+$ means that the turtle just turns clockwise by 90 degrees around its center, whereas $-$ means a counter-clockwise turn by 90 degrees.

What is the graphical interpretation of the production rule $x \mapsto x|w$? It means that you have a turtle trace graphics defined by x and then replace every straight line in this graphics by the graphics defined by w, drawn in the direction of that line, together with the shrinking factor k such that the total length of the graphics remains constant, i.e., 1 in our case. This is also why L-systems are labeled "rewriting systems". Observe that in contrast to F, the action G is not rewritten, it is a kind of "terminal" entity.

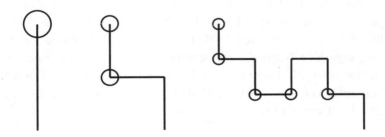

Fig. 19.3. The graphical interpretation of the first three words of the L-system featured in example 76.

An important class of production grammars is this:

Definition 131 *We are given a finite alphabet* $\mathcal{A} = \mathcal{T} \cup \mathcal{N}$*, which is the disjoint union of two subsets of letters: the* (lower case) *terminal symbols* $t \in \mathcal{T}$ *and the* (upper case) *nonterminal symbols* $X \in \mathcal{N}$*. We are also given a* start symbol $S \in \mathcal{N}$ *and a relation* $R \subset (Word(\mathcal{A}) - Word(\mathcal{T})) \times Word(\mathcal{A})$*. The production grammar* $f_{\mathcal{T},\mathcal{N},S,R}$ *is defined by the quadruple* $(\mathcal{T}, \mathcal{N}, S, R)$ *on words* $x \in Word(\mathcal{A})$ *as follows:*

$$f_{\mathcal{T},\mathcal{N},S,R}(x) = \{x\}$$

if there is no $u \in pr_1(R), a, b \in Word(\mathcal{A})$ *with* $x = aub$

$$f_{\mathcal{T},\mathcal{N},S,R}(x) = \{y \mid \text{ there are words } a, b \text{ and } (u, v) \in R$$
$$\text{such that } x = aub, y = avb\}$$

otherwise.

The quadruple $(\mathcal{T}, \mathcal{N}, S, R)$*, together with the production grammar it defines, is called a* phrase structure grammar*. In this context, the language* $(S : Word(\mathcal{T}))$ *is called the* language generated by the phrase structure *grammar* $(\mathcal{T}, \mathcal{N}, S, R)$*. If* $y \in f_{\mathcal{T},\mathcal{N},S,R}(x)$*, one also writes* $x \to y$ *and says that* y *is obtained from* x *by application of the rules* R*. This applies in particular if* $(x, y) \in R$*, then one says for* $x \to y$ *that* x *is the* pattern *for the replacement* y*. If* $f(x)$ *is a finite set* $\{y_1, y_2, \ldots y_r\}$*, then one also writes* $x \to y_1 | y_2 | \ldots y_r$*.*

Remark 23 We should add a remark here concerning the question when two languages L and L' are identical. By definition, there are two alphabets $\mathcal{A}$ and $\mathcal{A}'$ such that $L \subset Word(\mathcal{A})$ and $L' \subset Word(\mathcal{A}')$. Saying that the sets L and L' are the same means that their elements coincide, i.e., the words in L and in L' are the same, and this means that they are the same sequences of letters from $\mathcal{A}$ and $\mathcal{A}'$ respectively. In other words, there is a common subset $\mathcal{A}'' \subset \mathcal{A} \cap \mathcal{A}'$ such that $L \subset Word(\mathcal{A}'')$ and $L' \subset Word(\mathcal{A}'')$ and that these subsets are equal. In the above definition of a language $(S : Word(\mathcal{T}))$ generated by a phrase grammar, this applies in the sense that neither the set of nonterminals nor the total set of terminals is relevant to the definition of $(S : Word(\mathcal{T}))$, it is only the set-theoretic identification which counts.

19.2.1 The Chomsky Hierarchy

We now discuss the four-fold hierarchy

$$\text{type } 3 \subset \text{type } 2 \subset \text{type } 1 \subset \text{type } 0$$

of successively more comprising language types introduced by Noam Chomsky. *In the following discussion of Chomsky types, we suppose that in a phrase structure grammar* $(\mathcal{T}, \mathcal{N}, S, R)$, *the sets* $\mathcal{T}$, $\mathcal{N}$ *and R are all finite.* To begin with, we look at the innermost set of languages, those of type 3.

Definition 132 *If for a phrase structure grammar* $(\mathcal{T}, \mathcal{N}, S, R)$, *every rule* $x \to y$ *in R has the shape* $X \to Yt$ $(X \to tY)$ *or* $X \to s$ *for nonterminal letters X and Y and terminal letters s and t, the grammar is called* left linear (right linear). *If a rule* $X \to \varepsilon$ *is also admitted in a left linear (right linear) phrase structure grammar, it is called* left (right) regular.

Proposition 157 *For a language L the following four properties are equivalent:*

(i) *There is a left linear phrase structure grammar which generates the language* $L - \{\varepsilon\}$.

(ii) *There is a right linear phrase structure grammar which generates the language* $L - \{\varepsilon\}$.

(iii) *There is a left regular phrase structure grammar which generates the language L.*

(iv) *There is a right regular phrase structure grammar which generates the language L.*

Proof (i) implies (iii): If the left linear phrase structure grammar $G_l = (\mathcal{T}, \mathcal{N}, S, R)$ generates $L - \{\varepsilon\}$ (it cannot generate ε, by the nature of its rules), then add a new nonterminal element S_0 and the rules $S \to S_0$ and $S_0 \to \varepsilon$, and the new left regular phrase structure grammar does the job. (iii) implies (i): If the left regular phrase structure grammar $(\mathcal{T}, \mathcal{N}, S, R)$ generates L, and $\varepsilon \in L$, then we successively reduce the number of nonterminals X which have the rule $X \to \varepsilon$ until they have disappeared. The point is the case where we are given a single X with $X \to \varepsilon$. However, if we omit this rule, we not only prevent ε from being generated, but all the words stemming from a rule $Y \to xX$ followed by $X \to \varepsilon$ are in danger. Therefore, we have to add the rule $Y \to x$ to each $Y \to xX$, then we can omit $X \to \varepsilon$, and are done.

The proof for the right linear cases ((ii) iff (iv)) works the same way, therefore we omit it. We are left with the equivalence of left linear and right linear generation of languages, i.e., (i) iff (ii). We show that (i) implies (ii), the converse follows by exchanging left and right. To begin with, we may choose a new start symbol S'

and add to every rule $S \to$? a rule $S' \to$?. Then the new grammar never has its start symbol on the right hand side of any rule and is of course equivalent to the original. So we may suppose wlog that S never appears on the right side of a rule. We then construct a right linear grammar $G_r = G_l^*$ which is equivalent to G_l. But we construct more: The rules of G_r are such that the same operator $*$, when applied to G_r, with left and right exchanged, yields $*G_r = G_l$. This is the new rule set R^*, the alphabet being unaltered:

1. The rules $S \to t$ are left unchanged.
2. A rule $S \to Xs$ is replaced by the rule $X \to s$.
3. A rule $X \to s$ with $X \neq S$, is replaced by the rule $S \to sX$.
4. A rule $X \to Yt$ with $X \neq S$, is replaced by the rule $Y \to tX$.

We now show that $(S : R : \mathcal{T}) \subset (S : R^* : \mathcal{T})$. The converse is true by exchanging the roles of left and right and by the remark that the star operator, when applied to G_r gives us back G_l. The proof is by induction on the length n of a path $S \to Xs \to \ldots w$ with $w \in (S : R : \mathcal{T})$. For $n = 1$ this is rule 1. If $S \to Xs \to \ldots vu = w$ in G_l has length $n + 1$, where the path $Xs \to \ldots vu$ has length n and stems from the length n path $X \to \ldots v$, then we show that we have a path $S \to \ldots vX \to vu$ in G_r. We show by induction on path length, that if $X \to \ldots v$ in G_l has length m, then there is a path $S \to vX$ of length m in G_r. For $m = 1$, this is rule 3. In general, we have $X \to Yx \to \ldots yx = v$, where the rule $X \to Yx$ in R is converted into the rule $Y \to xX$ in R^* according to rule 4 above. By induction hypothesis we now have this new path: $S \to \ldots yY \to yxX = vX$, the first part being implied from $Y \to \ldots y$ to the right of $X \to Yx \to \ldots yx$, and we are done. $\square$

Definition 133 *A language which shares the equivalent properties of proposition 157 is called* regular *or of type 3.*

The crucial fact about type 3 languages is this:

Proposition 158 *If $L, L' \in Lang(\mathcal{A})$ are of type 3 (i.e., regular), then so are*

$$L \cup L', L \cap L', L^*, LL', Word(\mathcal{A}) - L.$$

Languages of type 3 are closed under all boolean operations as well as the Kleene operator and the product of languages.

Proof The proof idea is exemplified for the statement of $L \cup L'$ being of type 3 if L and L' are so. Take two phrase structure grammars $G = (\mathcal{T}, \mathcal{N}, S, R)$ and $G' = (\mathcal{T}, \mathcal{N}', S', R')$ which generate L and L', respectively. It is clear that one may suppose that the nonterminal sets N and N' are disjoint. From the proof of proposition 157, we also may assume that the start symbols S and S' are never on the right side of a rule. But then we create a new set $N^* = N \cup N' \cup \{S^*\}$

with a new start symbol S^* not in N and N', whereas the old rules are inherited, except the rules $S \to w, S' \to w'$, which we replace by the rules $S^* \to s, S^* \to w'$, and we are done. We refer to [43] for a complete proof. □

Example 77 Let $\mathcal{T} = \{a, b, c\}$ and $\mathcal{N} = \{S, A, B, C\}$ and consider the language $L_1 = \{a^l b^m c^n \mid l > 0, m > 0, n > 0\}$. A right linear grammar for this language consists of the rules $R_1 = \{S \to aA, A \to aA, A \to bB, B \to bB, B \to cC, B \to c, C \to cC, C \to c\}$. A right *regular* grammar can be expressed with the (simpler) set of rules $R_1^R = \{S \to aA, A \to aA, A \to bB, B \to bB, B \to cC, C \to cC, C \to \varepsilon\}$. Note the different handling of the end of words. A regular grammar can also be found for the language $L_2 = \{a^l b^m c^n \mid l \geq 0, m \geq 0, n \geq 0\}$, with the rules $R_2 = \{S \to A, A \to aA, A \to B, B \to bB, B \to C, C \to cC, C \to \varepsilon\}$. For L_2 no *linear* grammar exists.

A *left* linear grammar for L_1 is given by the rules $R_1^L = \{S \to Bc, S \to Cc, C \to Cc, C \to Bc, B \to Bb, B \to Ab, A \to Aa, A \to a\}$. Using this grammar a derivation for the word $abcc$ is given as follows:

$$S \to Cc$$
$$\to Bcc$$
$$\to Abcc$$
$$\to abcc$$

A useful property of languages of type 3 is embodied by the following lemma:

Lemma 159 (Type 3 Pumping Lemma) *Let $G = (\mathcal{T}, \mathcal{N}, S, R)$ be a linear grammar and L the language generated by G. Let $x \in L$ where $l(x) > card(\mathcal{N})$. Then there exist words $x, z, w \in Lang(\mathcal{T})$, $z \neq \varepsilon$, such that $x = yzw$ and $yz^k w \in L$ for $k = 0, 1, \ldots$.*

Proof Consider a word x of length $|x| > |\mathcal{N}|$. Then a derivation consists of $|x|$ steps. Since the number of non-terminals is less than $|x|$, there must be a subderivation of length at most $|\mathcal{N}|$ that begins with a non-terminal, say A, and ends with the same non-terminal A, e.g. $S \to \ldots yA \to \ldots yzA \to \ldots yzw$. But the subderivation $A \to \ldots zA$, can be substituted for the second A, thus yielding $yzzw$, and again, $yzzzw$, and so on. The subderivation can be left out entirely, yielding yw. □

The languages of type 2 are the famous context free languages which may be used to describe programming languages, mainly in its widespread Backus-Naur form (BNF), and more standardized as augmented

BNF (ABNF) or the extended BNF (EBNF) (standard ISO 14977); see 19.2.2 below for this type of grammars.

Definition 134 *A phrase structure grammar* $(\mathcal{T}, \mathcal{N}, S, R)$, *with alphabet* $\mathcal{A} = \mathcal{T} \cup \mathcal{N}$ *is said to be*

   (i) context free *if its rules are all of shape* $X \to x$ *with* $X \in \mathcal{N}$ *and* $x \in Word(\mathcal{A})$;

  (ii) reduced *if it is context free and for each nonterminal A different from the start symbol S, there is a rule* $A \to t$ *with* $t \in \mathcal{T}$ *terminal, and for each nonterminal* $A \in \mathcal{N}$, *there is a rule* $S \to vAw$ *with* $v, w \in Word(\mathcal{A})$;

 (iii) *in* Chomsky normal form *if its rules are of shape* $X \to t$ *or* $X \to AB$ *for* $X, A, B \in \mathcal{N}$ *and* $t \in \mathcal{T}$;

 (iv) *in* Greibach normal form *if its rules are of shape* $X \to xw$ *with* $X \in \mathcal{N}, x \in \mathcal{T}$, *and* $w \in Word(\mathcal{N})$.

Proposition 160 *For a language L the following four properties are equivalent:*

   (i) *There is a context free phrase structure grammar which generates the language L.*

  (ii) *There is a reduced context free phrase structure grammar which generates the language L.*

 (iii) *There is a phrase structure grammar in Chomsky normal form which generates the language* $L - \{\varepsilon\}$.

 (iv) *There is a phrase structure grammar in Greibach normal form which generates the language* $L - \{\varepsilon\}$.

Proof We have given a proof of proposition 157. The proof of this proposition is however too long for our context, therefore we refer to [43].   □

Definition 135 *A language which shares the equivalent properties of proposition 160 is called* context free *or of type 2.*

In virtue of the first criterion in definition 134, a language of type 3 is evidently of type 2. The crucial fact about type 2 languages is this:

Proposition 161 *If* $L, L' \in Lang(\mathcal{A})$ *are of type 2 (i.e., context free), then so are*

$$L \cup L', L^* \text{ and } LL'.$$

Proof Again, we give an idea of the proof for the union $L \cup L'$. Take the definition 134, (i), for two type 2 languages L and L'. Let two phrase structure grammars $G = (\mathcal{T}, \mathcal{N}, S, R)$ and $G' = (\mathcal{T}, \mathcal{N}', S', R')$ generate languages L and L', respectively. One may again assume that the sets N and N' of nonterminals are disjoint. Then just take the union of N and N' and add a new start symbol S^*, together with the two rules $S^* \rightarrow S$ and $S^* \rightarrow S'$, which solves the problem. We refer to [43] for a complete proof. □

There exists also a pumping lemma for languages of type 2:

Lemma 162 (Type 2 Pumping Lemma) *Let L be context free. Then there exists n such that for every $x \in L$ with $l(x) \geq n$ there exist words u, v, y, z, w where $v \neq \varepsilon$ or $z \neq \varepsilon$ such that $x = uvyzw$ and $uv^k yz^k w \in L$ for $k = 0, 1, \ldots$*

Proof See [28] for a proof. □

Example 78 We can use a context free grammar to describe expressions of elementary arithmetic. Expressions of this type are common in the syntax of programming languages. Let $\mathcal{N} = \{S, E, F, T\}$ and take $\mathcal{T} = \{+, *, (,), x, y, z\}$ where the letters x, y and z denote variables in the programming language. The rules are given by $R = \{E \rightarrow T+E, E \rightarrow T, T \rightarrow F*T, T \rightarrow F, F \rightarrow (E), F \rightarrow x|y|z\}$.

A derivation of the expression x + y $*$ (z + y) is given by:

$$
\begin{aligned}
E & \rightarrow T + E \\
& \rightarrow F + E \\
& \rightarrow x + E \\
& \rightarrow x + T \\
& \rightarrow x + F * T \\
& \rightarrow x + y * T \\
& \rightarrow x + y * F \\
& \rightarrow x + y * (E) \\
& \rightarrow x + y * (T + E) \\
& \rightarrow x + y * (F + E) \\
& \rightarrow x + y * (z + E) \\
& \rightarrow x + y * (z + T) \\
& \rightarrow x + y * (z + F) \\
& \rightarrow x + y * (z + y)
\end{aligned}
$$

Note how the rules model the usual precedence rules of the operators $+$ and $*$. This is of great practical value when implementing a parser for an actual programming language.

Figure 19.4 shows the derivation in form of a *syntax tree*. Each node of the tree is an application of a rule, the resulting expression can be read off the leaves of the tree in left-to-right order.

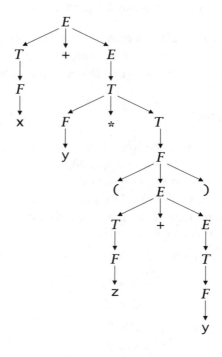

Fig. 19.4. The syntax tree for $x + y * (z + y)$.

Exercise 89 Not every language of type 2 is of type 3, i.e., the inclusion type 3 $\subset$ type 2 is proper. Construct a context free grammar for the language $L = \{a^n b^n \mid n \geq 1\}$ over $\mathcal{A} = \{a, b\}$, then use the pumping lemma for type 3 languages to show that there is no regular grammar for L.

Definition 136 *A phrase structure grammar* $(\mathcal{T}, \mathcal{N}, S, R)$, *with alphabet* $\mathcal{A} = \mathcal{T} \cup \mathcal{N}$ *is said to be* context sensitive *iff for every rule* $x \to y$ *in* R, *we have* $l(x) \leq l(y)$. *A language is called* context sensitive *or of type 1 iff it is generated by a context sensitive phrase structure grammar.*

Evidently, the characterization of context free languages by Chomsky normal form grammars implies that type 2 is a subset of type 1. The crucial fact about type 1 languages is this:

Proposition 163 *If $L \in Lang(\mathcal{A})$ are of type 1 (i.e., context sensitive), then so is its complement language $Word(\mathcal{A}) - L$.*

Proof We refer to [43] for a proof. □

Example 79 Let $\mathcal{T} = \{a, b, c\}$ and $\mathcal{N} = \{S, A, B, C, D, E\}$. The language $L = \{a^k b^k c^k \mid k > 0\}$ is of type 1. The rather complicated grammar is given by the following set of rules: $R = \{S \to Abc, A \to a, A \to aB, B \to aC, Cb \to bC, Cc \to Dc, D \to bc, D \to Ebc, bE \to Eb, aE \to aB\}$. Let us see how this works on the example of the word $aaabbbccc$:

$$
\begin{aligned}
S \;&\to\; Abc \\
&\to\; aBbc \\
&\to\; aaCbc \\
&\to\; aabCc \\
&\to\; aabDc \\
&\to\; aabEbcc \\
&\to\; aaEbbcc \\
&\to\; aaBbbcc \\
&\to\; aaaCbbcc \\
&\to\; aaabCbcc \\
&\to\; aaabbCcc \\
&\to\; aaabbDcc \\
&\to\; aaabbbccc
\end{aligned}
$$

We have yet to prove that there is no grammar of type 2 generating this language. We do this by invoking the type 2 pumping lemma 162. Suppose that L is context free. The lemma assures, that there is a number n such that the properties of the lemma will be fulfilled for words of length $\geq n$. Let us choose the word $a^n b^n c^n \in L$ which is certainly longer that n. The lemma tells us that this word must have a structure $uvyzw$ such that $uv^k yz^k w \in L$ for $k = 0, 1, \ldots$. But however we choose two subwords in $a^n b^n c^n$, the resulting "pumped-up" word will *not* be in L. Either the equal number of as, bs and cs will not be maintained, or the order of the letters will not be respected, as can be easily checked. Thus L cannot be context free.

Exercise 90 Prove, by finding a counterexample, that the intersection of two context free languages need not be context free.

The last type 0 is that of completely general phrase structure grammars:

Definition 137 *A phrase structure grammar $(\mathcal{T}, \mathcal{N}, S, R)$, with alphabet $\mathcal{A} = \mathcal{T} \cup \mathcal{N}$ is called*

(i) general *if there are no further conditions,*

(ii) separated *if each of its rules $x \to y$ has one of the following shapes:*

    a) $x \in Word(\mathcal{N}) - \{\varepsilon\}$ with $y \in Word(\mathcal{N})$,

    b) $x \in \mathcal{N}$ with $y \in \mathcal{T}$, or

    c) $x \in \mathcal{N}$ with $y = \varepsilon$,

(iii) normal *if each of its rules $x \to y$ has one of the following shapes:*

    a) $x \in \mathcal{N}$ with $y \in \mathcal{T}$,

    b) $x \in \mathcal{N}$ with $y = \varepsilon$,

    c) $x \in \mathcal{N}$ with $y \in \mathcal{N}^2$, or

    d) $x, y \in \mathcal{N}^2$.

It should be stressed that the definition (i) "general phrase structure grammar" is completely superfluous, but has been used in the computer community as a synonym of "phrase structure grammar". So the following proposition effectively is a statement about phrase structure grammars without any further attribute.

Proposition 164 *For a language L the following four properties are equivalent:*

(i) *There is a (general) phrase structure grammar generating L.*

(ii) *There is a separated phrase structure grammar generating L.*

(iii) *There is a normal phrase structure grammar generating L.*

Proof We refer to [43] for a proof.   □

Definition 138 *A language L which shares the equivalent properties of proposition 164 is called* recursively enumerable *or of type 0.*

19.2.2 Backus-Naur Normal Forms

The syntax of most programming languages, e.g., Algol 60, Extended Pascal, Minimal BASIC, or C, can be described by context free grammars. Originally, BNF was used by Peter Naur for the description of Algol 60 in an adaptation of a notation developed by John Backus.

The idea was to set up a standardized formal procedure to create terminal and nonterminal symbols and to describe the rules. Recall that we have one determined type of rules in context free grammars, i.e., $X \to w$, where $w \in Word(\mathcal{A})$ and X is a nonterminal symbol. To begin with, the arrow "$\to$" in a rule is replaced by the sign "$::=$" derived from the mathematical symbol "$:=$" meaning that x is defined by y in the definition $x := y$. The alternative $x \to y_1 | y_2 | \dots y_n$ is denoted in the same way, i.e., by $x ::= y_1 | y_2 | \dots y_n$.

The more important contribution of BNF is that the terminal and nonterminal symbols are provided by a standard construction from a given character set CH, in the ASCII encoding, say, i.e., "$CH = ASCII$". The procedure is very simple: Terminals are just single characters from CH. Nonterminals are all words of shape $\langle w \rangle, w \in Word(CH)$. The start symbol is mostly clear from the given rule system. For example, the Algol 60 specification of a floating point constant is called "unsigned number", and this is the start symbol, which defines the language of floating-point constants as follows:

```
<unsigned integer> ::= <digit> | <unsigned integer> <digit>
<integer> ::= <unsigned integer> | + <unsigned integer> |
              - <unsigned integer>
<decimal fraction> ::= . <unsigned integer>
<exponent part> ::= _10_ <integer>
<decimal number> ::= <unsigned integer> | <decimal fraction> |
                     <unsigned integer> <decimal fraction>
<unsigned number> ::= <decimal number> | <exponent part> |
                      <decimal number> <exponent part>
<digit> ::= 0|1|2|3|4|5|6|7|8|9
```

Here, the start symbol is $S = $ <unsigned number>. In contrast, the Extended BNF notation (EBNF) of the same grammar is this:

```
unsigned integer = digit | unsigned integer, digit;
integer = unsigned integer | "+", unsigned integer |
```

```
                  "-", unsigned integer;
decimal fraction = ".", unsigned integer;
exponent part = "_10_", integer;
decimal number = unsigned integer | decimal fraction |
                 unsigned integer, decimal fraction;
unsigned number = decimal number | exponent part |
                 decimal number, exponent part;
digit = "0"|"1"|"2"|"3"|"4"|"5"|"6"|"7"|"8"|"9";
```

The EBNF notation also includes some extensions to BNF, which improve readability and conciseness, e.g., the Kleene cross for a sequence of one or more elements of the class so marked, for example

```
unsigned integer = digit+;
```

The changes are evident, the idea behind this standardization is clear and is usually learned by doing some examples.[3]

It is also customary to represent a BNF grammar by use of syntax diagrams, i.e., groups of flow charts, where the alternatives in a rule $X ::= y_1 | y_2 | \ldots y_n$ are flow ramifications starting from the block of X and terminating at the leftmost symbol in the target replacement y_i. Here, the leaves are the terminal symbols. Figure 19.5 is a syntax diagram; such diagrams were used for the first time by Jensen and Wirth.

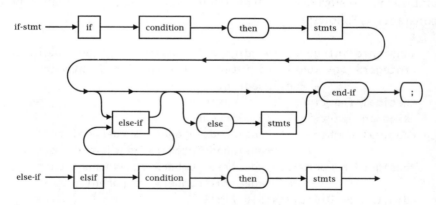

Fig. 19.5. A syntax diagram as used by Kathleen Jensen and Niklaus Wirth in [29].

---

[3] See http://www.cl.cam.ac.uk/~mgk25/iso-14977-paper.pdf for a more complete description of the BNF standard.

Exercise 91 Define (the fragments of) an alphabet and write the BNF rules corresponding to the flow charts shown in figure 19.5.

19.3 Automata and Acceptors

In this section, we shall establish a systematic relation between phrase structure languages and abstract machines, as they are axiomatically described by sequential machines, automata and acceptors. It will turn out that the languages of different Chomsky type are precisely those which may be defined by determined types of machines, such as Turing machines, for instance.

We had defined sequential machines in preliminary contexts earlier. Now it is time to give the full-fledged setup of those concepts.

Definition 139 *Given a finite alphabet set $\mathcal{A}$, a finite set S of "states", and an "initial" state $i \in S$ an* automaton *over $\mathcal{A}$ with initial state i is a pair (M, i) where M is a set map*

$$M : 2^S \times \mathcal{A} \to 2^S$$

such that M commutes with unions, i.e., for all families $(U_i)_{i \in I}$ of state sets $U_i \in 2^S$, and for all $a \in \mathcal{A}$, we have $M(\bigcup_i U_i, a) = \bigcup_i M(U_i, a)$; if M is clear from the context, one writes $U \cdot a$ instead of $M(U, a)$. In particular, we always have $\varnothing \cdot a = \varnothing$. The map M is completely determined by its values on singletons $\{s\} \in 2^S$, i.e., on single states s. As with production grammars, we therefore also write $M(\{s\}, a) = s \cdot a$. The corresponding map $S \times \mathcal{A} \to 2^S$ (now without any further properties), is also denoted by M and serves as an alternate definition for an automaton, much as this was done for production grammars.

An automaton is called deterministic *if its images $s \cdot a$ on singletons are always singletons $s \cdot a = \{x\}$ (attention: in particular, the images $s \cdot a$ are never empty!), we then also write $s \cdot a = x$, and correspondingly $M : S \times \mathcal{A} \to S$. A* nondeterministic *automaton is one which is not deterministic.*

The elementary graph *of an automaton is the digraph $\Gamma_M : Arr(M) \to S^2$ the arrow set of which is $Arr(M) = \{(s, a, x) \mid a \in \mathcal{A}, s \in S, x \in s \cdot a\}$ with $\Gamma_M((s, a, x)) = (s, x)$. The initial state i is rephrased as a morphism of digraphs $i : 1 \to \Gamma_M$ pointing to the vertex i. A path*

$$p = s_1 \xrightarrow{(s_1, a_1, s_2)} s_2 \xrightarrow{(s_2, a_2, s_3)} s_3 \cdots \xrightarrow{(s_{m-1}, a_{m-1}, s_m)} s_m$$

in Γ_M starting at the initial state i (i.e., with $s_1 = i$) is called a state sequence of the automaton associated with the word $W_p = a_1 a_2 \ldots a_m$. *The lazy path is associated with the empty word.*

Evidently, every automaton (M, i) determines its alphabet $\mathcal{A}$ and state set S, so we do not need to denote them explicitly.

Any automaton defines the *associated power automaton* (M', i'), which, evidently, is always deterministic, with these data: We replace S by $S' = 2^S$ and i by $i' = \{i\}$. Then the same map $M' = M$ defines an automaton, but we take the alternate definition this time! So in the first definition setup, we would switch to 2^{2^S}. This is nothing more than a trick of switching from the first definition to the second one. The point is that, however, not every deterministic automaton is of this type, since its state set need not be a powerset, and the commutation with unions need not work.

Although the elementary graph of an automaton is customary in computer science, it has some serious structural drawbacks which enforce a second digraph associated with an automaton: the *power graph* Γ^M *of an automaton* (i, M) is the elementary graph of the associated power automaton, $\Gamma^M = \Gamma_{M'}$. Since the power automaton is a priori deterministic, the power graph may be described more economically (it is in fact the Moore graph of the underlying sequential machine, check this out) as follows: vertexes are the sets of states, its arrows are the pairs $(s, a) \in 2^S \times \mathcal{A}$, which are mapped to the state set pairs $(s, s \cdot a)$, and the initial state pointer is $i' : 1 \to \Gamma^M$.

As to the graph representation $i : 1 \to \Gamma_M$ of the automaton (M, i), the representation of the associated deterministic automaton $i' : 1 \to \Gamma_{M'}$ contains the entire original information, whereas the original graph $i : 1 \to \Gamma_M$ need not in case where each set $s \cdot a$ is empty.

Definition 140 *An* acceptor *is a triple* (M, i, F), *where* $(M : S \times \mathcal{A} \to 2^S, i)$ *is an automaton and where* $F \subset S$ *is a subset of "terminal" or "accepting" states. By definition, the* language $(i : M : F)$, *or* $(i : F)$ *if M is clear, accepted by the acceptor* (M, i, F) *is the set of words W_p associated with state sequences p, which start at i and terminate in an element of F. In particular, the empty word W_i associated with the lazy path i is accepted iff $i \in F$. If the automaton is given by the first definition, i.e., $(M : 2^S \times \mathcal{A} \to 2^S, i)$, then a word W_p is accepted iff its path p starts in $\{i\}$ and ends in a* set s_n *of states such that* $s_n \cap F \neq \varnothing$. *Two acceptors are called* equivalent *if they accept the same language.*

Observe that an automaton (M, i) can be identified with the acceptor $(M, i, \varnothing)$. This is why we mostly talk about acceptors from now on, thereby including automata as special cases.

Proposition 165 *Every acceptor* $(M : 2^S \times \mathcal{A} \to 2^S, i, F)$ *is equivalent to the deterministic power acceptor* (M', i', F') *with* $M' = M, i' = \{i\}, F' = \{s \in 2^S \mid s \cap F \neq \varnothing\}$.

Exercise 92 The proof is left as an exercise.

Like automata, acceptors (M, i, F) are also represented in the form of digraphs. The *elementary graph of an acceptor* is the elementary graph Γ_M of the underlying automaton (M, i), together with the initial state pointer $i : 1 \to \Gamma_M$, and with the set $f : 1 \to \Gamma_M$ of digraph morphisms pointing to the elements $f \in F$. Also, the *power graph of an acceptor* (M, i, F) is the power graph Γ^M of the underlying automaton, together with the initial pointer $i' : 1 \to \Gamma^M$ and the final set pointer $F : 1 \to \Gamma^M$ targeting at the element $F \in 2^S$.

Example 80 Figure 19.6 shows the elementary graph for an acceptor with states $S = \{A, B, C, D\}$ and alphabet $\{a, b, c, d\}$. Its initial state $i = A$ is drawn as a square, the terminal states, given by $F = \{C, D\}$, as double circles. This acceptor is nondeterministic: from state A for example there are two transitions for the letter a.

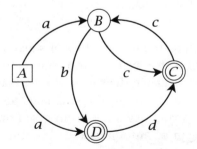

Fig. 19.6. The elementary graph for a nondeterministic acceptor.

The associated power graph is shown in Figure 19.8. To compute its transition function, it is best to draw up a table of the combined states, as in figure 19.7.

By inspection we see that the only states reachable from the initial state $I = \{A\}$ are $X = \{B\}$, $Y = \{C\}$, $Z = \{D\}$, $U = \{B, D\}$ and $V = \varnothing$, where we have renamed the combined states for easier reference. The new terminal states are $F' = \{U, Y, Z\}$.

	a	b	c	d
$\{A\}$	$\{B,D\}$	$\varnothing$	$\varnothing$	$\varnothing$
$\{B\}$	$\varnothing$	$\{D\}$	$\{C\}$	$\varnothing$
$\{C\}$	$\varnothing$	$\varnothing$	$\{B\}$	$\varnothing$
$\{D\}$	$\varnothing$	$\varnothing$	$\varnothing$	$\{C\}$
$\{A,B\}$	$\{B,D\}$	$\{D\}$	$\{C\}$	$\varnothing$
$\{A,C\}$	$\{B,D\}$	$\varnothing$	$\{B,C\}$	$\varnothing$
$\{A,D\}$	$\{B,D\}$	$\varnothing$	$\varnothing$	$\{C\}$
$\{B,C\}$	$\varnothing$	$\{D\}$	$\{B,C\}$	$\varnothing$
$\{B,D\}$	$\varnothing$	$\{D\}$	$\{C\}$	$\{C\}$
$\{C,D\}$	$\varnothing$	$\varnothing$	$\{B\}$	$\{C\}$
$\{A,B,C\}$	$\{B,D\}$	$\{D\}$	$\{B,C\}$	$\varnothing$
$\{B,C,D\}$	$\varnothing$	$\{D\}$	$\{B,C\}$	$\{C\}$
$\{A,C,D\}$	$\{B,D\}$	$\varnothing$	$\{B\}$	$\{C\}$
$\{A,B,D\}$	$\{B,D\}$	$\{D\}$	$\{C\}$	$\{C\}$
$\{A,B,C,D\}$	$\{B,D\}$	$\{D\}$	$\{B,C\}$	$\{C\}$
$\varnothing$	$\varnothing$	$\varnothing$	$\varnothing$	$\varnothing$

Fig. 19.7. The combined states of the power graph.

Clearly, one is not interested in acceptors of a given language which have superfluous ingredients, for example too many letters a which are never used in that language because their output $s \cdot a$ is always empty, or else states which do not contribute to the language. So we need to compare acceptors and to construct new ones from given ones. This leads to the concept of a morphism of automata and acceptors. To this end we use the following general construction on set maps known from set theory: If $f : X \rightarrow Y$ is a set map, then we have an associated set map $2^f : 2^X \rightarrow 2^Y :$ $U \mapsto f(U)$, moreover, if $g : Y \rightarrow Z$ is a second map, then $2^{g \circ f} = 2^g \circ 2^f$.

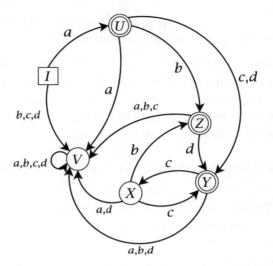

Fig. 19.8. The power graph for the nondeterministic acceptor of figure 19.6. Note that non-accessible states have been removed.

Definition 141 *If* $(i, M : S \times \mathcal{A} \to 2^S)$ *and* $(j, N : T \times \mathcal{B} \to 2^T)$ *are two automata, a morphism* $(\sigma, \alpha) : (i, M) \to (j, N)$ *is a pair of set maps* $\sigma : S \to T$ *and* $\alpha : \mathcal{A} \to \mathcal{B}$ *such that*

(i) *initial states are conserved:* $\sigma(i) = j$, *and*

(ii) *for any* $(s, a) \in S \times \mathcal{A}$, *we have* $2^\sigma(s \cdot a) = \sigma(s) \cdot \alpha(a)$, *where the products must be taken in the respective automata, i.e., one has the following diagram*

$$
\begin{array}{ccc}
S \times \mathcal{A} & \xrightarrow{\ M\ } & 2^S \\
{\scriptstyle \sigma \times \alpha}\big\downarrow & & \big\downarrow{\scriptstyle 2^\sigma} \\
T \times \mathcal{B} & \xrightarrow{\ N\ } & 2^T
\end{array}
$$

If (i, M, F) *and* (j, N, G) *are two acceptors, a morphism of the underlying automata* $(\sigma, \alpha) : (i, M) \to (j, N)$ *is a morphism of acceptors iff the terminal sets are also respected, i.e., if* $\sigma(F) \subset G$.

Let $(\sigma, \alpha) : (i, M : S \times \mathcal{A} \to 2^S) \to (j, N : T \times \mathcal{B} \to 2^T)$ *and* $(\tau, \beta) : (j, N : T \times \mathcal{B} \to 2^T) \to (k, O : U \times C \to 2^U)$ *be two morphisms of automata, then their composition* $(\tau, \beta) \circ (\sigma, \alpha)$ *is defined by* $(\tau, \beta) \circ (\sigma, \alpha) = (\tau \circ \sigma, \beta \circ \alpha)$. *The same definition is given for the composition of morphisms of acceptors.*

For an automaton $(i, M : S \times \mathcal{A} \rightarrow 2^S)$, *denote by* $Id_{(i,M)}$ *the morphism* $(Id_S, Id_\mathcal{A})$. *The notation for the identity of acceptors is correspondingly* $Id_{(i,M,F)}$. *The morphism* (σ, α) *is called an* isomorphism *of automata/acceptors iff there is a morphism* $(\tau : T \rightarrow S, \beta : \mathcal{B} \rightarrow \mathcal{A})$ *such that* $(\tau, \beta) \circ (\sigma, \alpha) = (Id_S, Id_\mathcal{A})$ *and* $(\sigma, \alpha) \circ (\tau, \beta) = (Id_T, Id_\mathcal{B})$.

Sorite 166 *Morphisms of automata/acceptors have these standard properties:*

(i) *(Associativity) Whenever the composition* $(\mu, \gamma) \circ ((\tau, \beta) \circ (\sigma, \alpha))$ *is defined, it is equal to* $((\mu, \gamma) \circ (\tau, \beta)) \circ (\sigma, \alpha)$ *and is denoted by* $(\mu, \gamma) \circ (\tau, \beta) \circ (\sigma, \alpha)$.

(ii) *(Identity) For any morphism* $(\sigma : S \rightarrow T, \ \alpha : \mathcal{A} \rightarrow \mathcal{B})$ *of automata, or acceptors, the identities* $(Id_S, Id_\mathcal{A})$, *respectively* $(Id_T, Id_\mathcal{B})$, *are right, respectively left, neutral, i.e.,*

$$(Id_T, Id_\mathcal{B}) \circ (\sigma, \alpha) = (\sigma, \alpha) = (\sigma, \alpha) \circ (Id_S, Id_\mathcal{A}).$$

(iii) *(Isomorphisms) A morphism* (σ, α) *is iso iff* σ *and* α *are both bijections of sets.*

Proof This is an easy exercise left to the reader. ☐

Exercise 93 Let $(\sigma, \alpha) : (i, M : S \times \mathcal{A} \rightarrow 2^S) \rightarrow (j, N : T \times \mathcal{B} \rightarrow 2^T)$ be a morphism of automata, show that the map $(s, a, x) \mapsto (\sigma(s), \alpha(a), \sigma(x))$ on arrows and σ on vertexes defines a morphism $\Gamma_{(\sigma,\alpha)} : \Gamma_M \rightarrow \Gamma_N$ which also maps the initial pointers into each other. Also, if we have final state sets and therefore acceptors, the corresponding final state pointers are preserved. Show that on power graphs Γ^M and Γ^N we have a corresponding morphism $\Gamma^{(\sigma,\alpha)} : \Gamma^M \rightarrow \Gamma^N$ defined by $2^\sigma \times \alpha$ on arrows and 2^σ on vertexes. Show that for two morphisms (τ, β) and (σ, α) which can be composed to $(\tau, \beta) \circ (\sigma, \alpha)$, we have $\Gamma_{(\tau,\beta)\circ(\sigma,\alpha)} = \Gamma_{(\tau,\beta)} \circ \Gamma_{(\sigma,\alpha)}$ and $\Gamma^{(\tau,\beta)\circ(\sigma,\alpha)} = \Gamma^{(\tau,\beta)} \circ \Gamma^{(\sigma,\alpha)}$. This kind of passage from one type of objects (automata) to another type (digraphs), which also preserves morphisms and their composition, is called *functorial* and will be discussed extensively in a later chapter of this course dedicated to so-called *categories*. Categories are a fundamental subject in modern mathematics and are becoming more and more important in computer science.

Proposition 167 *If* $(\sigma : S \to T, \alpha : \mathcal{A} \to \mathcal{B})$ *is a morphism of acceptors* $(\sigma, \alpha) : (i, M, F) \to (j, N, G)$, *then the induced homomorphism of the language set* $Word(\alpha) : Word(\mathcal{A}) \to Word(\mathcal{B})$ *maps* $(i : M : F)$ *into* $(j : N : G)$. *If in particular* (σ, α) *is an isomorphism, then* $Word(\alpha)$ *induces a bijection* $(i : M : F) \stackrel{\sim}{\to} (j : N : G)$. *More specifically, if also* $\alpha = Id_\mathcal{A}$, *then* $(i : M : F) = (j : N : G)$.

Exercise 94 Use the elementary graph of an acceptor and the morphism $\Gamma_{(\sigma,\alpha)}$ to give a proof of proposition 167.

Definition 142 *If* $(\sigma, \alpha) : (i, M, F) \to (j, N, G)$ *is a morphism of acceptors over the alphabets* $\mathcal{A}$ *and* $\mathcal{B}$ *respectively, such that* $\alpha : \mathcal{A} \subset \mathcal{B}$ *and* $\sigma : S \subset T$ *are subset inclusions, then we say that* (i, M, F) *is a subacceptor of* (j, N, G).

Corollary 168 *If* (i, M, F) *is a subacceptor of* (j, N, G), *then* $(i : M : F)$ *is a sublanguage of* $(j : N : G)$.

Proof This follows immediately from proposition 167. □

Definition 143 *An acceptor* (i, M, F) *is* simple *iff every state* $s \in S$ *is a vertex of a state sequence* p *from* i *to* F.

Proposition 169 *Every acceptor has an equivalent simple subacceptor.*

Proof Since the states not appearing in the path vertex of any state sequence have no meaning for the generated words, one obtains the same language when omitting those states. □

Besides the simplification procedure for an acceptor, we may also need to look for subacceptors which are present in multiple "copies", i.e., which play the same role with respect to the language they accept. We now want to eliminate such multiplicities, since it is not reasonable to have machines with equivalent functional units in multiple instantiations.

Definition 144 *If* M *is a sequential machine and* $F \subset S$ *a set of final states, then two states* $s, t \in S$ *are called* equivalent *if* $(s : M : F) = (t : M : F)$. *If for an acceptor* (i, M, F) *any two different states* $s \neq t$ *on state sequences from* i *to* F *are not equivalent, the acceptor is called* reduced.

We now discuss the construction of a reduced acceptor from a given deterministic acceptor $(i, M : S \times \mathcal{A} \to S, F)$. To begin with, we need a generic sequential machine associated with the alphabet $\mathcal{A}$.

Definition 145 *For the alphabet* $\mathcal{A}$ *the sequential machine* $LangMachine_{\mathcal{A}}$ *of* $\mathcal{A}$ *is defined by the map*

$$LangMachine_{\mathcal{A}} : Lang(\mathcal{A}) \times \mathcal{A} \to Lang(\mathcal{A})$$
$$(L, a) \mapsto L/a = \{x \in Word(\mathcal{A}) \mid a \cdot x \in L\}.$$

If (i, M, F) *is an acceptor, the* associated generic acceptor *is defined by* $(i_{\mathcal{A}}, LangMachine_{\mathcal{A}}, F_{\mathcal{A}})$, *where*

(i) $i_{\mathcal{A}} = (i : M : F)$

(ii) *and* $F_{\mathcal{A}} = \{(f : M : F) \mid f \in F\}$.

Exercise 95 If one defines more generally $L/w = \{x \in Word(\mathcal{A}) \mid w \cdot x \in L\}$ for $w \in Word(\mathcal{A})$, then $v, w \in Word(\mathcal{A})$ implies $(L/v)/w = L/vw$.

Proposition 170 *For a deterministic acceptor* $(i, M : S \times \mathcal{A} \to S, F)$, *consider the morphism*

$$(\sigma, Id_{\mathcal{A}}) : (i, M, F) \to (i_{\mathcal{A}}, LangMachine_{\mathcal{A}}, F_{\mathcal{A}})$$

of acceptors given by $\sigma(s) = (s : M : F)$. *Then the image* $(i, M, F)_{\mathcal{A}}$ *of* $(\sigma, Id_{\mathcal{A}})$ *is an equivalent reduced deterministic acceptor, more precisely, for each state* $s \in S$ *we have*

$$(s : M : F) = ((s : M : F) : LangMachine_{\mathcal{A}} : F_{\mathcal{A}}).$$

(i) *If* (i, M, F) *is simple, then so is* $(i, M, F)_{\mathcal{A}}$.

(ii) *If* (i, M, F) *is reduced, then* $(\sigma, Id_{\mathcal{A}}) : (i, M, F) \to (i, M, F)_{\mathcal{A}}$ *is an isomorphism.*

(iii) *If* (i, M, F) *and* (j, N, G) *are reduced, simple, and equivalent, then* $(i, M, F)_{\mathcal{A}} = (j, N, G)_{\mathcal{A}}$.

Proof To begin with, we have to check whether $(\sigma, Id_{\mathcal{A}})$ is a morphism. This readily follows from the fact that for a pair $(s, a) \in S \times \mathcal{A}$, we have $(s \cdot a, M, F) = (s, M, F)/a$. To show that

$$(s : M : F) = ((s : M : F) : LangMachine_{\mathcal{A}} : F_{\mathcal{A}}),$$

let $w \in (s : M : F)$. Then $s \cdot w \in F$. Therefore, $(s : M : F) \cdot w = (s \cdot w : M : F) \in F_{\mathcal{A}}$, whence $w \in ((s : M : F) : LangMachine_{\mathcal{A}} : F_{\mathcal{A}})$, i.e., $(s : M : F) \subset ((s : M : F) : LangMachine_{\mathcal{A}} : F_{\mathcal{A}})$. Conversely, if $w \in ((s : M : F) : LangMachine_{\mathcal{A}} : F_{\mathcal{A}})$, then we have $(s : M : F)/w = (f : M : F), f \in F$. In other words, $s \cdot w \cdot v \in F$ iff $f \cdot v \in F$, for any word v. In particular, $v = \varepsilon$, the empty word gives us

$f \cdot \varepsilon = f \in F$ whence $s \cdot w \cdot \varepsilon = s \cdot w \in F$, so $w \in (s : M : F)$, whence $(s : M : F) \supset ((s : M : F) : LangMachine_{\mathcal{A}} : F_{\mathcal{A}})$, and equality holds.

The new acceptor is deterministic by construction, and it is reduced since the language we obtain in the image $((s : M : F) : LangMachine_{\mathcal{A}} : F_{\mathcal{A}})$ is exactly the starting state $(s : M : F)$, so no two different starting states can produce the same language. If $(s : M : F)$ is simple, every state t is visited by a state sequence starting at s in a word w. But then the state $(s \cdot w, M, F)$ is reached by w from the starting point (s, M, F), whence claim (i). As to claim (ii), observe that the fiber of a state (s, M, F) is the set of states t which generate the same language (t, M, F), i.e., the equivalent states of the (possibly) not reduced acceptor. Therefore, if all fibers are singletons, the map on states is a bijection, and we are done. To prove (iii), observe that the initial states $(i : M : F)$ and $(j, N : G)$ are equal by hypothesis. But the composition rule is the same in $LangMachine_{\mathcal{A}}$, so the terminal states of these acceptors are the same, i.e., the images of the common initial state under the common language. $\square$

Definition 146 *An acceptor over the alphabet $\mathcal{A}$ which has a minimal number of states for a given language $L \in Lang(\mathcal{A})$ is called minimal.*

Corollary 171 (Theorem of Myhill-Nerode) *Any two minimal acceptors (i, M, F) and (j, N, G) of a given language $L \in Lang(\mathcal{A})$ are isomorphic. In fact, we have $(i, M, F)_{\mathcal{A}} = (j, N, G)_{\mathcal{A}}$.*

Proof In fact, a minimal acceptor is reduced and simple, whence the claim by proposition 170. $\square$

We now have a central theorem which relates acceptors and languages:

Proposition 172 *Let $L \in Lang(\mathcal{A})$ be a language over the alphabet $\mathcal{A}$. Then the following statements are equivalent.*

  (i) *The language L is of Chomsky type 3, i.e., regular.*

 (ii) *There is a minimal (and therefore deterministic, reduced, simple) acceptor (i, M, F) such that $L = (i : M : F)$.*

(iii) *There is an acceptor (i, M, F) such that $L = (i : M : F)$.*

Proof The equivalence of statements (ii) and (iii) is evident from the above theory. For the other equivalence, see [43]. $\square$

19.3.1 Stack Acceptors

Stack acceptors are a special type of acceptors. The point is that their state space is not finite and is composed in a specific way of three kinds of entities: a (finite) set S of elementary states, a (finite) elementary alphabet $\mathcal{A}$ and a stack alphabet K. The practical relevance of a stack acceptor's stack set K is illustrated by the typical example: We are given a stack of plates in a cafeteria. Each time when a plate is taken away (the so-called *pop* action) from the top of the stack, a spring moves the stack of the remaining plates upwards in order to make available a new plate to the service personnel. When, on the contrary, a cleaned plate is put onto the existing stack (the so-called *push* action), the augmented stack moves down one unit to receive the new top plate. The theoretical relevance of such stack acceptors is given by the fact that the languages which are accepted by this type of acceptors are precisely the context free, or type 2, languages.

In the context of stack automata, one starts from three finite sets $S, \mathcal{A}, K$, the elements of which are called the *elementary states, input letters, and stack elements,* respectively. We then consider the Cartesian product $Word(S, \mathcal{A}, K) = Word(S) \times Word(\mathcal{A}) \times Word(K)$ of word monoids which is a monoid by factorwise multiplication, i.e., $(u, v, w) \cdot (x, y, z) = (ux, vy, wz)$. If $X \subset Word(S, \mathcal{A}, K)$ and $x \in Word(S, \mathcal{A}, K)$ we write X/x for the set of elements y such that $x \cdot y \in X$. This is a construction we already know from the theory of generic acceptors. We further need the set $\mathcal{A}_\varepsilon = \mathcal{A} \cup \{\varepsilon\} \subset Word(\mathcal{A})$, ε being the neutral (empty) word. This is all we need to introduce stack acceptors. Observe that as acceptors, stack automata are deterministic. But attention: in theoretical computer science a slightly different terminology is customary, we shall explain this below.

Definition 147 *Given three sets $S, \mathcal{A}, K$ of* elementary states, input letters, and stack elements, *respectively, a* stack acceptor over S, $\mathcal{A}$, K *(also:* push down acceptor*) consists of*

(i) *the state space $2^{\mathfrak{s}}$ for the set of* configurations $\mathfrak{s} = S \times Word(\mathcal{A}) \times Word(K) \subset Word(S, \mathcal{A}, K)$,

(ii) *the stack alphabet $Alpha = Alpha(S, \mathcal{A}, K) = S \times \mathcal{A}_\varepsilon \times K$,*

(iii) *a* state transition function $\mu : Alpha \to 2^{S \times Word(K)}$, *which defines the following operation on states $M : 2^{\mathfrak{s}} \times Alpha \to 2^{\mathfrak{s}}$. Let $x \in Alpha$ and $X \subset \mathfrak{s}$. Then we define*

$$
\begin{aligned}
M(X, x) &= X \cdot x \\
&= \mu(x) \cdot (X/x) \\
&= \{(z, \varepsilon, k) \cdot y \mid y \in X/x, (z, k) \in \mu(x)\}
\end{aligned}
$$

(iv) *the initial element (in $2^{\mathbb{S}}$) is defined by two elementary initial elements $i \in S$ and $k \in K$ and is given by $I_{i,k} = \{i\} \times Word(\mathcal{A}) \times \{k\}$,*

(v) *the final set is defined by an elementary final set $E \subset S$ via*

$$\mathcal{F}_E = \{Y \times \{\varepsilon\} \times \{\varepsilon\} \subset \mathbb{S} \mid Y \cap E \neq \varnothing\} \subset 2^{\mathbb{S}}.$$

Such a stack acceptor is symbolized by $Stack(i, \mu, E)$, and again, the elementary sets $S, \mathcal{A}, K$ are determined by μ. Contrary to the strict terminology a stack acceptor is traditionally coined *deterministic* iff (1) all $\mu(u, v, w)$ are empty or singletons, and (2), $\mu(u, \varepsilon, w) = \varnothing$ implies that all $\mu(u, v, w), v \in \mathcal{A}$ are singletons; vice versa, if $\mu(u, \varepsilon, w)$ is a singleton, then $\mu(u, v, w) = \varnothing$ for all $v \in \mathcal{A}$. This means in particular that $X \cdot x$ is a singleton or empty if X is a singleton. It is coined *nondeterministic* if it is not deterministic. In order to distinguish these confusing wordings, we mark this traditional terminology by the prefix "stack", i.e., saying "*stack* deterministic/nondeterministic" if a confusion is likely.

Definition 148 *The* stack language which is accepted by $Stack(i, \mu, E)$ *will be denoted by* $Stack(i : \mu : E)$ *and consists by definition of all words* $w \in Word(\mathcal{A})$ *such that there is a word* $W_p = (z_1, a_1, k_1)(z_2, a_2, k_2) \ldots$ (z_n, a_n, k_n) *of a state sequence in* $(I_{i,k} : \mu : \mathcal{F}_E)$ *with* $w = a_1 \cdot a_2 \cdot \ldots a_n$.

Exercise 96 The initial element $I_{i,k}$ has the property that with any of its elements (s, w, t), it contains all elements $(s, v, t), v \in Word(\mathcal{A})$. Show that this property is inherited under the state operation M. More precisely, if $X \subset \mathbb{S}$ is such that with any of its configurations (s, w, t), it contains all configurations $(s, v, t), v \in Word(\mathcal{A})$, then so is $X \cdot x$ for any $x \in Alpha$. In other words, for any state sequence in $(I_{i,k} : \mu : \mathcal{F}_E)$, all its states share the property that any input words are admitted in the middle coordinate. We may therefore adopt the saying that when calculating a state sequence, we may "read a word, or letter, v from the input alphabet". This means that we are given any triple $(s, w, t) \in X$ and then choose any v, take (s, v, t), which is also in X, and look for configurations in $X \cdot x$ deduced from (s, v, t). Therefore, in a more sloppy denotation, one also forgets about the middle coordinate and only denotes the elementary state and stack coordinates. This is what will be done in example 81.

And here is the long awaited proposition about languages and stack automata:

Proposition 173 *A language L over an alphabet $\mathcal{A}$ is context free, i.e., of type 2, iff it is the language $Stack(i : \mu : E)$ of a stack acceptor over the input alphabet $\mathcal{A}$.*

Proof For the lengthy proof of this proposition, we refer to [43].   □

Example 81 Reprising our earlier example 78 of the context free language generating arithmetical expressions, we endeavour to construct a stack acceptor $e = Stack(i_e, \mu_e, E_e)$ for this language. The elementary states, input letters and stack elements are defined as $S_e = \{i, f\}$, $\mathcal{A}_e = \{+, *, (,), x, y, z\}$ and $K_e = \mathcal{A} \cup \{E, T, F, k\} = \{+, *, (,), x, y, z, E, T, F, k\}$, respectively. The final set $E_e \subset S$ is $\{f\}$ and the elementary initial elements are $i_e = i$ and $k_e = k$.

Table 19.9 describes the state transition function μ_e.

$S_e \times \mathcal{A}_{e\varepsilon} \times K_e \quad \longrightarrow_{\mu_e}$	$2^{S \times Word(K)}$
(i, ε, k)	$\{(f, E)\}$
(f, ε, E)	$\{(f, T + E), (f, T)\}$
(f, ε, T)	$\{(f, F * E), (f, F)\}$
(f, ε, F)	$\{(f, (E)), (f, x), (f, y), (f, z)\}$
(f, x, x)	$\{(f, \varepsilon)\}$
(f, y, y)	$\{(f, \varepsilon)\}$
(f, z, z)	$\{(f, \varepsilon)\}$
$(f, +, +)$	$\{(f, \varepsilon)\}$
$(f, *, *)$	$\{(f, \varepsilon)\}$
$(f, (, ()$	$\{(f, \varepsilon)\}$
$(f,),))$	$\{(f, \varepsilon)\}$

Fig. 19.9. The transition map μ_e for the stack acceptor of example 81.

We now write down a derivation of the word $x + y$. Note that the sets of states get rather large, therefore we abbreviate the sets involved and leave out the states that will play no further role. We begin with the set of states $\{(i, k)\}$. Reading the empty word, i.e., reading nothing at all, we reach the set $\{(f, E)\}$. At each step we apply every matching rule of the transition map μ_e to all states in the current state set. In the following, $\rightarrow_a$ indicates a transition by reading letter a from the input word.

$$\{(i,k)\} \;\rightarrow_\varepsilon\; \{(f,E)\}$$
$$\rightarrow_\varepsilon\; \{(f,T+E),(f,T)\}$$
$$\rightarrow_\varepsilon\; \{(f,F*E+E),(f,F*E),(f,F+E),(f,F)\}$$
$$\rightarrow_\varepsilon\; \{\ldots,(f,x+E),\ldots\}$$
$$\rightarrow_x\; \{\ldots,(f,+E),\ldots\}$$
$$\rightarrow_+\; \{\ldots,(f,E),\ldots\}$$
$$\rightarrow_\varepsilon\; \{\ldots,(f,T+E),(f,T),\ldots\}$$
$$\rightarrow_\varepsilon\; \{\ldots,(f,F*E+E),(f,F*E),(f,F+E),(f,F),\ldots\}$$
$$\rightarrow_\varepsilon\; \{\ldots,(f,y),\ldots\}$$
$$\rightarrow_y\; \{\ldots,(f,\varepsilon),\ldots\}$$

In the final line, the current set of states includes a state (f,ε) which satisfies the conditions of a final state, E_e being $\{f\}$ and the stack word being empty.

Note how the form of a rule $(s,a,k) \rightarrow (s,k_1 k)$ justifies the image of "push", k_1 being pushed on top of k. In the same way a rule $(s,a,k) \rightarrow (s,\varepsilon)$ "pop"s off k.

We leave it to the reader to compare the map μ_e with the set of grammar rules in example 78 and to find out how they relate.

19.3.2 Turing Machines

Intuitively, Turing machines—named after the mathematician Alan Turing (1912-1954), one of the founders of computer science—are finite automata which are provided with an infinite tape memory and a sequential access mode by a read/write head. We shall give a formal definition below, but we should make a point concerning the general formalism of automata and the concrete technical setup where determined types of automata are realized. We had in fact already learned above that stack acceptors are acceptors which are defined by auxiliary operations on complex spaces. The same is true for Turing machines.

For Turing machines, one is essentially given a state set S, a tape alphabet B which includes a special "empty" sign # for a "blank" entry on a specific place on the tape. The tape is thought to be infinite to the left and to the right of the read/write head. On the tape, we may have any elements of B, except that only finitely many differ from #. The set of tape states is therefore described by the subset $B^{(\mathbb{Z})}$ of $B^{\mathbb{Z}}$ of those sequences $t = (t_i), t_i \in B$ with $t_i = \#$ for all but finitely many indexes $i \in \mathbb{Z}$ (mathematicians often say in this case: "for *almost all* indexes"). The read/write head position is by definition the one with index $i = 0$. With this vocabulary, a Turing machine, as we shall define shortly, yields a map

$$\tau : S \times B^{(\mathbb{Z})} \to 2^{S \times B^{(\mathbb{Z})}},$$

which describes the transition from one couple "state s of automaton plus tape state t" to a set of possible new couples s' and t' of the same type. The machine goes on with this update of states until it halts. This looks somewhat different from the definition of an automaton, which requires a map $S \times \mathcal{A} \to S$. However, this formalism is also present in the previous setup, one must in fact use the natural adjunction of set maps discussed in proposition 59: The sets $Set(a \times b, c)$ and $Set(a, c^b)$ are in a natural bijection. Therefore

$$\mathcal{A}d : Set(S \times B^{(\mathbb{Z})}, 2^{S \times B^{(\mathbb{Z})}}) \xrightarrow{\sim} Set(S \times B^{(\mathbb{Z})} \times B^{(\mathbb{Z})}, 2^S)$$

which means that the Turing automaton map τ corresponds to a map

$$\mathcal{A}d(\tau) : S \times (B^{(\mathbb{Z})} \times B^{(\mathbb{Z})}) \to 2^S,$$

and this is precisely the type of maps we need for automata, i.e., the alphabet is $\mathcal{A} = B^{(\mathbb{Z})} \times B^{(\mathbb{Z})}$. We call the bijection $\mathcal{A}d$ the *Turing adjunction*.[4] The alphabet $\mathcal{A}$ happens to be infinite, but the formalism is the required one. The meaning of the correct writing $\mathcal{A}d(\tau)$ is this: We are given a present state $s \in S$ and a couple (t, t') of tape states, t is the present tape state, whereas t' is one of the possible successor tape state. The set $\mathcal{A}d(\tau)(s, t, t')$ is precisely the set of those successor states $s' \in S$ such that $(s', t') \in \tau(s, t)$, i.e., which correspond to the successor tape state t'.

But we shall stick to the map τ in order to maintain the intuitive setup. Here is the formal definition of Turing machines:

Definition 149 *A Turing machine is given by*

    (i) *a finite state set S, an initial state $i \in S$, and a special halt state s_H being specified,*

---

[4] Observe the omnipresence of universal constructions of mathematics in computer science.

(ii) *a finite tape alphabet B, containing a special* empty *sign #, together with an input alphabet* $\mathcal{A} \subset B$,

(iii) *three symbols* H, L, R *not in B, one writes* $B_{HLR} = \{H, L, R\} \cup B$,

(iv) *a state transition map* $tr : S \times B \rightarrow 2^{S \times B_{HLR}}$ *with* $tr(s, b) \neq \emptyset$ *for all* $(s, b) \in S \times B$ *and such that only pairs* $(s_H, H) \in tr(s, b)$ *may appear with second coordinate H.*

A *Turing machine is* deterministic *iff every set* $tr(s, b)$ *is a singleton, otherwise it is called* nondeterministic. *The Turing machine is also denoted by* $Turing(i, tr, s_H)$, *according to our general notation of acceptors.*

This definition generates the following map τ. To begin with, the element $q \in B_{HLR}$ of the extended tape alphabet operates on tape states $t \in B^{(\mathbb{Z})}$ as follows:

1. if $q \in B$, then $(q \cdot t)_i = t_i$ for $i \neq 0$, and $(q \cdot t)_0 = q$, i.e., the zero position on the tape is replaced by q;

2. if $q = H$, nothing happens: $q \cdot t = t$;

3. if $q = R$, the tape moves one unit to the right, i.e., $(q \cdot t)_i = t_{i-1}$;

4. if $q = L$, the tape moves one unit to the left, i.e., $(q \cdot t)_i = t_{i+1}$.

Then we have this map:

$$\tau : S \times B^{(\mathbb{Z})} \rightarrow 2^{S \times B^{(\mathbb{Z})}}$$
$$\tau(s, t) = \{(s', b' \cdot t) \mid (s', b') \in tr(s, t_0)\}$$

and the idea is that the change of states should go on until the halt state s_H is obtained. More precisely,

Definition 150 *With the above notations, a* state sequence *of a Turing machine* $Turing(i, tr, s_H)$ *is a sequence* $(s^{(0)}, t^{(0)}), (s^{(1)}, t^{(1)}), \ldots (s^{(n)}, t^{(n)})$ *such that* $(s^{i+1}, t^{i+1}) \in \tau(s^i, t^i)$ *for all indexes* $i = 0, 1, \ldots n - 1$.

A *word* $w = w_1 w_2 \ldots w_k \in Word(\mathcal{A})$ *is* accepted *by* $Turing(i, tr, s_H)$ *iff there is a state sequence* $(s^{(0)}, t^{(0)}), (s^{(1)}, t^{(1)}), \ldots (s^{(n)}, t^{(n)})$ *which terminates at the halt state, i.e.,* $s^{(n)} = s_H$, *and starts at state* $s^{(0)} = i$ *and a tape state* $t^{(0)}$ *with* $t_i^{(0)} = \#$, *for* $i \leq 0$ *and* $i > k$, *and* $t_i^{(0)} = w_i$ *for* $i = 1, 2, \ldots k$. *The language of words accepted by the Turing machine* $Turing(i, tr, s_H)$ *is denoted by* $Turing(i : tr : s_H)$. *Languages of type* $Turing(i, tr, s_H)$ *for given Turing machines are also called* semi-decidable.

And here is the rewarding proposition relating Turing machines and type 0 languages:

Proposition 174 *A language* $L \in Lang(\mathcal{A})$ *is of type 0 (i.e., recursively enumerable) iff it is semi-decidable, i.e., iff there is a Turing machine* $Turing(i, tr, s_H)$ *with input alphabet* $\mathcal{A}$ *such that* $L = Turing(i : tr : s_H)$.

Proof For the complicated proof of this proposition, we refer to [43]. □

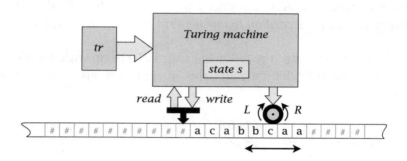

Fig. 19.10. A schematic illustration of a Turing machine, with the transition map ("program") *tr*, current state *s*, input alphabet $\mathcal{A} = \{a, b, c\}$. The arrows show the flow of information into and out of the machine.

Example 82 A simple example of an actual "program" for a Turing machine will illustrate the principles of the foregoing discussion, and show that Turing machines are capable of doing "useful" work, in this case the incrementation of a natural number represented in binary representation, and encoded using the symbols 0 and 1 on the initial tape state, where the least-significant bit of the number is on the right of the tape. Thus the number 151 will appear on the tape as $\ldots \#\#\#\overline{\#}10010111\#\#\#\#\ldots$ At the beginning, the machine's head will be at the position indicated by $\overline{\#}$. The final tape state when the machine terminates in the halt state will be $\ldots\#\#\#\overline{\#}10011000\#\#\#\#\ldots$, i.e., 152. Note that we use the Turing machine in a way a little different than previously described. The object here is not to accept an initial word, but to transform the initial word into the result word.

Now we proceed to describe the required data for the Turing machine T_I: the state set $S_I = \{i, A, B, C, D, E, F, s_H\}$ and the tape alphabet $B_I = \{\#, 1, 0\}$. The state transition map tr_I is shown in table 19.11. We assume that, initially, the tape is not empty, and omit any kind of error handling. The rudimentary character of Turing machines wouldn't allow much of anything in this direction anyhow. Observe that T_I is essentially deter-

$S_I \times B_I$ $\longrightarrow_{tr_I}$	$2^{S_I \times B_{I_{HLR}}}$
$(i, \#)$	$\{(A, L)\}$
$(A, 0)$	$\{(A, L)\}$
$(A, 1)$	$\{(A, L)\}$
$(A, \#)$	$\{(B, R)\}$
$(B, 1)$	$\{(D, 0)\}$
$(B, 0)$	$\{(E, 1)\}$
$(B, \#)$	$\{(E, 1)\}$
$(D, 0)$	$\{(B, R)\}$
$(D, 1)$	$\{(B, R)\}$
$(E, 0)$	$\{(C, R)\}$
$(E, 1)$	$\{(C, R)\}$
$(C, 0)$	$\{(E, 0)\}$
$(C, 1)$	$\{(E, 1)\}$
$(C, \#)$	$\{(F, L)\}$
$(F, 0)$	$\{(s_H, H)\}$
$(F, 1)$	$\{(s_H, H)\}$

Fig. 19.11. The transition map tr_I for the Turing machine of example 82.

ministic, since the values of the transition map are all singletons, except that we left out the definition of tr_I for some couples (s, b) that wouldn't occur anyway.

The reader is invited to simulate this machine on a few input words.

Concluding this short section on Turing machines, we should notice that the length of a state sequence which accepts a word w as a function of the word length $l(w)$ may be exorbitant, and one is therefore interested in those languages, where the lengths of accepting state sequences for their words are not too long. Here is the precise definition.

Definition 151 *A language $L \in Lang(\mathcal{A})$ is called* of (polynomial) com-plexity class P *if there is a polynomial $P(X)$ and a deterministic Turing machine $Turing(i, tr, s_H)$ with input alphabet $\mathcal{A}$ and $L = Turing(i, tr, s_H)$,* *such that each word $w \in L$ is accepted by a state sequence of length at most $P(l(w))$.*

A language $L \in Lang(\mathcal{A})$ is called of (nondeterministic polynomial) complexity class NP *if there is a polynomial $P(X)$ and a nondeterministic Turing machine $Turing(i, tr, s_H)$ with input alphabet $\mathcal{A}$ and $L = Turing(i, tr, s_H)$, such that each word $w \in L$ is accepted by a state sequence of length at most $P(l(w))$.*

A final remark: It is one of the deepest unsolved problems of computer science to understand the relation between class P and class NP. In particular, it is not currently known whether "P = NP". Recently, it has been shown that there is a deterministic algorithm that decides for every natural number n in $(\log(n))^k$ steps, k being a fixed natural exponent, whether n is prime or not. The logarithm is proportional to the length of n in its binary representation, so it represents the length of the word needed to write down n. This is what is meant, when the result is stated as "PRIME is in P". See [11] for a lucid exposition.

A comprehensive treatment of automata theory and languages is [28]. For more on the subject of computability, P and NP, see [30].

CHAPTER 20

Categories of Matrixes

The present chapter opens a field of mathematics which is basic to all applications, be it in equation solving, geometry, optimization, calculus, or numerics. This field is called linear algebra. It is the linear part of algebra, which we do control best—in contrast to the non-linear algebra, also called algebraic geometry, which is far more difficult than the linear part. Linear algebra deals with the structure theory of vectors, and the geometry they describe. We shall however see later in the chapters on calculus that even non-linear, or—worse than that—non-algebraic phenomena of continuous and infinitesimal phenomena can in a first approximation be dealt with by linear algebra. So here we enter one of the most powerful domains of mathematics, and also one where algorithms and corresponding computer programs are most developed.

The structure of linear algebra rests on three pillars, which complement each other, but are of equal importance for a real grasp of the subject: Matrix theory, axiomatic vector space theory, and linear geometry. Already this seemingly divergent triad makes it plausible that linear algebra provides a very powerful access to problem solving.

First, *matrix theory* is the calculatory backbone, it is nothing less than the mathematical theory of tables, a subject which is at the core of any operational programming methodology, database theory or listing strategy. Without knowing the essentials about matrixes, any understanding of arrays, lists, or vectors becomes illusory, and no real, concrete calculation is possible. Strangely enough it turns out that the system of matrixes is at the same time the most concrete and the most abstract structure of this field. This is hidden behind the attribute "category" in our title, a specification which will become more and more clear with the evolution of the student's understanding of fundamental structures in mathematics. Recall that we have already added this attribute in the context of graphs. It is remarkable that things seemingly so distant as graphs and tables turn out to be core instances of a common substance: the structure of a category. Presently, this is a philosophical allusion rather than hard mathematics. But it is a hint at the beauty of mathematics any reader should learn to feel in the course of this curriculum.

Second, *axiomatic vector space theory* is the account to the fact that the truths which are "hard coded" in matrix theory can be encountered in much less concrete situations, which are seemingly unrelated to matrixes, but, when analyzed in view of their structural substance reveal a fantastic kind of conceptual generalization of matrix calculus, an effect which gives back very operational perspectives to seemingly abstract situations.

Last, *linear geometry* is a very traditional branch of geometry, essentially related to Descartes and his analytical geometry. If one rephrases the structural substance of points, lines, surfaces, together with their operations, metrical properties, and transformational behavior, then it turns out that this returns what is condensed in vector space theory. And this gives the dry calculations of matrix theory and the abstract manipulations of vector space theory a sensorial perspective which is not only nice to have, but in turn helps to understand abstract phenomena of matrix theory in a revealing environment of geometric intuition. We should however point out that the geometric intuition about abstract truths is not the only one: the auditory intuition as cultivated by music is another sensorial image of abstract mathematical truths which for a number of problems is even better suited to yield evidence to "abstract artifacts" (or even what some scientific businessmen call "abstract nonsense").

20.1 What Matrixes Are

In this and the following sections of this chapter, the zero and unit of a ring R will be denoted by 0 and 1, respectively, if the context is clear. For any natural number n, we denote by $[1, n]$ the set of natural numbers between 1 and n, including the extremal values 1 and n. For $n = 0$, the set $[1, 0]$ is the empty set.

Definition 152 *Given a ring R and two natural numbers m and n, a* matrix *of size $m \times n$, or $m \times n$-matrix, with coefficients in R is a triple $(m, n, M : [1, m] \times [1, n] \to R)$. For non-zero m and n, the value $M(i, j)$, which is also written as M_{ij}, is called the* coefficient *of M at index pair ij; its component i is called the* row index, *while j is called the* column index. *If the number of rows m and the number of columns n are clear, M is also denoted by (M_{ij}). For a row number i the* i-th row matrix *in M is the matrix $M_{i\bullet}$ of size $1 \times n$ with $(M_{i\bullet})_{1j} = M_{ij}$. For a column number j the* j-th column matrix *in M is the matrix $M_{\bullet j}$ of size $m \times 1$ with $(M_{\bullet j})_{i1} = M_{ij}$.*

The set of $m \times n$-matrixes over R is denoted by $\mathbb{M}_{m,n}(R)$, while the set of all matrixes, i.e., the disjoint union of all sets $\mathbb{M}_{m,n}(R)$, is denoted by $\mathbb{M}(R)$. Clearly, every ring homomorphism $f : R \to S$ between rings gives rise to a map $\mathbb{M}(f) : \mathbb{M}(R) \to \mathbb{M}(S)$, which sends a matrix $M = (M_{ij}) \in \mathbb{M}(R)$ to the matrix $f(M) = f \circ M \in \mathbb{M}(S)$ with $(f(M)_{ij}) = (f(M_{ij}))$, and which therefore also sends $\mathbb{M}_{m,n}(R)$ into $\mathbb{M}_{m,n}(S)$.

Example 83 There is a number of special matrixes which we want to present now: The unique matrix for either m or n equal to 0 is denoted by $0 \square n$, $m \square 0$, or $0 \square 0$, respectively. If $m = n$, the matrix is called a *square* matrix. The following convention is very useful in matrix calculus: If i and j are two natural numbers, the *Kronecker delta symbol* is the number

$$\delta_{ij} = \begin{cases} 1 \in R & \text{if } i = j, \\ 0 \in R & \text{if } i \neq j. \end{cases}$$

Then the quadratic $n \times n$-matrix defined by $E_n = (\delta_{ij})$ is called the *unit matrix* of rank n—including the unique matrix $0 \square 0$ as a "degenerate" special case. Also important are the so-called *elementary* matrixes. Given an index pair ij, the number of rows m and the number of columns n, the elementary $m \times n$-matrix for this datum is the matrix $E(i\ j)$ such that $(E(i\ j))_{uv} = 0$ except for $uv = ij$, where we have $(E(i\ j))_{ij} = 1$.

Very often, a matrix M is represented as a rectangular table, where the entry on row i and column j is the matrix coefficient M_{ij}. Here is the tabular representation of one example of a 2×3-matrix M over the ring $R = \mathbb{Z}$ of integer numbers:

$$M = \begin{pmatrix} -2 & 5 & 0 \\ 0 & 26 & 3 \end{pmatrix}$$

If we have the canonical ring homomorphism $f = \text{mod}_7 : \mathbb{Z} \to \mathbb{Z}_7$, then the image $f(M)$ is equal to this matrix over $\mathbb{Z}_7$:

$$f(M) = \begin{pmatrix} \mathrm{mod}_7(-2) & \mathrm{mod}_7(5) & \mathrm{mod}_7(0) \\ \mathrm{mod}_7(0) & \mathrm{mod}_7(26) & \mathrm{mod}_7(3) \end{pmatrix}$$

One also writes $M \equiv N \mod d$ if $M, N \in \mathbb{M}_{m,n}(\mathbb{Z})$ and their images under the canonical homomorphism $\mathrm{mod}_d : \mathbb{Z} \to \mathbb{Z}_d$ coincide; so for example, we have

$$\begin{pmatrix} -2 & 5 & 0 \\ 0 & 26 & 3 \end{pmatrix} \equiv \begin{pmatrix} 5 & 5 & 0 \\ 0 & 5 & 3 \end{pmatrix} \quad \mod 7.$$

Or else, consider the 4×3-matrix

$$M = \begin{pmatrix} 3.5 + i \cdot 4 & -i & 4 + i \cdot \sqrt{5} \\ -0.5 & 0 & 4 - i \cdot 2 \\ i \cdot 20 & 1 & 3.78 - i \\ 0 & -i & -5 - i \cdot \sqrt{3} \end{pmatrix}$$

with complex coefficients. We have the field automorphism of conjugation $f(z) = \bar{z}$, which gives us the *conjugate matrix*

$$f(M) = \overline{M} = \begin{pmatrix} \overline{3.5 + i \cdot 4} & \overline{-i} & \overline{4 + i \cdot \sqrt{5}} \\ \overline{-0.5} & \overline{0} & \overline{4 - i \cdot 2} \\ \overline{i \cdot 20} & \overline{1} & \overline{3.78 - i} \\ \overline{0} & \overline{-i} & \overline{-5 - i \cdot \sqrt{3}} \end{pmatrix}$$

$$= \begin{pmatrix} 3.5 - i \cdot 4 & i & 4 - i \cdot \sqrt{5} \\ -0.5 & 0 & 4 + i \cdot 2 \\ -i \cdot 20 & 1 & 3.78 + i \\ 0 & i & -5 + i \cdot \sqrt{3} \end{pmatrix}.$$

A third example is more in the vein of tables for common usage: word processing environments. Suppose that we are given an alphabet C, which, to be concrete, denotes the letters in the Courier font, whereas S denotes the alphabet of letters in the Old German Schwabacher font. In a common word processing software, a text may be converted from Courier to Schwabacher, more formally, we have a map $y : C \to S$. By the universal property of monoids (proposition 111) and monoid algebras (proposition 120), this induces a ring homomorphism $f = Id_{\mathbb{Z}}\langle Word(y)\rangle :$ $\mathbb{Z}\langle Word(C)\rangle \to \mathbb{Z}\langle Word(S)\rangle$, which we may now apply to a table with Courier-typed text, i.e., elements from the ring $\mathbb{Z}\langle Word(C)\rangle$, in order to obtain a table with text in the Schwabacher font, i.e., elements from the ring $\mathbb{Z}\langle Word(S)\rangle$. For example, if

$$M = \begin{pmatrix} \texttt{Author: Shakespeare} \\ \texttt{Work: Hamlet} \end{pmatrix}$$

then

$$f(M) = \left(\begin{array}{ll} \text{Author:} & \text{Shakespeare} \\ \text{Work:} & \text{Hamlet} \end{array} \right)$$

Definition 153 *For any ring R, the* transposition *is the map*

$$?^\tau : \mathbb{M}(R) \to \mathbb{M}(R)$$

defined by $(M^\tau)_{ij} = M_{ji}$ *for all index pairs* i, j, *thereby transforming a* $m \times n$-*matrix M into a* $n \times m$-*matrix* M^τ. *A matrix M is called* symmetric *if* $M^\tau = M$.

Exercise 97 For any natural number n, the identity matrix E_n is symmetric. If $E(i\ j)$ is elementary, then $E(i\ j)^\tau = E(j\ i)$. Show that for any matrix M, we have

$$(M^\tau)^\tau = M.$$

In particular, matrix transposition is a bijection on the set $\mathbb{M}(R)$.

$$E_3 = \begin{pmatrix} 1 & 0 & 0 \\ 0 & 1 & 0 \\ 0 & 0 & 1 \end{pmatrix} \quad E(2\ 3) = \begin{pmatrix} 0 & 0 & 0 \\ 0 & 0 & 1 \\ 0 & 0 & 0 \end{pmatrix} \quad E(3\ 2) = \begin{pmatrix} 0 & 0 & 0 \\ 0 & 0 & 0 \\ 0 & 1 & 0 \end{pmatrix}$$

Fig. 20.1. The unit matrix and the elementary matrixes $E(2\ 3)$ and $E(3\ 2) = E(2\ 3)^\tau$ in $\mathbb{M}_{3,3}$.

20.2 Standard Operations on Matrixes

We now proceed to the algebraic standard operations on matrixes. We now tacitly assume that the underlying ring R for $\mathbb{M}(R)$ is commutative, and we shall stress any deviation of this general assumption.

Definition 154 *Given two matrixes* $M, N \in \mathbb{M}_{m,n}(R)$, *their sum* $M + N \in \mathbb{M}_{m,n}(R)$ *is defined as follows: If one of the numbers* m *or* n *is* 0, *then there is only one matrix in* $\mathbb{M}_{m,n}(R)$, *and we set* $M + N = m\square n$. *Else, we set* $(M + N)_{ij} = M_{ij} + N_{ij}$.

Sorite 175 *With the addition defined in definition 154, the set* $\mathbb{M}_{m,n}(R)$ *becomes an abelian group. For m or n equal to* 0, *this is the trivial (zero) group. In the general case, the neutral element of* $\mathbb{M}_{m,n}(R)$ *is the zero matrix* $0 = (0)$, *whereas the additive inverse* $-M$ *is defined by* $(-M)_{ij} = -M_{ij}$.

Exercise 98 Give a proof of sorite 175.

Definition 155 *Given a matrix* $M \in \mathbb{M}_{m,n}(R)$, *and a scalar* $\lambda \in R$, *the scalar multiplication* $\lambda \cdot M$ *is defined as follows: If one of the numbers m or n is* 0, *then there is only one matrix in* $\mathbb{M}_{m,n}(R)$, *and we set* $\lambda \cdot M = m \square n$. *Otherwise we set* $(\lambda \cdot M)_{ij} = \lambda \cdot M_{ij}$. *The matrix* $\lambda \cdot M$ *is also called "M scaled by* λ."

In other words, as is the case with ring homomorphisms and addition, scalar multiplication proceeds coefficient-wise, i.e., by operating on each coefficient of the involved matrixes.

Sorite 176 *With the definitions 154 and 155 of addition and scalar multiplication, we have these properties for any* $\lambda, \mu \in R$ *and* $M, N \in \mathbb{M}_{m,n}(R)$:

(i) *Scalar multiplication is* homogeneous, *i.e., we have* $\lambda \cdot (\mu \cdot M) = (\lambda \cdot \mu) \cdot M$, *therefore we may write* $\lambda \cdot \mu \cdot M$ *for this expression.*

(ii) *Scalar multiplication is* distributive, *i.e., we have*

$$(\lambda + \mu) \cdot M = \lambda \cdot M + \mu \cdot M$$

and

$$\lambda \cdot (M + N) = \lambda \cdot M + \lambda \cdot N.$$

(iii) *Scalar multiplication and transposition commute:* $(\lambda \cdot M)^\tau = \lambda \cdot M^\tau$.

Exercise 99 Give a proof of sorite 176.

Proposition 177 *Given positive row and column numbers m and n, every matrix* $M \in \mathbb{M}_{m,n}(R)$ *can be uniquely represented as a sum*

$$M = \sum_{\substack{i=1,\dots m \\ j=1,\dots n}} M_{ij} \cdot E(i\,j)$$

of scaled $m \times n$ *elementary matrixes* $E(i\,j)$. *I.e., if we have any representation*

$$M = \sum_{\substack{i=1,\dots m \\ j=1,\dots n}} \mu_{ij} \cdot E(i\ j)$$

with $\mu_{ij} \in R$, then $\mu_{ij} = M_{ij}$.

Proof The sum $\sum_{i=1,\dots m, j=1,\dots n} M_{ij} \cdot E(i\ j)$, when evaluated at an index pair uv, yields $\sum_{i=1,\dots m, j=1,\dots n} M_{ij} \cdot (E(i\ j)(uv)) = M_{uv}$, since all elementary matrixes vanish, except for $ij = uv$. If any representation $M = \sum_{i=1,\dots m, j=1,\dots n} \mu_{ij} \cdot E(i\ j)$ of M is given, then $M_{uv} = \mu_{uv}$, and we are done. $\qquad \square$

The next operation is the most important in matrix theory: the product of two matrixes. It is crucial to learn this operation by heart.

Definition 156 Let $M \in \mathbb{M}_{m,n}(R)$ and $N \in \mathbb{M}_{n,l}(R)$ be two matrixes, then a matrix $M \cdot N \in \mathbb{M}_{m,l}(R)$, the product of M and N, is defined as follows: If one of the outer numbers, m or l is 0, then $M \cdot N = m \square l \in \mathbb{M}_{m,l}(R)$ is the unique matrix in $\mathbb{M}_{m,l}(R)$. Otherwise, if the middle number $n = 0$, then we set $M \cdot N = 0 \in \mathbb{M}_{m,l}(R)$, the $m \times l$-matrix with zeros at every position. In the general case (no number m, n, l vanishes), one sets

$$(M \cdot N)_{ij} = \sum_{k=1,\dots n} M_{ik} N_{kj}$$

for every index pair ij.

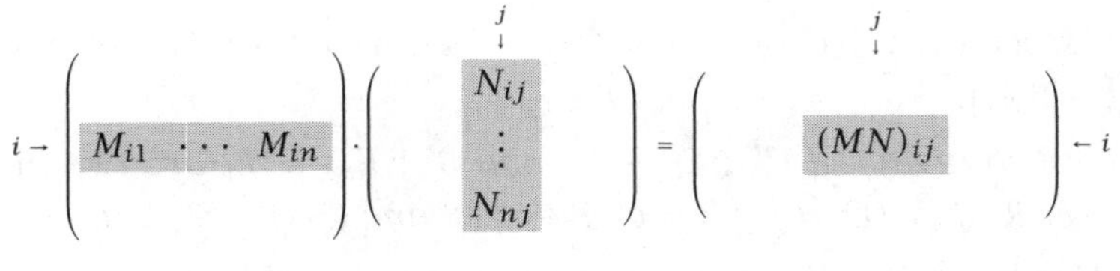

Fig. 20.2. Multiplication of matrixes $M \in \mathbb{M}_{m,n}$ and $N \in \mathbb{M}_{n,l}$, with the result $MN \in \mathbb{M}_{m,l}$.

So pay attention: The product of matrixes $M \in \mathbb{M}_{m,n}(R)$ and $N \in \mathbb{M}_{n',l}(R)$ is never defined if $n \neq n'$. To make this restriction really evident, we set up a more telling representation of matrixes. If $M \in \mathbb{M}_{m,n}(R)$, we shall also write

$$M : E_n \to E_m \text{ or else } E_n \xrightarrow{M} E_m$$

in order to indicate the possibilities to compose matrixes as if they were set maps! In fact, the product of $E_l \xrightarrow{N} E_{n'}$ and $E_n \xrightarrow{M} E_m$ is only possible if $n = n'$, i.e., the "codomain" $E_{n'}$ of N equals the "domain" E_n of M, and then gives us the matrix product $E_l \xrightarrow{M \cdot N} E_m$ which we may visualize by a commutative diagram

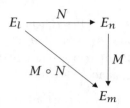

What looks somehow mysterious here is a very modern point of view of maps: Matrixes look like maps without domains or codomains. We just have composition of maps, and the domains and codomains must be reinvented by a trick. Let us take this approach as it is and postpone its deeper signification to a more advanced discussion of category theory in the second volume of this book. *The only point to retain here is the very useful notational advantage which makes immediately evident which matrixes may be multiplied with each other.* What should however be expected from this arrow formalism is, that matrix products are associative whenever defined. This is of course true:

Sorite 178 *Let $A : E_n \to E_m, B : E_m \to E_l$ and $C : E_l \to E_k$ be three matrixes over R. Then*

(i) *(Associativity) $(C \cdot B) \cdot A = C \cdot (B \cdot A)$, which we therefore write as $C \cdot B \cdot A$.*

(ii) *(Distributivity) If $C' : E_l \to E_k$ and $B' : E_m \to E_l$ are two matrixes over R, then $(C + C') \cdot B = C \cdot B + C' \cdot B$ and $C \cdot (B + B') = C \cdot B + C \cdot B'$.*

(iii) *(Homogeneity) If $\lambda \in R$ is a scalar, then $\lambda \cdot (C \cdot B) = (\lambda \cdot C) \cdot B = C \cdot (\lambda \cdot B)$, which we therefore write as $\lambda \cdot C \cdot B$.*

(iv) *(Neutrality of identity matrixes) We have $A \cdot E_n = E_m \cdot A = A$.*

(v) *$(C \cdot B)^\top = B^\top \cdot C^\top$.*

Proof Let $A = (A_{tu})$, $B = (B_{st})$ and $C = (C_{rs})$, with $1 \le r \le k$, $1 \le s \le m$ and $1 \le t \le n$. Then $((C \cdot B) \cdot A)_{ru} = \sum_t (C \cdot B)_{rt} A_{tu} = \sum_t (\sum_s (C_{rs} \cdot B_{st})) A_{tu} = \sum_t \sum_s (C_{rs} B_{st}) A_{tu} = \sum_s \sum_t C_{rs} (B_{st} A_{tu}) = \sum_s C_{rs} (\sum_t B_{st} A_{tu}) = \sum_s (C_{rs} \cdot (B \cdot A)_{su}) = (C \cdot (B \cdot A))_{ru}$, whence (i).

Then $((C + C') \cdot B)_{rt} = \sum_s (C + C')_{rs} B_{st} = \sum_s (C_{rs} + C'_{rs}) B_{st} = \sum_s (C_{rs} B_{st} + C'_{rs} B_{st}) = \sum_s C_{rs} B_{st} + \sum_s C'_{rs} B_{st} = (C \cdot B)_{rt} + (C \cdot B')_{rt}$, whence (ii).

Further, $(\lambda \cdot (C \cdot B))_{rt} = \lambda (C \cdot B)_{rt} = \lambda \sum_s C_{rs} B_{st} = \sum_s (\lambda C_{rs}) B_{st} = ((\lambda \cdot C) \cdot B)_{rt}$, and similarly $\sum_s (\lambda C_{rs}) B_{st} = \sum_s C_{rs} (\lambda B_{st}) = (C \cdot (\lambda \cdot B))_{rt}$, whence (iii).

Claim (iv) is left to the reader.

Finally, $((C \cdot B)^\top)_{tr} = (C \cdot B)_{rt} = \sum_s C_{rs} B_{st} = \sum_s B_{ts}^\top C_{sr}^\top = (B^\top \cdot C^\top)_{tr}$. □

We now urgently need first justifications of the matrix product definition from more practical points of view. Here are some examples.

Example 84 From high school and practical experience it is well known that linear equations are very important. Here is one such equation, which we set up in its concrete shape

$$
\begin{array}{rcrcrcrcr}
3.7 & = & 23x_1 & - & 4.5x_2 & & & + & 45x_4 \\
-8 & = & 0.9x_1 & + & 9.6x_2 & + & x_3 & - & x_4 \\
0 & = & & & 20x_2 & - & x_3 & + & x_4 \\
1 & = & & & 3x_2 & + & x_3 & - & 2x_4
\end{array}
$$

in order to let the student recognize the common situation. This system of equations is in fact an equation among matrixes: On the left hand side of the equation, we have a 4×1-matrix which is the product

$$
\begin{pmatrix} 3.7 \\ -8 \\ 0 \\ 1 \end{pmatrix} = \begin{pmatrix} 23 & -1 & 0 & 45 \\ 0.9 & 9.6 & 1 & -1 \\ 0 & 20 & -1 & 1 \\ 0 & 3 & 1 & -2 \end{pmatrix} \cdot \begin{pmatrix} x_1 \\ x_2 \\ x_3 \\ x_4 \end{pmatrix}
$$

of a 4×4-matrix and a 4×1-matrix, the latter being the matrix of the unknowns $x_1, \ldots x_4$. Hence the theory of matrix products should (and will) provide us tools for finding solutions of linear equations. The ideal thing would be to construct a kind of inverse to the 4×4-matrix of the equation's coefficients and then multiply both sides with this inverse. This is indeed what is done, but we are not yet ready for such a construction and need more theory.

Example 85 This example is taken from graph theory and uses the adjacency matrix introduced in definition 67. Given a digraph $\Gamma : A \to V^2$, and fixing a bijection $c : [1, n] \to V, n = card(V)$, we obtain a $n \times n$-matrix $Adj_c = (Adj_c(i, j))$, where $Adj_c(i, j)$ is the number of arrows from the vertex $c(i)$ to the vertex $c(j)$.

What is the role of matrix products in this graph-theoretical context? The entry at index ij of the adjacency matrix is the number of arrows from vertex $c(i)$ to vertex $c(j)$, i.e., the number of paths of length one! We contend that the square Adj_c^2 of the adjacency matrix has as entry at ij the number of paths of length 2. In fact, any such path must reach $c(j)$ from $c(i)$ through an intermediate vertex, which runs through $c(1), \ldots c(n)$. Now, for each such intermediate vertex $c(k)$, the paths which cross it are one arrow $c(i) \to c(k)$, composed with one arrow $c(k) \to c(j)$, and this yields the product $Adj_c(i, k) \cdot Adj_c(k, j)$, therefore the total number of paths of length 2 is the coefficient at ij of the square Adj_c^2. More generally, the numbers of paths of length r are the coefficients in the r-th power Adj_c^r of the adjacency matrix.

We consider again the adjacency matrix of the graph Γ from example 34 in section 10:

$$Adj_c(\Gamma) = \begin{pmatrix} 0 & 0 & 0 & 0 & 0 & 0 \\ 1 & 0 & 1 & 0 & 2 & 0 \\ 0 & 0 & 0 & 0 & 0 & 2 \\ 0 & 0 & 0 & 1 & 0 & 0 \\ 0 & 0 & 0 & 0 & 0 & 0 \\ 2 & 0 & 0 & 0 & 0 & 1 \end{pmatrix}$$

The square of this matrix is:

$$Adj_c^2(\Gamma) = \begin{pmatrix} 0 & 0 & 0 & 0 & 0 & 0 \\ 0 & 0 & 0 & 0 & 0 & 2 \\ 4 & 0 & 0 & 0 & 0 & 2 \\ 0 & 0 & 0 & 1 & 0 & 0 \\ 0 & 0 & 0 & 0 & 0 & 0 \\ 2 & 0 & 0 & 0 & 0 & 1 \end{pmatrix}$$

As an illustration, figure 20.3 shows the four paths from vertex 2 to vertex 0 as indicated by entry $(Adj_c^2(\Gamma))_{3,1}$. Remember that vertex number i is associated with matrix index $i + 1$.

Exercise 100 What does it mean for the adjacency matrix of a digraph if the digraph has three connected components? Try to find a reasonable indexing of the vertexes. How do the powers of such a matrix look like?

20.3 Square Matrixes and their Determinant

The most important matrixes are the square matrixes M of positive size n, i.e., $M \in \mathbb{M}_{n,n}(R)$. We also tacitly assume $R \neq 0$. We have this general result to start with:

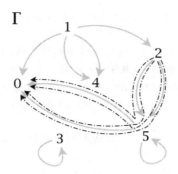

Fig. 20.3. The four paths of length 2 from vertex 2 to vertex 0.

Proposition 179 *Let n be a positive natural number and R a (commutative, non-zero) ring. Then the set $\mathbb{M}_{n,n}(R)$, together with the sum and product of matrixes is a ring which is not commutative except for $n = 1$. The homomorphism of rings $\Delta : R \to \mathbb{M}_{n,n}(R) : r \mapsto r \cdot E_n$ identifies R with the* diagonal *matrixes of size n with zeros off the diagonal indexes ii and with $M_{ii} = r$. We therefore also identify $r \in R$ with its diagonal matrix $r \cdot E_n$ if no confusion is likely. Then, we have $r \cdot M = M \cdot r$ for all $M \in \mathbb{M}_{n,n}(R)$; we say that R commutes with all matrixes of size n. No other matrixes commute with all of $\mathbb{M}_{n,n}(R)$.*

Proof The proposition immediately follows from sorite 178, except for the last statement. Suppose that a $n \times n$-matrix N commutes with all of $\mathbb{M}_{n,n}(R)$. Then it must commute with all elementary matrixes $E(i\ j)$. But $N \cdot E(i\ j)$ has zeros except in column j which is $N_{\bullet i}$, whereas the product $E(i\ j) \cdot N$ has zeros except in row i which is $N_{j\bullet}$. So, at the intersection of this row and this column, we have the equation $N_{ii} = N_{jj}$, whereas all other coefficients must vanish, whence $N = \lambda = \lambda \cdot E_n$. □

Definition 157 *The group $\mathbb{M}_{n,n}(R)^*$ of invertible matrixes of size n is called the* general linear group of size n *and denoted by $GL_n(R)$. An invertible matrix is also called* regular.

In order to control $GL_n(R)$, one needs a special function, the *determinant* of a square matrix. Here is its construction:

Definition 158 *If $M = (M_{ij}) \in \mathbb{M}_{n,n}(R)$, then $\det(M) \in R$ is defined by*

$$\det(M) = \sum_{\pi \in S_n} (-1)^{sig(\pi)} \prod_{j=1\ldots n} M_{\pi(j)j}$$

where S_n is the symmetric group of rank n defined in section 15.2.

Observe that, a priori, this function has $n!$ summands and will therefore require special tools to handle with reasonable effort, especially in computerized implementations. Before delving into the many beautiful properties of this strange formula, let us give some easy examples:

Example 86 For $n = 1$, we have $\det(M) = M_{11}$, and for $n = 2$, we have the formula

$$\det(M) = M_{11}M_{22} - M_{21}M_{12}$$

which is well known from high school. For $n = 3$, we have

$$\det(M) = M_{11}M_{22}M_{33} - M_{11}M_{32}M_{23} - M_{21}M_{12}M_{33}$$
$$+ M_{21}M_{32}M_{13} + M_{31}M_{12}M_{23} - M_{31}M_{22}M_{13}$$

Exercise 101 Calculate the determinants of these matrixes, the first over $\mathbb{C}$, the second over $\mathbb{Z}[X]$:

$$\begin{pmatrix} 2 - i \cdot 3 & \sqrt{3} + i \\ 3 & 1 - i \end{pmatrix} \qquad \begin{pmatrix} X^2 + 2 & 3X - 1 & 0 \\ -X & X^3 + 5 & 4 \\ 0 & X + 12 & -8 \end{pmatrix}$$

Show that for any positive n, $\det(E_n) = 1$.

Proposition 180 *For a ring R and a positive natural number n, we have these properties:*

(i) *If $M \in \mathbb{M}_{n,n}(R)$, then $\det(M) = \det(M^\top)$.*

(ii) *(Column Additivity) If for a column $M_{\bullet j}$ of $M \in \mathbb{M}_{n,n}(R)$, we have $M_{\bullet j} = N + L$, and if $M|N$ is the matrix obtained from M after replacing $M_{\bullet j}$ by N, while $M|L$ is the matrix obtained from M after replacing $M_{\bullet j}$ by L, then $\det(M) = \det(M|N) + \det(M|L)$.*

(iii) *(Row Additivity) If for a row $M_{i\bullet}$ of $M \in \mathbb{M}_{n,n}(R)$, we have $M_{i\bullet} = N + L$, and if $M|N$ is the matrix obtained from M after replacing $M_{i\bullet}$ by N, while $M|L$ is the matrix obtained from M after replacing $M_{i\bullet}$ by L, then $\det(M) = \det(M|N) + \det(M|L)$.*

(iv) *(Column Homogeneity) If for a column $M_{\bullet j}$ of $M \in \mathbb{M}_{n,n}(R)$, we have $M_{\bullet j} = \lambda \cdot N$, and if $M|N$ is the matrix obtained from M after replacing $M_{\bullet j}$ by N, then $\det(M) = \lambda \cdot \det(M|N)$.*

(v) *(Row Homogeneity) If for a row $M_{i\bullet}$ of $M \in \mathbb{M}_{n,n}(R)$, we have $M_{i\bullet} = \lambda \cdot N$, and if $M|N$ is the matrix obtained from M after replacing $M_{i\bullet}$ by N, then $\det(M) = \lambda \cdot \det(M|N)$.*

(vi) *(Column Skew Symmetry) If M' is obtained from $M \in \mathbb{M}_{n,n}(R)$ by exchanging two columns $M_{\bullet j}, M_{\bullet k}$, with $j \neq k$, then $\det(M') = -\det(M)$.*

(vii) *(Row Skew Symmetry) If M' is obtained from $M \in \mathbb{M}_{n,n}(R)$ by exchanging two rows $M_{i\bullet}, M_{k\bullet}$, with $i \neq k$, then $\det(M') = -\det(M)$.*

(viii) *(Column Equality Annihilation) If in $M \in \mathbb{M}_{n,n}(R)$, we have two equal columns $M_{\bullet j} = M_{\bullet k}$ with $j \neq k$, then $\det(M) = 0$.*

(ix) *(Row Equality Annihilation) If in $M \in \mathbb{M}_{n,n}(R)$, we have two equal rows $M_{i\bullet} = M_{k\bullet}$ with $i \neq k$, then $\det(M) = 0$.*

(x) *(Uniqueness of Determinant) Any function $D : \mathbb{M}_{n,n}(R) \to R$ with properties (ii), (iv), (viii) is uniquely determined by its value $D(E_n)$, and then we have $D(M) = D(E_n) \cdot \det(M)$.*

(xi) *(Product Rule for Determinants) If $M, N \in \mathbb{M}_{n,n}(R)$, then*

$$\det(M \cdot N) = \det(M) \cdot \det(N).$$

(xii) *(General Linear Group Homomorphism) The determinant function induces a homomorphism*

$$\det : GL_n(R) \to R^*$$

onto the multiplicative group R^* of invertible elements of R. Its kernel is denoted by $SL_n(R)$ and called the special linear group of size *n, whence $GL_n(R)/SL_n(R) \overset{\sim}{\to} R^*$.*

(xiii) *(Invariance under Conjugation) If $M \in \mathbb{M}_{n,n}(R)$ and $C \in GL_n(R)$, then*

$$\det(C \cdot M \cdot C^{-1}) = \det(M),$$

the matrix $C \cdot M \cdot C^{-1}$ being called the C-conjugate of M.[1]

---

[1] Do not confuse matrix conjugation with the conjugation of a complex number.

Proof We have

$$
\begin{aligned}
\det(M) &= \sum_{\pi \in S_n} (-1)^{sig(\pi)} \prod_{j=1...n} M_{\pi(j)j} \\
&= \sum_{\pi \in S_n} (-1)^{sig(\pi)} \prod_{\pi(j)=1...n} M_{\pi(j)\pi^{-1}(\pi(j))} \\
&= \sum_{\pi \in S_n} (-1)^{sig(\pi)} \prod_{j=1...n} M_{j\pi^{-1}(j)} \\
&= \sum_{\pi \in S_n} (-1)^{sig(\pi)} \prod_{j=1...n} M_{j\pi(j)} \\
&= \sum_{\pi \in S_n} (-1)^{sig(\pi)} \prod_{j=1...n} M^{\tau}_{\pi(j)j} \\
&= \det(M^{\tau}),
\end{aligned}
$$

whence (i).

If we have $M_{ij} = N_i + L_i$, for all i, then for each product in the determinant function, we have

$$
\begin{aligned}
\prod_{t=1...n} M_{\pi(t)t} &= M_{\pi(j)j} \prod_{t \neq j} M_{\pi(t)t} \\
&= N_{\pi}(j) \prod_{t \neq j} M_{\pi(t)t} + L_{\pi}(j) \prod_{t \neq j} M_{\pi(t)t} \\
&= (M|N)_{\pi(j)j} \prod_{t \neq j} M_{\pi(t)t} + (M|L)_{\pi(j)j} \prod_{t \neq j} M_{\pi(t)t} \\
&= (M|N)_{\pi(j)j} \prod_{t \neq j} (M|N)_{\pi(t)t} + (M|L)_{\pi(j)j} \prod_{t \neq j} (M|N)_{\pi(t)t} \\
&= \prod_{t} (M|N)_{\pi(t)t} + \prod_{t} (M|L)_{\pi(t)t}
\end{aligned}
$$

whence (ii).

For (iii), we observe that by (i), we can use (ii) to tackle the problem.

For column homogeneity (iv), we observe that

$$
\begin{aligned}
\prod_{t} M_{\pi(t)t} &= M_{\pi(j)j} \prod_{t \neq j} M_{\pi(t)t} \\
&= \lambda(N_{\pi(j)} \prod_{t \neq j} M_{\pi(t)t}) \\
&= \prod_{t} (M|N)_{\pi(t)t}.
\end{aligned}
$$

Row homogeneity (v) follows from (iv) and (i).

Suppose that (viii) is true. Then (vi) follows immediately. In fact, take a matrix M and two column indexes k and j. Take the new matrix M' which is derived from M by adding column $M_{\bullet k}$ to column $M_{\bullet j}$ and adding $M_{\bullet j}$ to column $M_{\bullet k}$. Then by (viii), $\det(M') = 0$. But by (iii), $0 = \det(M') = \det(M|k|k) + \det(M|j|j) +$

$\det(M|j|k) + \det(M)$, where $M|k|k$ is the matrix derived from M, where we have the k-th column $M_{\bullet k}$ at column positions k, j, and $M|j|j$ is the matrix derived from M, where we have the j-th column $M_{\bullet j}$ at column positions k and j, while $M|j|k$ is the matrix where the k-th and j-th columns of M have been exchanged. But by (viii)$\det(M|k|k) = \det(M|j|j) = 0$, whence $\det(M|j|k) = -\det(M)$. Also, by (i) and (vi), (vii) follows.

To prove (viii), recall that the number of even permutations is $n!/2$ and that for each even permutation π there is an odd permutation $\pi^* = (\pi(k), \pi(j)) \circ \pi$. This gives us a bijection from even to odd permutations. But then, the product $\prod_t M_{\pi(t)t}$ is equal to the product $\prod_t M_{\pi^*(t)t}$ since both columns at positions k and j are equal. So by the change of signs, i.e., $(-1)^{\pi^*} = -(-1)^{\pi}$, these products neutralize each other. Further, by (i) and (viii), (ix) follows.

Claim (x) is demonstrated as follows: Each column is the sum of special columns N, where only one coefficient N_i is possibly different from zero. So the function $D(M)$ is determined on matrixes having only one coefficient possibly different from zero. But such a column N is also the scaling $N = N_i \cdot N'$, where N' has coefficient 1 instead of N_i, and zeros else. So by homogeneity in columns, the function D is determined by its values on matrixes which have only columns with a 1 and zero coefficients else. Now, by (viii), the value of our function D must vanish if two columns are equal. If not, the matrix M results from a permutation π of the columns of E_n, and the value must be $D(M) = (-1)^{sig(\pi)}D(E_n)$. Now, since $D(E_n)\cdot\det(M)$ has all the properties (ii),(iv),(viii) of D, and $D(E_n)\cdot\det(E_n) = D(E_n)$ we must have $D(M) = D(E_n)\det(M)$.

Suppose for (xi) that $M, N \in \mathbb{M}_{n,n}(R)$. Then, fixing M, the function $D(N) = \det(M \cdot N)$ evidently has the properties (ii), (iv), and (viii) by the laws of matrix multiplication. Since its value for $N = E_n$ is $\det(M)$, we are done by (x). Further, since by (xi) the determinant commutes with products of matrixes, it sends the product $E_n = M^{-1} \cdot M$ to $1 = \det(M^{-1}) \cdot \det(M)$, i.e., we have a group homomorphism $GL_n(R) \to R^*$, which is surjective, since the matrix $M = (\lambda - 1)E(1\,1) + E_n$ has $\det(M) = \lambda$. Claim (xiii) is evident from claim (xii). $\qquad \square$

The calculation of the inverse of an invertible square matrix uses the determinant function in a rather complex way. We first establish the necessary auxiliary structures.

Definition 159 *For positive n, we denote by $_iM^j$ the $(n-1) \times (n-1)$-matrix derived from $M \in \mathbb{M}_{n,n}(R)$ by the cancellation of its i-th row and j-th column. For $n > 1$, the determinant $\det(_iM^j)$ is called the ij-minor of M. The number $cof(M)_{ij} = (-1)^{i+j} \det(_jM^i)$ is called the ij-cofactor of M. The matrix $Ad(M) = (cof(M)_{ij}) \in \mathbb{M}(R)$ is called the adjoint of M. If $n = 1$, we set $Ad(M) = E_1$.*

Lemma 181 *For a matrix $M \in \mathbb{M}_{n,n}(R)$, of size $n > 1$, if $1 \le i \le n$ is a row index, then*

$$\det(M) = \sum_{j=1,\dots n} M_{ij} cof(M)_{ji}.$$

If $1 \le j \le n$ is a column index, then

$$\det(M) = \sum_{i=1,\dots n} M_{ij} cof(M)_{ji}.$$

Proof We have

$$\det(M) = \sum_{\pi \in S_n} (-1)^{sig(\pi)} \prod_{t=1\dots n} M_{\pi(t)t}$$

$$= \sum_{i=1,\dots n} M_{ij} \sum_{\pi \in S_n, \pi(j) \neq i} (-1)^{sig(\pi)} \prod_{t \neq j} M_{\pi(t)t}.$$

The factor $\sum_{\pi \in S_n, \pi(j) \neq i} (-1)^{sig(\pi)} \prod_{t \neq j} M_{\pi(t)t}$ is easily seen to be $cof(M)_{ji}$, whence the second formula. The first follows from the invariance of the determinant under transposition. □

Proposition 182 (Cramer's Rule) *For a matrix $M \in \mathbb{M}_{n,n}(R)$, of positive size n, we have the following equation:*

$$M \cdot Ad(M) = \det(M) \cdot E_n.$$

Proof Cramer's rule is an immediate consequence of the lemma 181. If we take the formula $\sum_{j=1,\dots n} M_{ij} cof(M)_{ji}$ and change the coefficient i to $k \neq i$, then this is the same formula for the matrix deduced from M, where the row at index i is replaced by the row at index k. But such a matrix has determinant zero by row equality annihilation (ix) in proposition 180. So the Cramer formula results. □

Proposition 183 *A matrix $M \in \mathbb{M}_{n,n}(R)$, of positive size n is invertible iff $\det(M) \in R^*$. In particular, if R is a field, this means that $\det(M) \neq 0$. If M is invertible, then the inverse is given by this formula:*

$$M^{-1} = \frac{1}{\det(M)} Ad(M).$$

Proof If M is invertible, we know that $\det(M) \in R^*$. Conversely, if $\det(M) \in R^*$, then Cramer's formula yields that the inverse is given by $M^{-1} = \frac{1}{\det(M)} Ad(M)$. □

Exercise 102 Decide whether the matrix

$$M = \begin{pmatrix} 25 & -1 \\ 12 & 5 \end{pmatrix}$$

over $\mathbb{Z}$ is invertible. Is its image M mod 12 over $\mathbb{Z}_{12}$ invertible? Try to calculate the adjoint and, if M is invertible, the inverse matrix of M.

Exercise 103 If $M \in \mathbb{M}_{n,n}(R)$ is an upper triangular matrix, i.e., $M_{ij} = 0$ for all $i > j$, then $\det(M) = \prod_{i=1...n} M_{ii}$.

Exercise 104 As a special case of matrix multiplications, we have already mentioned linear equations, such as shown in example 84 above. We are now in a position to solve such an equation

$$\begin{pmatrix} 3.7 \\ -8 \\ 0 \\ 1 \end{pmatrix} = \begin{pmatrix} 23 & -1 & 0 & 45 \\ 0.9 & 9.6 & 1 & -1 \\ 0 & 20 & -1 & 1 \\ 0 & 3 & 1 & -2 \end{pmatrix} \cdot \begin{pmatrix} x_1 \\ x_2 \\ x_3 \\ x_4 \end{pmatrix}.$$

In fact, if the coefficient matrix

$$\begin{pmatrix} 23 & -1 & 0 & 45 \\ 0.9 & 9.6 & 1 & -1 \\ 0 & 20 & -1 & 1 \\ 0 & 3 & 1 & -2 \end{pmatrix}$$

is invertible, then the solution is

$$\begin{pmatrix} x_1 \\ x_2 \\ x_3 \\ x_4 \end{pmatrix} = \begin{pmatrix} 23 & -1 & 0 & 45 \\ 0.9 & 9.6 & 1 & -1 \\ 0 & 20 & -1 & 1 \\ 0 & 3 & 1 & -2 \end{pmatrix}^{-1} \cdot \begin{pmatrix} 3.7 \\ -8 \\ 0 \\ 1 \end{pmatrix}.$$

Proposition 184 *If $f : R \to S$ is a ring homomorphism of commutative rings, then for a matrix $M \in \mathbb{M}_{n,n}(R)$ of positive size $n > 0$, we have*

$$\det(f(M)) = f(\det(M)).$$

Proof We have $\det(f(M)) = \det(f((M_{ij})))$, but the determinant is a sum of products of the images $f(M_{ij})$, and since f is a ring homomorphism, we have $\det(f((M_{ij}))) = f(\det((M_{ij})))$. □

In particular, let $R = T[X]$ be the polynomial algebra over the commutative ring T with the indeterminate X. By the universal property of the polynomial algebra, if $N \in \mathbb{M}_{n,n}(T)$, we have a unique ring homomorphism $f : R \to \mathbb{M}_{n,n}(T)$ defined by $X \mapsto N$. Its image ring S is commutative, since the polynomial ring $T[X]$ is so. The ring S consists of all polynomials in N with coefficients in T. Therefore, by proposition 184, $\det(f(M)) = f(\det(M))$ for any matrix $M \in \mathbb{M}_{n,n}(T[X])$. In particular, if $M = N - X \cdot E_n$ for $N \in \mathbb{M}_{n,n}(T)$ for the unit matrix E_n in $\mathbb{M}_{n,n}(T)$, then $\det(M) = P(X) \in T[X]$, and we have $f(P(X)) = P(N)$. But also $f(P(X)) = f(\det(M)) = \det(f(M)) = \det(f(N - X \cdot E_n)) = \det(N - N) = 0$. Therefore,

$$P(N) = 0$$

which means that N is a solution of the polynomial equation $P(X) = 0$. This somewhat tricky, but immediate corollary of proposition 184 is the famous equation of Cayley-Hamilton, together with a number of evident corollaries from the preceding results:

Corollary 185 *If T is a (commutative) ring, and if $N \in \mathbb{M}_{n,n}(T)$ for positive n, the polynomial $\chi_N = \det(N - X \cdot E_n)$, $E_n \in \mathbb{M}_{n,n}(T)$ being the unit matrix of size n over T, and X an indeterminate, is called the* characteristic *polynomial of N. Then we have the Cayley-Hamilton equation*

$$\chi_N(N) = 0,$$

moreover,

$$\chi_N(0) = \det(N).$$

The coefficients of the characteristic polynomial are invariant under conjugation, i.e., if $C \in GL_n(T)$, then $\chi_N = \chi_{C \cdot N \cdot C^{-1}}$.

Proof Everything is already proved, except the invariance of the characteristic polynomial under conjugation. But $\chi_{C \cdot N \cdot C^{-1}} = \det(C \cdot N \cdot C^{-1} - X \cdot E_n) = \det(C \cdot N \cdot C^{-1} - X \cdot C \cdot C^{-1}) = \det(C \cdot (N - X \cdot E_n) \cdot C^{-1}) = \det(N - X \cdot E_n) = \chi_N$. □

Exercise 105 Calculate the characteristic polynomial χ_N for the matrix

$$N = \begin{pmatrix} -1 & 0 \\ 2 & 3 \end{pmatrix}$$

over the integers. Verify the Cayley-Hamilton equation $\chi_N(N) = 0$.

CHAPTER 21

Modules and Vector Spaces

If the matrixes are the backbones of linear algebra, here is the flesh: the axiomatic theory of modules and vector spaces, which encompasses a large variety of comparable structures; they will eventually be cast into matrixes in many interesting cases. The setup is drawn from the historic approach of René Descartes in his analytic geometry. *In this chapter, we shall again stick to commutative rings except when we explicitly express the contrary.*

Definition 160 *Let R be a ring, then a left R-module is a triple $(R, M, \mu : R \times M \to M)$, where M is an additively written abelian group of so-called* vectors, *and μ is the* scalar multiplication, *usually written as $\mu(r, m) = r \cdot m$ if μ is clear, with these properties:*

(i) *We have $1 \cdot m = m$, for all $m \in M$.*

(ii) *For all $r, s \in R$ and $m, n \in M$, we have $(r + s) \cdot m = r \cdot m + s \cdot m$, $r \cdot (m + n) = r \cdot m + r \cdot n$ and $r \cdot (s \cdot m) = (rs) \cdot m$.*

Given an R-module M, a subgroup $S \subset M$ such that for each $r \in R$ and $s \in S, r \cdot s \in S$, is called a submodule *of M; it is also an R-module on its own right. If the ring R is a field, the module is called a R-vector space.*

Exercise 106 Show that by statement (ii) of definition 160, one always has $0 \cdot m = r \cdot 0 = 0$ in a module.

Example 87 In the course of the last chapter, we have encountered plenty of modules: Each set $M = \mathbb{M}_{m,n}(R)$, together with the sum of matrixes and the scalar multiplication by ring elements defined in sorite 176 is an R module. In particular, we have zero modules $\mathbb{M}_{0,n}(R)$, $\mathbb{M}_{m,0}(R)$, $\mathbb{M}_{0,0}(R)$, each consisting of a zero group and the uniquely possible scalar multiplication.

But the fundamental idea of defining matrixes as R-valued functions on certain domains ($[1, m] \times [1, n]$ for matrixes) can easily be generalized: Take any set D and consider the set R^D of functions on D with values in R. Then the addition $f + g$ of $f, g \in R^D$ defined by $(f + g)(d) = f(d) + g(d)$ and the scalar multiplication $(r \cdot f)(d) = r \cdot f(d)$ for $r \in R$ defines a module, which generalizes the idea for matrixes.

An important subset is also a module under the same addition and scalar multiplication: Take the set $R^{(D)}$ of functions $f : D \to R$ such that $f(d) = 0$ except for at most a finite number of arguments. For example, if we consider the monoid algebra $R\langle M \rangle$ of a monoid M, this is precisely $R^{(M)}$, and the sum of its elements is the one we just defined. Moreover, the identification of R-elements r with the element $1 \cdot e_M$ yields the scalar multiplication $r \cdot f$ for elements $f \in R\langle M \rangle$.

This idea generalizes as follows: We may view R as an R-module over itself: The vectors are the elements of R, the sum is the given sum in R, and the scalar multiplication is the ring multiplication. This module is of course "isomorphic" to $\mathbb{M}_{1,1}(R)$ (we shall define what a module isomorphism is in a few lines from here). It is called the *free R-module of dimension one* and denoted by $_R R$, or also by R if the context is clear. With this special module in mind, suppose we are given a module M over R. Then for any set D, we have the module $M^{(D)}$ whose elements $f : D \to M$ vanish except for a finite set of arguments d, and where sum and scalar multiplication are again defined point-wise, i.e., $(f + g)(d) = f(d) + g(d), (r \cdot f)(d) = r \cdot f(d)$. This module is called the *direct sum of D copies of M* and usually denoted by $M^{\oplus D}$; in the special case where $D = n$ is a natural number, $M^{\oplus n}$ is also written as M^n. We now recognize part of the structure of complex numbers $\mathbb{C}$ as being the module $\mathbb{R}^2$. In fact, addition of complex numbers is the vector addition on $\mathbb{R}^2$, whereas the multiplication of a real number r with a complex number z plays the role of the scalar multiplication $r \cdot z$. The special case $M = {}_R R$ has been introduced above in the module $R^{(D)}$! We also recognize that the matrix module $\mathbb{M}_{m,n}(R)$ identifies with $R^{\oplus [1,m] \times [1,n]}$.

If we are given a finite family $(M_i)_{i=1,\ldots n}$ of R-modules, the Cartesian product $M_1 \times \ldots M_n$ is given a module structure as follows: Vector addition and scalar multiplication are defined component-wise, i.e.,

$$(m_1, \ldots m_n) + (m'_1, \ldots m'_n) = (m_1 + m'_1, \ldots m_n + m'_n),$$

and

$$r \cdot (m_1, \ldots m_n) = (r \cdot m_1, \ldots r \cdot m_n).$$

This module is denoted by $\bigoplus_{i=1,\ldots n} M_i$ and is called the direct sum of the modules $M_1, \ldots M_n$.

The following example is a very comfortable restatement of abelian groups in terms of modules: Each abelian group M is automatically a $\mathbb{Z}$-module by the following scalar multiplication: One defines $z \cdot m = m + m + \ldots m$, z times, for $z > 0$, $-((-z) \cdot m)$ for $z < 0$, and $0 \cdot m = 0$. Check that this construction satisfies the module axioms.

Already after these first examples one recognizes that many modules essentially are manifestations of the same structure. And this is why we once more invoke the "catechism" of morphisms, which was already performed for sets, digraphs, rings, and automata:

Definition 161 *If $(R, M, \mu : R \times M \to M)$ and $(R, N, \nu : R \times N \to N)$ are two R-modules, an R-linear homomorphism $f : M \to N$ is a group homomorphism such that for all $(r, m) \in R \times M$, $f(\mu(r, m)) = \nu(r, f(m))$. If no ambiguity about the scalar multiplications in M and in N is likely, one uses the dot notation, and then linearity reads as $f(r \cdot m) = r \cdot f(n)$.*

The set of R-linear homomorphisms $f : M \to N$ is denoted by $Lin_R(M, N)$. By point-wise addition and scalar multiplication, $Lin_R(M, N)$ is also provided with the structure of an R-module, which we henceforth tacitly assume.

If L is a third R-module, the composition $g \circ f$ of two R-linear homomorphisms $f \in Lin_R(M, N)$ and $g \in Lin(N, L)$ is defined by their set-theoretic composition; it is again R-linear.

For $M = N$, one writes $End_R(M) = Lin_R(M, M)$ and calls its elements R-module endomorphisms. In particular, the identity $Id_M : M \to M$ is an R-module endomorphism on M.

An R-linear homomorphism $f : M \to N$ is called an isomorphism if it has an inverse $g : N \to M$ such that $g \circ f = Id_M$ and $f \circ g = Id_N$. Evidently, this is the case iff the underlying group homomorphism is a group isomorphism; the inverse is uniquely determined and denoted by $g = f^{-1}$. If, moreover, $M = N$, an isomorphism is called automorphism.

Exercise 107 Show that the matrix R-module $\mathbb{M}_{m,n}(R)$ is isomorphic to R^{mn}. In particular, the column matrix module $\mathbb{M}_{m,1}(R)$ and the row matrix module $\mathbb{M}_{1,m}(R)$ are both isomorphic to R^m, and all modules $\mathbb{M}_{0,n}(R)$, $\mathbb{M}_{m,0}(R)$, $\mathbb{M}_{0,0}(R)$ and R^0 are isomorphic, i.e., they are trivial R-modules.

Exercise 108 Show that the ring $Lin_R(R,R)$ is isomorphic to R, in particular, observe that therefore these objects are also isomorphic as R-modules.

Sorite 186 *If* $f \in Lin_R(N,L)$ *and* $g \in Lin_R(M,N)$, *then*

(i) *if* $f = f_1 + f_2$, *then* $f \circ g = (f_1 + f_2) \circ g = f_1 \circ g + f_2 \circ g$, *whereas for* $g = g_1 + g_2$, $f \circ g = f \circ (g_1 + g_2) = f \circ g_1 + f \circ g_2$.

(ii) *If* $r \in R$, *then* $r \cdot (f \circ g) = (r \cdot f) \circ g = f \circ (r \cdot g)$.

(iii) *With the addition and composition of linear endomorphisms on M, the set $End_R(M)$ is a (generally not commutative) ring. If $M \neq 0$, R identifies with the subring $R \cdot Id_M$ by the ring isomorphism $R \xrightarrow{\sim} R \cdot Id_M : r \mapsto r \cdot Id_M$, and it commutes with every endomorphism. The group of invertible elements in $End_R(M)$ is denoted by $\mathrm{GL}(M)$, it is called the* general linear group *of M.*

Proof Let $g : M \to N$ and $f : N \to L$ be R-linear homomorphisms. If $f = f_1 + f_2$, then for $x \in M$, $((f_1 + f_2) \circ g)(x) = (f_1 + f_2)(g(x)) = f_1(g(x)) + f_2(g(x)) = (f_1 \circ g)(x) + (f_2 \circ g)(x) = (f_1 \circ g) + f_2 \circ g)(x)$. If $g = g_1 + g_2$, then $(f \circ (g_1 + g_2))(x) = f((g_1 + g_2)(x)) = f(g_1(x) + g_2(x)) = f(g_1(x)) + f(g_2(x)) = (f \circ g_1)(x) + (f \circ g_2)(x) = (f \circ g_1 + f \circ g_2)(x)$, whence (i).

If $r \in R$, then $(r \cdot (f \circ g))(x) = r \cdot ((f \circ g)(x)) = r \cdot (f(g(x)) = r \cdot (f(g(x))$. But also $((r \cdot f) \circ g)(x) = ((r \cdot f)(g(x)) = r \cdot f(g(x))$. Finally $(f \circ (r \cdot g))(x) = f((r \cdot g)(x)) = f(r \cdot g(x)) = r \cdot f(g(x))$, whence (ii).

Claim (iii) follows now immediately from (i) and (ii).   □

Example 88 Consider the R modules $\mathbb{M}_{n,1}(R)$ and $\mathbb{M}_{m,1}(R)$ of n- and m-element columns. Then a $m \times n$-matrix $M : E_n \to E_m$ defines a map $f_M : \mathbb{M}_{n,1}(R) \to \mathbb{M}_{m,1}(R)$ by the matrix multiplication $f_M(X) = M \cdot X$. This gives us an a posteriori justification of the functional notation of a matrix. The general laws of matrix multiplication stated in sorite 178 imply that this map is R-linear. Moreover, applying f_M to the elementary column $E(i\ 1) \in \mathbb{M}_{n,1}(R)$ gives us the column $M_{i\bullet}$ of M. Therefore the map $M \mapsto f_M$ is injective. We shall soon see that the map is often also surjective, i.e., the matrixes are essentially the same thing as linear homomorphisms!

This is not always the case, but for a large class of rings, the fields, and therefore for all vector spaces, this is true.

As for groups, one can also build quotient and image modules as follows:

Proposition 187 *Given an R-linear homomorphism $f : M \to N$, the image group $Im(f)$ is a submodule of N. The kernel $Ker(f)$ of the underlying group homomorphism is a submodule of M.*

If $S \subset M$ is a submodule of the R-module M, then the quotient group M/S is also an R-module by the scalar multiplication $r \cdot (m + S) = r \cdot m + S$. This will be called the quotient R-module.

Proof The fact that $Im(f)$ and $Ker(f)$ are submodules is immediate and left to the reader. If $S \subset M$ is a submodule, then the scalar multiplication is well defined, in fact, $m + S = m' + S$ iff $m - m' \in S$ But then $r \cdot m + S = r \cdot m' + S$ since $r \cdot m - r \cdot m' = r(m - m') \in S$. That this scalar multiplication verifies the module axioms, and that p is R-linear, is then immediate. □

And here is the universal property of quotient modules:

Proposition 188 *Let M be an R-module and $S \subset M$ a sub-R-module of M. Then for every R-module N, we have a bijection*

$$Lin_R(M/S, N) \xrightarrow{\sim} \{f \mid f \in Lin_R(M, N) \text{ and } S \subset Ker(f)\}.$$

Proof Let $p : M \to M/S$ be the canonical projection. If $g : M/S \to N$ is R-linear, then the composition $g \circ p : M \to N$ is in the set on the right hand side. Since p is surjective, the map $g \mapsto g \circ p$ is injective (see the characterization of surjections of sets in sorite 16). Conversely, if $f : M \to N$ is such that $S \subset Ker(f)$, then we may define a map $g : M/S \to N : m + S \mapsto f(m)$. Is this map well defined? If $m + S = m' + S$, then $m - m' \in S$, hence $f(m) - f(m') = f(m - m') = 0$. It is evidently R-linear, and we are done, since $f = g \circ p$. □

Here is the (double) universal property of the finite direct sum:

Proposition 189 *If $M_1, \ldots M_n$ is a finite family of R-modules M_i, we have R-linear injections $\iota_j : M_j \to \bigoplus_i M_i : m \mapsto (0, \ldots 0, m, 0, \ldots 0)$ for each $j = 1, \ldots n$, which map an element $m \in M_j$ to the n-tuple having zeros except at the j-th position, where the element m is placed. For any R-module N, this defines a bijection*

$$Lin_R(\bigoplus_i M_i, N) \xrightarrow{\sim} \bigoplus_i Lin_R(M_i, N)$$

between sets of homomorphisms defined by $f \mapsto (f \circ \iota_i)_i$.

Dually, we have R-linear projections $\pi_j : \bigoplus_i M_i \to M_i : (m_i)_i \mapsto m_j$, *for each* $j = 1, \ldots n$. *For any R-module N, this defines a bijection*

$$Lin_R(N, \bigoplus_i M_i) \xrightarrow{\sim} \bigoplus_i Lin_R(N, M_i)$$

between sets of homomorphisms defined by $f \mapsto (\pi_i \circ f)_i$.

Proof The isomorphisms $f \mapsto (f \circ \iota_i)_i$ and $f \mapsto (\pi_i \circ f)_i$ indicated in the proposition allow an immediate verification of the claims. We leave the details to the reader. $\qquad\Box$

Exercise 109 Suppose we are given two subspaces $U, V \subset M$ and consider the homomorphism $f : U \oplus V \to M$ guaranteed by proposition 189 and these two inclusions. Then show that f is an isomorphism iff (i) $U \cap V = 0$ and (ii) $M = U + V$, which means that every $m \in M$ is a sum $m = u + v$ of $u \in U$ and $v \in V$. In this case M is also called the *inner direct sum of U and V*, and U and V are said to be *complements* to each other.

Remark 24 If we consider the two injections $i_M : M \to M \oplus N$ and $i_N : N \to M \oplus N$, an element $m \in M$ is mapped to $i_M(m) = (m, 0)$, while an element $n \in N$ is mapped to $i_N(n) = (0, n)$. If it is clear from the context that m belongs to the direct summand M and n to N, then one also identifies m with $i_M(m)$ and n to $i_N(n)$. With this identification, one may add m to n and write $m + n$ instead of $i_M(m) + i_N(n)$. In the sequel, this comfortable and economic notation will often be used without special emphasis.

This proposition is the germ of the reduction of linear algebra to matrix algebra for a class of modules called *free*:

Definition 162 For a finite number n, an R-module M is called free of dimension n if it is isomorphic to R^n.

Attention: for general rings, a module is not necessarily free. A very simple example is a finite abelian group, such as $\mathbb{Z}_n$, which, as a $\mathbb{Z}$-module, cannot be free since any free non-zero $\mathbb{Z}$ module is infinite. At present, you cannot know whether the dimension of a free module is uniquely determined. That it is in fact unique is shown by this result:

Proposition 190 *If an R-module M is free of dimension n and of dimension m, then n = m, and this uniquely defined dimension is also called the* $dim(M)$.

Proof This follows from the properties of the determinant. Let $f : R^n \to R^m$ be an R-linear isomorphism with $n < m$. Any element $x = (x_1, \ldots x_n) \in R^n$ can be written as $x = \sum_{i=1,\ldots n} x_i e_i$, where $e_i = (0, \ldots, 0, 1, 0, \ldots 0)$ has 1 at the i-th position. Then since every element $y \in R^m$ is an image under f, it can be written as $y = \sum_i x_i f(e_i)$. Now, consider the unit matrix E_m, for which we have $\det(E_m) = 1$. By the above, we may write each row $E(1\ i) \in R^m$ as $E(1\ j) = \sum_{i=1,\ldots n} x(j)_i f(e_i), x(j)_i \in R$, i.e., as a combination of less than m row vectors $f(e_i), i = 1, \ldots n$. Therefore, by the properties of the determinant, especially equal row annihilation, the determinant must vanish, a contradiction, therefore $n \geq m$, and a symmetric argument shows $m \geq n$, whence the claim. $\square$

It will be shown in the course of the next chapter that every vector space has a dimension.

Proposition 191 *If M is a free R-module of dimension n, and if N is a free R-module of dimension m, then the R-module $Lin_R(M, N)$ is isomorphic to* $\mathbb{M}_{m,n}(R)$, *i.e., free of dimension mn.*

Proof We may wlog suppose that $M = R^n$ and $N = R^m$. Then by proposition 189, we have an R-linear isomorphism $Lin_R(M, N) \overset{\sim}{\to} \bigoplus_{i=1,\ldots m, j=1,\ldots n} Lin_R(R, R)$, with $Lin_R(R, R) \overset{\sim}{\to} R$, and therefore $dim(Lin_R(M, N)) = n \cdot m$. One now maps the homomorphism f defined by the sequence $(m_{ij})_{i=1\ldots m, j=1,\ldots n} \in \bigoplus_{i=1,\ldots m, j=1,\ldots n} R$ to the matrix M_f with $(M_f)_{ij} = m_{ij}$. This map is evidently a linear bijection. $\square$

We are now in the position to define the determinant of any linear endomorphism of a module M of dimension n by the following observation: If we have a free R-module M of dimension n and a linear endomorphism $f : M \to M$, then, if $u : M \to R^n$ is an isomorphism, we may consider the linear homomorphism $u \circ f \circ u^{-1} : R^n \to R^n$. This corresponds to a matrix $M_{f,u} \in \mathbb{M}_{n,n}(R)$. If we take another isomorphism $u' : M \to R^n$, then we have the corresponding matrix $M_{f,u'} \in \mathbb{M}_{n,n}(R)$, and it easily follows that

$$M_{f,u'} = (u'u^{-1}) \cdot M_{f,u} \cdot (u'u^{-1})^{-1}.$$

Therefore, the determinant of f, if defined by

$$\det(f) = \det(M_{f,u})$$

is well defined by our previous result on conjugation of matrixes, see (xiii) of theorem 180.

Free modules therefore are fairly transparent, since their theory seems to reduce to matrix theory. However, we still have some unknown situations even with this easy type of modules: For example, if $f : R^n \to R^m$ is an R-linear map, what is the structure of $Ker(f)$ or $Im(f)$? Are these modules free? In general, they are not, but again, for vector spaces, this is true. This will be shown in the next chapter.

CHAPTER 22

Linear Dependence, Bases, and Dimension

In practice, modules often do not occur as free constructions, but as subspaces, more precisely: kernels or even quotient spaces related to linear homomorphisms. For example, if we are given a matrix $M \in \mathbb{M}_{m,n}(R)$, the corresponding linear homomorphism $f_M : R^n \to R^m$ has a kernel $Ker(f_M)$ which plays a crucial role in the theory of linear equations. Here is the relation. A linear equation is a matrix equation of this type:

$$\begin{pmatrix} y_1 \\ y_2 \\ \vdots \\ y_m \end{pmatrix} = \begin{pmatrix} a_{11} & a_{12} & \cdots & a_{1n} \\ a_{11} & a_{12} & \cdots & a_{1n} \\ \vdots & \vdots & & \vdots \\ a_{m1} & a_{m2} & \cdots & a_{mn} \end{pmatrix} \cdot \begin{pmatrix} x_1 \\ x_2 \\ \vdots \\ x_n \end{pmatrix}$$

where the matrixes (y_i) and $M = (a_{ij})$ are given and one looks for the unknown column matrix (x_i). One reinterprets this equation by the linear homomorphism $f_M : R^n \to R^m$ associated with M. We are given an element $y = (y_1, y_2, \ldots y_m) \in R^m$ and look for the inverse image $f_M^{-1}(y)$ of y under f_M. The solutions of the above equations are by definition the elements $x \in f_M^{-1}(y)$. So this solution space is structured as follows: If x_0 is a special solution, the solution space is

$$f_M^{-1}(y) = x_0 + Ker(f_M).$$

This means that we have to deal with these two problems: (1) Decide if there is at least one special solution and possibly find it. (2) Describe the kernel of f_M.

22.1 Bases in Vector Spaces

To tackle these problems, *from now on until the end of the linear algebra subject of this part we shall restrict our theory to vector spaces, i.e., to modules over fields R.*

Definition 163 *A finite sequence* $(x_i) = (x_1, x_2, \ldots x_k), k \geq 1,$ *of elements* $x_i \in M$ *of an R-vector space M is called* linearly independent *if one of the equivalent properties hold:*

(i) *A linear combination* $\sum_{i=1,\ldots k} \lambda_i x_i$ *of the vectors* x_i *equals 0, iff we have* $\lambda_i = 0$ *for all scalars* λ_i.

(ii) *The R-linear homomorphism* $f : R^k \to M$ *defined by* $f(\lambda_1 \ldots \lambda_k) = \sum_{i=1,\ldots k} \lambda_i x_i$ *is injective, i.e.,* $Ker(f) = 0$.

If (x_i) *is not linearly independent, the sequence is called* linearly dependent.

Exercise 110 Give a proof of the equivalence of the properties in definition 163.

Exercise 111 A linearly independent sequence (x_i) cannot contain the zero vector, nor can it contain the same vector twice. If (x_i) is linearly independent, then so is every permutation of this sequence. This means that linear independence is essentially a property of the underlying set $\{x_i, i = 1, \ldots k\}$ of vectors. However, there are many reasons to keep the sequential setup in this and also in the following definitions of generators and bases.

Exercise 112 Show that in the real vector space $\mathbb{R}^{(\mathbb{N})}$, every sequence $(e_i)_{i=0,1,2,\ldots k}, k \geq 1$ with $e_i = (0, \ldots 0, 1, 0, \ldots)$ having a 1 exactly in position $i \in \mathbb{N}$ and 0 else, is linearly independent.

Exercise 113 Show that in a free vector space R^n of dimension n, the sequence $(x_i)_{i=1,2,\ldots n}$ of the vectors $x = (1, 1, \ldots, 1, 0, 0 \ldots 0)$ which have 1 entries up to and including the i-th coordinate, and 0 after, is linearly independent.

Exercise 114 Consider the $\mathbb{Q}$-vector space $\mathbb{R}$, defined by the usual addition of "vectors", i.e., real numbers, and the usual scalar multiplication, but restricted to rational scalars. Show that the two vectors 1 and $\sqrt{2}$

are linearly independent over $\mathbb{Q}$. Use our results in exercise 68 about the irrationality of $\sqrt{2}$.

Definition 164 *A finite sequence* $(x_i) = (x_1, x_2, \ldots x_k), k \geq 1$, *of elements* $x_i \in M$ *of an* R-*vector space* M *is said to generate* M *if one of the equivalent properties hold:*

(i) *The vector space* M *equals the subspace of all linear combinations* $\sum_{i=1,\ldots k} \lambda_i x_i$ *of the vectors* x_i *(also called the space generated by* (x_i)*).*

(ii) *The* R-*linear homomorphism* $f : R^k \to M$ *defined by* $f(\lambda_1 \ldots \lambda_k) = \sum_{i=1,\ldots k} \lambda_i x_i$ *is surjective, i.e.,* $Im(f) = M$.

A vector space M *is called finitely generated if there is a finite sequence* $(x_i) = (x_1, x_2, \ldots x_k), k \geq 1$ *of elements* $x_i \in M$ *which generates* M.

Exercise 115 Give a proof of the equivalence of the properties in definition 164.

Exercise 116 Consider the $\mathbb{R}$-vector space $M = \mathbb{R}[X, Y]/(X^{12}, Y^{12})$. Show that it is generated by the images of $X^i Y^j, 0 \leq i, j, \leq 11$, in M.

Definition 165 *A finite sequence* $(x_i) = (x_1, x_2, \ldots x_k), k \geq 1$ *of elements* $x_i \in M$ *of an* R-*vector space* M *is called a* basis *of* M *iff it is linearly independent and generates* M. *Equivalently,* (x_i) *is a basis, iff the* R-*linear homomorphism* $f : R^k \to M : (\lambda_1 \ldots \lambda_k) \mapsto \sum_{i=1,\ldots k} \lambda_i x_i$ *is an isomorphism. Since by proposition 190, the dimension of* M *is uniquely determined, every basis of* M *must have the same number of elements, i.e.,* $k = dim(M)$.

Remark 25 We have excluded the zero vector spaces here, because in those no finite sequence $(x_1, \ldots x_k), k \geq 1$, can be linearly independent. To complete the general terminology, one also says that the empty sequence is linearly independent, and that it forms a basis for a zero space, but this is merely a convention.

Exercise 117 Show that in a free vector space R^n of dimension n, the sequence $(x_i)_{i=1,2,\ldots n}$ of the vectors $x = (1, 1, \ldots, 1, 0, 0 \ldots 0)$ which have 1 as entries up to and including the i-th coordinate, and 0 after, is a basis of R^n. Show that the elementary matrixes $E(i\ j)$ of $\mathbb{M}_{m,n}(R), m, n > 0$, form a basis of this vector space (R being any field).

Here is the guarantee that bases always exist:

Proposition 192 *A vector space which is finitely generated has a basis, more precisely, for every finite sequence* (x_i) *of generators, there is a subsequence which is a basis of* M.

Proof Let $(x_1, \ldots x_k), k \geq 1$, be a generating sequence, then consider the first $x_i \neq 0$ (if there is none, we have the case of a zero space, and the "empty basis" does the job). This sequence (x_{i_1}) is linearly independent since $\lambda \cdot x_{i_1} = 0, \lambda \neq 0$, implies $\lambda^{-1} \cdot \lambda \cdot x_{i_1} = x_{i_1} = 0$, a contradiction. Suppose we have found a subsequence $(x_{i_1}, x_{i_2}, \ldots x_{i_r}), i_1 < i_2 < \ldots i_t$ of linearly independent vectors of maximal length. Then this generates the space for the following reason. If i is an index $i_1 < i < i_r$, then x_i is linearly dependent of the vectors $x_{i_j}, i_j \leq i$, by construction. If $i > i_r$, then there is a non-trivial linear combination $0 = \mu \cdot x_i + \sum_{j=1,\ldots r} x - i_j$ by the maximality of our sequence. But then $\mu \neq 0$, otherwise, we would have linear dependence of the maximal sequence. Therefore x_i is contained in the space generated by the maximal sequence $(x_{i_1}, x_{i_2}, \ldots x_{i_r})$, and we are done.   □

And here is the famous Steinitz exchange theorem, which guarantees that sub-vector spaces are always embedded in a distinguished way:

Proposition 193 (Steinitz Exchange Theorem) *If* $(y_1, y_2, \ldots y_l)$ *is a sequence of linearly independent vectors in a finitely generated vector space* M, *and if* $(x_1, x_2, \ldots x_k)$ *is a basis of* M *(guaranteed by proposition 192), then* $l \leq k$, *and there is a (possibly empty) subsequence* $(x_{i_1}, x_{i_2}, \ldots x_{i_{k-l}})$ *of* $(x_1, x_2, \ldots x_k)$ *such that* $(y_1, y_2, \ldots y_l, x_{i_1}, x_{i_2}, \ldots x_{i_{k-l}})$ *is a basis of* M.

Proof There is a representation $y_1 = \sum_i \lambda_i x_i$. Since $y_1 \neq 0$, there is $\lambda_t \neq 0$. Then x_t is in the space generated by the sequence $(y_1, x_1, \ldots \hat{x}_t, \ldots x_r)$ (refer to the footnote of page 162 for the $\hat{}$ notation). But this is again a basis, since it generates the whole space and it is linearly independent. In fact, if $0 = \mu \cdot y_1 + \sum_{i=1,\ldots \hat{t},\ldots r} \lambda_i x_i$ is a non-trivial linear combination, then necessarily, $\mu \neq 0$, but then y also has a representation as a linear combination without x_t, so $0 = y - y$ would have a non-trivial representation by the basis, a contradiction! Therefore we have a new basis $(y_1, x_1, \ldots \hat{x}_t, \ldots x_r)$. Suppose now that we have found a new basis $(y_1, y_2, \ldots y_r, x_{i_1}, x_{i_2}, \ldots x_{i_{k-l}}), r \leq l$. If $r = l$ we are done, otherwise we may proceed as initially with y_1, however, we must show that we can still eliminate one of the remaining x_s. But if $y_{r+1} = \sum_{e=1,\ldots r} \mu_e y_e + \sum_{f=1,\ldots k-r} \lambda_f x_{i_f}$, then there must exist a $\lambda_{f_0} \neq 0$, otherwise, the y would be linearly dependent. So we may eliminate $x_{i_{f_0}}$ and we may proceed until all y are integrated in the basis.   □

Remark 26 The proof of the Steinitz theorem is quite algorithmic. Let us sketch the procedure: Suppose that we can find a linear combination

$y_j = \sum_{i=1,\ldots k} \lambda_{ji} x_i$ for each y_j. Then, starting with y_1, take the first non-vanishing coefficient $\lambda_{1i(1)}$ in the sequence (λ_{1i}), which exists since y_1 is not zero. Clearly, replacing $x_{i(1)}$ by y_1 in the basis (x_i) gives us a new basis. Now, suppose that we have already replaced some x_i by $y_1, \ldots y_r$ and still have a basis. Now, y_{r+1} (if such a vector is left) is also a linear combination of these new basis elements. However, it is impossible that all coefficients of the remaining x_i vanish since then the (y_j) would not be linearly independent. So we may go on as in the beginning and replace one such x_i whose coefficient is not zero. This procedure eventually yields the new basis which contains all y_j.

Corollary 194 *If N is a subspace of a vector space M of dimension $dim(M)$, then N is also finitely generated and $dim(N) \leq dim(M)$, equality holding iff $N = M$. Moreover, there is a complementary subspace $C \subset M$ to N, i.e., the homomorphism $N \oplus C \to M$ defined by the inclusions $C, N \subset M$ via the universal property of direct sums (proposition 189) is an isomorphism; in other words, M is the inner direct sum of N and C (see exercise 109).*

Proof We first show that N has a basis. If $N = 0$, we are done, so take a sequence $(y_1, \ldots y_r)$ of linearly independent vectors in N. Then $r \leq k$ by Steinitz. So take a maximal such sequence. This must generate N, otherwise, let $z \in N$ be a vector which is not a linear combination of $y_1, \ldots y_r$. Then evidently $(y_1, \ldots y_r, z)$ is linearly independent, a contradiction. So let $(y_1, \ldots y_l)$ be a basis of N, and $(x_1, \ldots x_k)$ a basis of M. Then by Steinitz, $l \leq k$. If $N = M$, then by uniqueness of dimension, $l = k$. If $l = k$, we may replace all of $(x_1, \ldots x_k)$ by the basis elements $y_1, \ldots y_r$, and therefore $N = M$.

To find a complement of N, take the space spanned by the $k - l$ elements x of the old basis in the basis $(y_1, y_2, \ldots y_l, x_{i_1}, x_{i_2}, \ldots x_{i_{k-l}})$. This is clearly a complement. $\qquad\square$

Remark 27 So we have this image in the case of finite-dimensional vector spaces over a fixed field R: If we fix an isomorphism $u_M : M \xrightarrow{\sim} R^{dim(M)}$ for each R-vector space M, we obtain an isomorphism

$$t_{u_M, u_N} : Lin_R(M, N) \xrightarrow{\sim} \mathbb{M}_{dim(N), dim(M)}(R)$$

defined by the conjugation $f : M \to N \mapsto u_N \circ f \circ (u_M)^{-1}$ and the canonical interpretation of linear maps $R^n \to R^m$ as matrixes. In this setup, if we are given a second linear map $g : N \to L$, $dim(L) = l$, then we have

$$t_{u_M, u_L}(g \circ f) = t_{u_N, u_L}(g) \cdot t_{u_M, u_N}(f) \text{ and } t_{u_M, u_M}(Id_M) = E_{dim(M)},$$

i.e., the matrix product commutes with the composition of linear maps. This may be visualized by commutative diagrams:

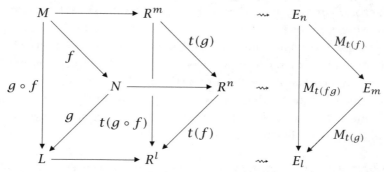

We therefore have restated vector space structures in terms of matrixes, as predicted. But what happens if we change the basis of a factor space M? Let us discuss this for endomorphisms of M which we transform into quadratic matrices in $\mathbb{M}_{n,n}(R)$ of size $n = dim(M)$. Suppose we are given two bases which induce the isomorphisms $u, v : M \overset{\sim}{\to} R^n$. Then $u \circ v^{-1} : R^n \overset{\sim}{\to} R^n$ defines a matrix X such that we have

$$t_{v,v}(f) = X^{-1} \cdot t_{u,u}(f) \cdot X,$$

i.e., conjugation with the base change matrix X gives us the second matrix of f. In particular, if $M = R^n$, and if $v = Id_{R^n}$, then this formula gives us the matrix of f when calculated in the matrix representation from the new basis, whose elements are the column vectors of X. In other words:

Corollary 195 *If we have a new basis (x_i) of the vector space R^n given in terms of a matrix X of columns $X_{\bullet i}$ which correspond to x_i, then the representation of a linear map matrix $f : R^n \to R^n$ in terms of the basis (x_i) is $X^{-1} \cdot f \cdot X$.*

We can now state the relation between the dimensions of the kernel and the image of a linear map.

Corollary 196 *Let $f : M \to N$ be a linear homomorphism defined on a finite-dimensional R-vector space M (the vector space N need not be finite-dimensional). Then we have*

$$dim(M) = dim(Im(f)) + dim(Ker(f)).$$

More precisely, there is a subspace $U \subset M$, which is a complement of $Ker(f)$, i.e., $M \overset{\sim}{\to} U \oplus Ker(f)$, and such that $f|_U : U \to Im(f)$ is an isomorphism.

Proof Let U be a complement of $Ker(f)$ in M. Then $dim(U) + dim(Ker(f)) = dim(M)$, by corollary 194. But the restriction $f|_U : U \to N$ is evidently a surjection onto $Im(f)$ since $Ker(f)$ is mapped to zero. Moreover, $Ker(f) \cap U = 0$ means that $Ker(f|_U) = 0$. Therefore $f|_U : U \xrightarrow{\sim} Im(f)$ is an isomorphism, and we are done. $\qquad\square$

Example 89 A simple example of this fact is illustrated in figure 22.1. Here we have a projection π of the 3-dimensional space $\mathbb{R}^3$ onto the 2-dimensional plane $\mathbb{R}^2$. A point x in $\mathbb{R}^3$ is mapped to a point $\pi(x)$ in the plane of dimension 2, which is $Im(\pi)$. The points on the line through the origin O parallel to the projection axis are all mapped to O, i.e., these are all the points y such that $\pi(y) = O$. Thus this line of dimension 1 is $Ker(\pi)$. As predicted by corollary 196, $dim(Im(\pi)) + dim(Ker(\pi)) = dim(\mathbb{R}^3)$, i.e., $2 + 1 = 3$.

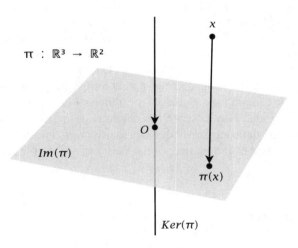

Fig. 22.1. The image and kernel of a projection.

Definition 166 *The dimension $dim(Im(f))$ of a linear homomorphism $f : M \to N$ of R-vector spaces is called the* rank *of f.*

Here is the numerical criterion which allows us to calculate the rank of f in terms of its matrix:

Definition 167 *The rank $rk(M)$ of a matrix $M \in \mathbb{M}_{m,n}(R)$, for positive m and n over a field R is the maximal r such that there is a quadratic*

$r \times r$-submatrix D of M with $\det(D) \neq 0$, $r = 0$ meaning that $M = 0$. By definition, such a submatrix is obtained by eliminating $m - r$ rows and $n - r$ columns from M.

Proposition 197 *The rank of a linear homomorphism $f : M \to N$ of non-trivial R-vector spaces coincides with the rank of an associated matrix $t(f) \in M_{dim(N),dim(M)}(R)$ with respect to selected bases.*

Proof We know that the matrix $t(f)$ of a linear homomorphism $f : M \to N$ with respect to selected bases of M and N is described as follows: The associated linear map $g : R^n \to R^m$ with $n = dim(M), m = dim(N)$, has the images $g(e_j)^\tau = t(f)_{\bullet j}$ for the canonical basis $e_i = E(1\ i), i = 1, \ldots n$, of R^n, and of course $rk(g) = rk(f)$. If we take the submatrix of $t(f)$ defined by selecting $q > rk(f)$ columns, then these q columns are linearly dependent. This remains true if we cancel all but q rows of this matrix. After canceling, we have a $q \times q$ submatrix with linearly dependent columns, which clearly has zero determinant. So the rank of $t(f)$ is at most $rk(f)$. Take $k = rk(f)$ columns, which generate a basis of $Im(g)$. Then, by Steinitz complete this basis by elementary columns $E(\pi(j)\ j), j = k + 1, \ldots m$. This yields a $m \times m$ matrix whose determinant does not vanish. But the determinant only changes its sign if we permute columns or rows. We may obviously exchange rows and columns such that the new matrix has as last $n - k$ columns the elementary columns $E(j + 1\ j + 1), \ldots E(m\ m)$, filling up the diagonal with 1s after the columns associated with the basis of $Im(g)$. But then the determinant is the necessarily non-zero determinant of the upper left $k \times k$-block of the matrix, so the rank of $t(f)$ is at least k and we are done. □

22.2 Equations

We may now decide upon the existence of a solution of the linear equation

$$
\begin{pmatrix} y_1 \\ y_2 \\ \vdots \\ y_m \end{pmatrix} = \begin{pmatrix} a_{11} & a_{12} & \ldots & a_{1n} \\ a_{11} & a_{12} & \ldots & a_{1n} \\ \vdots & \vdots & & \vdots \\ a_{m1} & a_{m2} & \ldots & a_{mn} \end{pmatrix} \cdot \begin{pmatrix} x_1 \\ x_2 \\ \vdots \\ x_n \end{pmatrix}
$$

with given matrixes $(y_i), M = (a_{ij})$ and unknown (x_i): Let (y, M) be the $m \times (n + 1)$-matrix obtained by prepending the column y to the left of M. Here is the somewhat redundant but useful list of cases for solutions of the system:

Proposition 198 *With the preceding notations, the linear equation* $y = M \cdot x$

(i) *has a solution iff* $rk(M) = rk((y, M))$;

(ii) *has at most one solution if* $rk(M) = n$;

(iii) *has exactly one solution iff* $rk((y, M)) = rk(M) = n$;

(iv) *in the case* $m = n = rk(M)$ *(so-called "regular equation") has the unique solution* $x = M^{-1} \cdot y$.

(v) *Given one solution* x_0, *and a basis* $(z_t)_{t=1,\dots s}$ *of the kernel of the linear map* f_M *associated with* M, *the solution space consists of the coset* $x_0 + Ker(f_M)$, *i.e., all the vectors* $x_0 + \sum_{t=1,\dots s} \lambda_t z_t$, $\lambda_t \in R$. *The elements of* $Ker(f_M)$ *are also called the solutions of the homogeneous equation* $0 = M \cdot x$ *associated with* $y = M \cdot x$.

Proof If the equation $y = M \cdot x$ has a solution, it is of the form $y = \sum_{j=1,\dots d} \lambda_i M_{\bullet j_i}$ for a basis $(M_{\bullet j_1}^{\tau}, \dots M_{\bullet j_d}^{\tau})$ of the image of the homomorphism $f_M : R^n \to R^m$ associated with M. But then the matrix (y, M) has no regular quadratic submatrix containing the column y, by the common column equality annihilation argument, whence $rk(M) = rk((y, M))$. Conversely, consider the linear map $h : R^n \oplus R \to R^m$ which on the first summand is f_M, and on the second summand just maps the basis 1 to y. Then since $rk(M) = rk((y, M))$, $dim(Im(h)) = dim(f_M)$, so the images are equal, and y is in the image of f_M, this proves (i). As to (ii), if $rk(M) = n$, then by corollary 196, $Ker(f_M) = 0$ and f_M is injective. If in (iii) we suppose $rk((y, M)) = rk(M) = n$, then by (i) and (ii), there is exactly one solution. Conversely, if there is exactly one solution, then there is a solution, and (i) shows $rk((y, M)) = rk(M)$, while if $rk(M) < n$ would yield a non-trivial kernel, and for each solution y, we get another solution $y + w$ for $w \in Ker(f_M) - \{0\}$. Therefore also $rk(M) = n$. Statements (iv) and (v) are clear, since any two solutions have their difference in the kernel, and any solution y, when changed to $y + w$, $w \in Ker(f_M)$, yields another solution. $\square$

In chapter 23, we shall present more algorithmic methods for finding solutions.

22.3 Affine Homomorphisms

It is sometimes customary to distinguish between vectors and points when dealing with elements of an R-vector space M. Why? Because vectors may play different roles within a space. So far we know that vectors are elements of a space which may be added and scaled. But all vectors play the same role. The representation of a vector $x \in M$ by means of its coordinate sequence $f(x) = (\lambda_1, \ldots \lambda_n)$ for a given basis (x_i) of M and the associated isomorphism $f : M \overset{\sim}{\to} R^n$ positions the vector in a coordinate system, whose axes are the 1-dimensional base spaces $R \cdot x_i$. This is the common image of traditional analytical geometry.

In "affine geometry" however, one adopts a slightly different point of view in that the addition $u + y$ of two vectors is restated as an operation of u upon y, the operation being denoted by $T^u : M \to M : y \mapsto T^u(y) = u+y$, it is called the *translation by* u. The exponential notation has its justification in the obvious formula $T^u \circ T^v = T^{u+v}$. In this understanding, y plays the role of a point which is shifted by the vector u to a new point $T^u(y) = u + y$. Clearly, we therefore have an injection $T : M \to Sym(M)$ of the additive group of M into the symmetric group of M; denote by T^M the image *group of translations on M*. This identification of a vector u with its translation T^u is what in affine geometry creates two kinds of vectors: the given ones, y, and the associated operators T^y. In this way, addition of vectors is externalized as an operator on the "point set M". Like linear algebra, affine algebra deals with modules and in particular vector spaces, but the morphisms between such spaces are a little more general, in fact they include also translations. Here is the formal definition.

Definition 168 *If M and N are R-vector spaces, then a map $f : M \to N$ is called an R-affine homomorphism, if there is a map $g \in Lin_R(M,N)$ and a vector $u \in N$ such that $f = T^u \circ g$. The homomorphism g is called the linear part, whereas T^u is called the translation part of f. The set of affine homomorphisms $f : M \to N$ is denoted by $Aff_R(M,N)$. The group of invertible elements in the ring $Aff_R(M,M)$ of affine endomorphisms of M is denoted by $GA(M)$ and called the general affine group of M.*

Since for the affine homomorphism $f = T^u \circ g$, $u = f(0)$, the linear part g is $g = T^{-u} \circ f$, so both linear and translation parts of f are uniquely determined by f, they are denoted by $u = \tau(f)$ and $g = \lambda(f)$, i.e.,

$$f = T^{\tau(f)} \circ \lambda(f).$$

Exercise 118 Show that together with the point-wise addition and scalar multiplication, $Aff_R(M,N)$ is also an R-vector space.

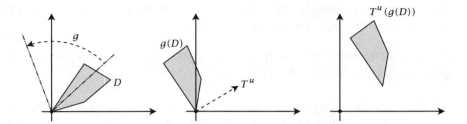

Fig. 22.2. An affine homomorphism $f = T^u \circ g$ on $\mathbb{R}^2$ is shown as a rotation g around the origin, followed by a translation T^u.

Lemma 199 *If $f : M \to N$ and $g : N \to L$ are R-affine homomorphisms, then their composition $g \circ f : M \to L$ is R-affine and we have this formula:*

$$g \circ f = (T^{\tau(g)} \circ \lambda(g)) \circ (T^{\tau(f)} \circ \lambda(f)) = T^{\tau(g)+\lambda(g)(\tau(f))} \circ (\lambda(g) \circ \lambda(f)).$$

The inverse of an affine isomorphism $f = T^{\tau(f)} \circ \lambda(f)$ (i.e., its linear part is an isomorphism) is given by the formula

$$f^{-1} = T^{-\lambda(f)^{-1}(\tau(f))} \circ \lambda(f)^{-1}.$$

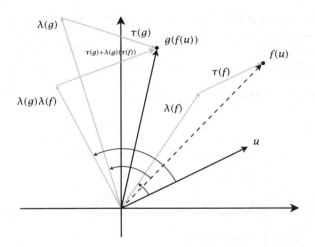

Fig. 22.3. Composition $g \circ f$ of affine homomorphisms g and f acting on a vector u.

Proof The lemma follows without any trick from the explicit formulas, which it presents. We therefore leave it to the reader.   □

Exercise 119 Show that the group T^M of translations is a normal subgroup of $GA(M)$. Show that $GA(M)/T^M \xrightarrow{\sim} GL(M)$.

We were happy to have linear maps represented by matrixes, so let us see how this can also be obtained for affine maps. Of course, the usual representation does not work, but there is a beautiful trick to interpret an affine map as if it would leave the origin fixed, which it really does not, in general. The trick consists in inventing a new origin and embedding the given vector space in a larger space, where we have a linear map which on the embedded space behaves like the given affine map. The new system is related to what is known as the method of "homogeneous coordinates". We have the affine injection

$$\eta_M : M \to M \oplus R : m \mapsto (m, 1)$$

of *homogenization*, i.e., $\eta_M = T^{(0,1)} \circ i_1$, where i_1 is the usual linear embedding of M as first summand of $M \oplus R$. Then, for a given affine homomorphism $f : M \to N$, we consider the linear homomorphism

$$\hat{f} : M \oplus R \to N \oplus R : (m, r) \mapsto (\lambda(f)(m), 0) + r(\tau(f), 1)$$

associated with an affine homomorphism $f : M \to N$. Clearly, $\hat{f}$ sends $\eta_M(M)$ to $\eta_N(N)$. More precisely, we have

$$\hat{f} \circ \eta_M = \eta_N \circ f,$$

which is best represented as a diagram

$$
\begin{array}{ccc}
M & \xrightarrow{\;\eta_M\;} & M \oplus R \\
\downarrow{\scriptstyle f} & & \downarrow{\scriptstyle \hat{f}} \\
N & \xrightarrow[\;\eta_N\;]{} & N \oplus R
\end{array}
$$

We may get back f by the formula

$$\pi_1 \circ \hat{f} \circ \eta_M = f$$

with the first projection $\pi_1 : N \oplus R \to N$, since $\pi_1 \circ \eta_N = Id_N$. If $\eta_M(m) = (m, 1)$ is in the image of M, then we have $\hat{f}(m, 1) = (f(m), 1)$. If M

is identified with R^n and N is identified with R^m by the choice of two bases, then the linear part $\lambda(f)$ identifies with a $m \times n$-matrix (a_{ij}), the translation vector $\tau(f)$ to a column vector (t_{i1}) and $\hat{f}$ identifies with the matrix

$$
\begin{pmatrix}
a_{11} & a_{12} & \cdots & a_{1n} & t_{11} \\
\vdots & \vdots & & \vdots & \vdots \\
a_{m1} & a_{m2} & \cdots & a_{mn} & t_{m1} \\
0 & 0 & \cdots & 0 & 1
\end{pmatrix}
$$

which evidently sends a column (y_{i1}) from the image $\eta_M(M)$ with last coordinate 1 to a column with the same last coordinate. The *homogeneous coordinates* of a column vector are the given coordinates plus the new last coordinate 1.

Exercise 120 Show that $\widehat{Id_M} = Id_{M \oplus R}$. Given two affine homomorphisms $f : M \to N$ and $g : N \to L$, show that $\widehat{g \circ f} = \hat{g} \circ \hat{f}$.

Example 90 In $M = \mathbb{R}^2$, the counter-clockwise rotation R_0 by 60 degrees around the origin $(0,0)$ is given by the matrix

$$
M_{R_0} = \begin{pmatrix} \frac{1}{2} & -\frac{\sqrt{3}}{2} \\ \frac{\sqrt{3}}{2} & \frac{1}{2} \end{pmatrix}
$$

where we anticipate the concept of angles and associated matrixes. The subject will be treated properly in section 24, for the moment, we may just recall the high school education in mathematics. We now consider the counter-clockwise rotation R_p of 60 degrees around any point $p = (x, y)$. Figure 22.4 illustrates the case of $p = (2, 1)$ and the value $R_p(a)$ of the point $a = (-\frac{1}{2}, -1)$. The rotation R_p is an affine transformation on $\mathbb{R}^2$ by the following argument: consider the composition $T^{-p} \circ R_p \circ T^p$. This map fixes the origin and is in fact the counter-clockwise rotation R_0 of 60 degrees around the origin $(0,0)$. Then the equation $R_0 = T^{-p} \circ R_p \circ T^p$ yields $R_p = T^p \circ R_0 \circ T^{-p} = T^{\Delta_p} \circ R_0$ with $\Delta_p = p - R_0(p)$. Let us calculate the numeric values and the 3×3-matrix of $\hat{R}_p$ in terms of homogeneous coordinates for the concrete vector $p = (2, 1)$. We have

$$
\Delta_p = \begin{pmatrix} 2 \\ 1 \end{pmatrix} - \begin{pmatrix} \frac{1}{2} & -\frac{\sqrt{3}}{2} \\ \frac{\sqrt{3}}{2} & \frac{1}{2} \end{pmatrix} \cdot \begin{pmatrix} 2 \\ 1 \end{pmatrix} = \begin{pmatrix} 1 + \frac{\sqrt{3}}{2} \\ \frac{1}{2} - \sqrt{3} \end{pmatrix}
$$

and therefore the matrix $M_{\hat{R}_p}$ of $\hat{R}_p$ is

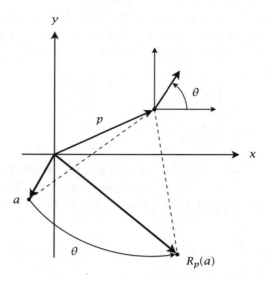

Fig. 22.4. The rotation $R_p(a)$ of the point $a = (-\frac{1}{2}, -1)$ by $\theta = 60$ degrees around the point $p = (2, 1)$.

$$M_{\hat{R}_p} = \begin{pmatrix} M_{R_0} & \Delta_p \\ 0 & 1 \end{pmatrix} = \begin{pmatrix} \frac{1}{2} & -\frac{\sqrt{3}}{2} & 1 + \frac{\sqrt{3}}{2} \\ \frac{\sqrt{3}}{2} & \frac{1}{2} & \frac{1}{2} - \sqrt{3} \\ 0 & 0 & 1 \end{pmatrix}.$$

The transformation corresponding to $M_{\hat{R}_p}$ is shown in figure 22.5. If applied to a vector $a = (-\frac{1}{2}, -1)$, rewritten in homogeneous coordinates $\hat{a} = (-\frac{1}{2}, -1, 1)$, we get the product

$$\hat{R}_p(\hat{a})^\top = \begin{pmatrix} \frac{1}{2} & -\frac{\sqrt{3}}{2} & 1 + \frac{\sqrt{3}}{2} \\ \frac{\sqrt{3}}{2} & \frac{1}{2} & \frac{1}{2} - \sqrt{3} \\ 0 & 0 & 1 \end{pmatrix} \cdot \begin{pmatrix} -\frac{1}{2} \\ -1 \\ 1 \end{pmatrix} = \begin{pmatrix} \frac{3}{4} + \sqrt{3} \\ -\frac{5\sqrt{3}}{4} \\ 1 \end{pmatrix}.$$

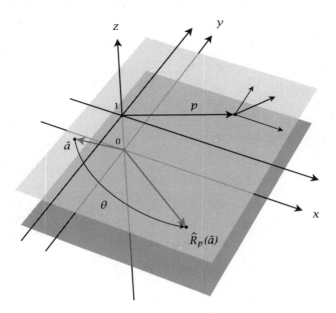

Fig. 22.5. The same transformation as in figure 22.4, but this time in homogeneous coordinates, i.e., $\widehat{R}_p(\widehat{a})$, where $\widehat{a} = (-\frac{1}{2}, -1, 1)$.

CHAPTER 23

Algorithms in Linear Algebra

In practice, one needs algorithmic methods to calculate matrixes which arise from problems in linear algebra or vector spaces. In particular, it is important to obtain solutions of linear equations, to calculate determinants and inverse matrixes. This is a vast field of numerical mathematics and of ongoing research since fast and reliable algorithms for special matrixes are of crucial importance in the entire computer-aided science, be it physics, economics, chemistry or biology. There are also very important applications of matrix calculations in special technological fields such as signal processing (e.g., Fast Fourier Transform, see later in the second volume of this book), which are particularly sensitive to fast methods because they pertain to real-time technologies. We shall only discuss some very elementary algorithms here in order to give a first idea of the subject. For object-oriented implementations of these and other algorithms, we refer to [6].

23.1 Gauss Elimination

The Gauss algorithm is based on two simple observations concerning solutions of linear equations:

1. If we are given a linear equation $y = f(x)$ for a R-linear map $f : M \to N$, the solutions x are unchanged if we compose f with a $g \in \mathrm{GL}(N)$ and thus obtain the new equation $g(y) = g(f(x)) = (g \circ f)(x)$.

2. If we have a well-known $h \in GL(M)$, then we may rewrite the equation as $y = f(x) = (f \circ h^{-1}) \circ h(x)$, and hope that the new unknown $h(x)$ is easier to manage than the original one.

In terms of matrixes, the first method looks as follows: We are given a system of linear equations

$$(y_i) = (a_{ij}) \cdot (x_j)$$

with m equations ($i = 1, \ldots m$) and n unknowns ($j = 1, \ldots n$), then the solution set is left unchanged if we multiply both sides by an invertible $m \times m$-matrix (b_{ik}), obtaining the new matrix equation

$$(y_i') = (b_{ik}) \cdot (y_i) = ((b_{ik}) \cdot (a_{ij})) \cdot (x_j) = (a_{ij}') \cdot (x_j).$$

The second method means that we take an invertible $n \times n$-matrix (c_{lj}) and rewrite

$$(y_i) = (a_{ij}) \cdot (x_j) = ((a_{ij})(c_{lj})^{-1}) \cdot ((c_{lj}) \cdot (x_j)) = (a_{ij}') \cdot (x_j')$$

with the new unknowns

$$(x_j') = (c_{lj}) \cdot (x_j).$$

Gauss elimination consist of a clever choice of a series of transformations (b_{ki}) and (c_{lj}) which successively changes the equation until it has a particularly simple shape. Mostly, Gauss elimination is applied in the case of a regular system of n equations with n unknowns, i.e., the coefficient matrix (a_{ij}) is supposed to be invertible, such that we have exactly one solution.

The idea of the simplifying procedure is to obtain an upper triangular coefficient matrix (a_{ij}), i.e., $a_{ij} = 0$ for $i > j$. This means that the diagonal elements $a_{ii} \neq 0$, because their product is the non-zero determinant. Then we may solve the equation system by *backward substitution*, which means that we first calculate x_n, then x_{n-1}, etc. until we obtain the solution of x_1. In fact, a triangular matrix yields these n equations:

$$y_i = \sum_{j=i,i+1,\ldots n} a_{ij} x_j$$

which can be solved recursively, beginning with

$$x_n = \frac{1}{a_{nn}} y_n$$

and yielding

$$x_i = \frac{y_i - \sum_{j=i+1,\ldots n} a_{ij} x_j}{a_{ii}}$$

from the calculation of $x_n, x_{n-1}, \ldots x_{i+1}$.

So we are left with the construction of an upper triangular coefficient matrix. Again, this is a recursive construction, running along the size n. To begin with, one would like to have $a_{11} \neq 0$. If this is not the case, we look for a first index $k > 1$ such that $a_{1k} \neq 0$. This exists since otherwise, the determinant of (a_{ij}) would vanish. Now we rename the unknowns: x_1 becomes x_k, and x_k becomes x_1. This is achieved by replacing (a_{ij}) with $(a'_{ij}) = (a_{ij}) \cdot P(1k)$ and the column (x_j) with $(x'_j) = P(1k) \cdot (x_j)$. The matrix $P(1k)$ corresponds to the transposition $(1k)$ in the symmetric group S_n and is defined by $P(1k) = E_n - E(1\ 1) - E(k\ k) + E(k\ 1) + E(1\ k)$. Multiplying (a_{ij}) from the right by $P(1k)$ exchanges column 1 and k of (a_{ij}), whereas multiplying (x_j) by $P(1k)$ from the left, exchanges x_1 and x_k, and this does not change anything in the product $(y_i) = (a'_{ij}) \cdot (x'_{ij})$, since $P(1k)^2 = E_n$. Therefore, up to renaming x_1 to x'_k, and x_k to x'_1, and leaving all other unknowns $x_j = x'_j$, we have achieved that $a_{11} \neq 0$. Next, we find an invertible (b_{ik}) such that $(a'_{ij}) = (b_{ik}) \cdot (a_{ij})$ has zeros for all index pairs $ji, j > i$. This is the matrix:

$$(b_{ik}) = E_n - \sum_{j=2,\ldots n} \frac{a_{j1}}{a_{11}} \cdot E(j\ 1).$$

It is invertible (the determinant is 1) and makes the product $(a'_{ij}) = (b_{ik}) \cdot (a_{ij})$ have zeros for index pairs $j1, j > 1$. Proceeding recursively, the situation is restricted to the equations $i = 2, \ldots n$ involving only the unknowns $x_2, \ldots x_n$. This settles the problem.

Example 91 Consider the system of equations

$$
\begin{array}{rrrrr}
2x_1 & + & 3x_2 & + & 4x_3 & = & 8 \\
4x_1 & + & 2x_2 & - & x_3 & = & -3 \\
2x_1 & - & 3x_2 & + & 3x_3 & = & 17
\end{array}
$$

This can be rewritten as

$$Ax = y$$

where

$$A = \begin{pmatrix} 2 & 3 & 4 \\ 4 & 2 & -1 \\ 2 & -3 & 3 \end{pmatrix} \text{ and } y = \begin{pmatrix} 8 \\ -3 \\ 17 \end{pmatrix}$$

To perform Gauss elimination, we first have to transform A into an upper triangular matrix. In a first step, we will multiply the equation with a matrix B that nullifies the second and third entries in the first column. After that, a matrix C will be used that nullifies the third entry in the second column, i.e.,

$$CBAx = CBy$$

where CBA will be an upper triangular matrix. As defined above, the matrix B is given by

$$B = E_3 - \sum_{j=2,3} \frac{a_{j1}}{a_{11}} \cdot E(j\,1)$$

i.e.,

$$B = \begin{pmatrix} 1 & 0 & 0 \\ 0 & 1 & 0 \\ 0 & 0 & 1 \end{pmatrix} - \left[\frac{4}{2} \cdot \begin{pmatrix} 0 & 0 & 0 \\ 1 & 0 & 0 \\ 0 & 0 & 0 \end{pmatrix} + \frac{2}{2} \cdot \begin{pmatrix} 0 & 0 & 0 \\ 1 & 0 & 0 \\ 0 & 0 & 0 \end{pmatrix} \right] = \begin{pmatrix} 1 & 0 & 0 \\ -2 & 1 & 0 \\ -1 & 0 & 1 \end{pmatrix}$$

A simple calculation shows

$$BA = \begin{pmatrix} 2 & 3 & 4 \\ 0 & -4 & -9 \\ 0 & -6 & -1 \end{pmatrix} \text{ and } By = \begin{pmatrix} 8 \\ -19 \\ 9 \end{pmatrix}$$

Now we have to nullify the third element in the second row of BA. For this, we look at the submatrix A' of A:

$$A' = \begin{pmatrix} -4 & -9 \\ -6 & -1 \end{pmatrix}$$

Using the same procedure as above, we get

$$C' = E_2 - \frac{a'_{21}}{a'_{11}} \cdot E(2\,1) = \begin{pmatrix} 1 & 0 \\ 0 & 1 \end{pmatrix} - \frac{3}{2} \cdot \begin{pmatrix} 0 & 0 \\ 1 & 0 \end{pmatrix} = \begin{pmatrix} 0 & 0 \\ -\frac{3}{2} & 0 \end{pmatrix}$$

This yields

$$C = \begin{pmatrix} 1 & 0 & 0 \\ 0 & 1 & 0 \\ 0 & -\frac{3}{2} & 1 \end{pmatrix}$$

hence

$$CBA = \begin{pmatrix} 2 & 3 & 4 \\ 0 & -4 & -9 \\ 0 & 0 & \frac{25}{2} \end{pmatrix} \text{ and } CBy = \begin{pmatrix} 8 \\ -19 \\ \frac{75}{2} \end{pmatrix}$$

Now we have modified our original equation to

$$Ax = y$$

where

$$A = \begin{pmatrix} 2 & 3 & 4 \\ 0 & -4 & -9 \\ 0 & 0 & \frac{25}{2} \end{pmatrix} \text{ and } y = \begin{pmatrix} 8 \\ -19 \\ \frac{75}{2} \end{pmatrix}$$

Going for the third unknown

$$x_3 = \frac{1}{a_{33}} y_3 = \frac{2}{25} \frac{75}{2} = 3$$

$$x_2 = \frac{y_2 - a_{23} x_3}{a_{22}} = \frac{-19 + 9 \cdot 3}{-4} = \frac{8}{-4} = -2$$

and finally

$$x_1 = \frac{y_3 - a_{12} x_2 - a_{13} x_3}{a_{11}} = \frac{8 - 3 \cdot (-2) - 4 \cdot 3}{2} = \frac{2}{2} = 1$$

It is left as an exercise for the reader to check that these values satisfy the original equation.

23.2 The LUP Decomposition

This algorithm computes a decomposition of a regular matrix $A = (a_{ij})$ which is also useful for calculating its determinant. The decomposition yields a product:

$$A = L \cdot U \cdot P$$

where the factors are as follows: $L = (l_{ij})$ is a lower triangular $n \times n$-matrix (i.e., $l_{ij} = 0$ for $i < j$), U is an upper triangular $n \times n$-matrix, and P is a permutation matrix (see figure 23.1). This means that $P = \sum_{i=1,\ldots n} \pi(i) E^i$ for a permutation $\pi \in S_n$.

Such a decomposition yields $\det(A)$ as a product of the diagonal coefficients of L, times the product of the diagonal elements of U, times the determinant of P, which is the signature $sig(\pi)$ of the given permutation.

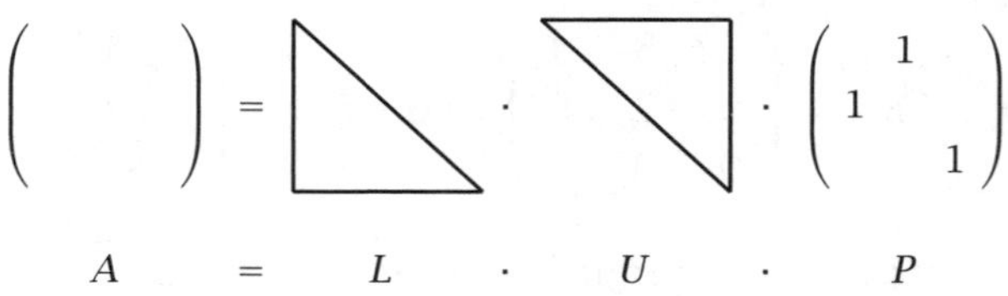

$$A \quad = \quad L \quad \cdot \quad U \quad \cdot \quad P$$

Fig. 23.1. The LUP decomposition of a matrix A.

The solution of an equation $(y_i) = A \cdot (x_i)$ proceeds in three steps: first one solves the auxiliary equation $(y_i) = L \cdot (z_i)$. The lower diagonal L allows this recursive calculation of z_i by *forward substitution*:

$$z_1 = \frac{1}{l_{11}} y_1$$

and producing

$$z_i = \frac{y_i - \sum_{j=1,\dots i-1} l_{ij} z_j}{l_{ii}}$$

from the calculation of $z_1, z_2, \dots z_{i-1}$. Then, we observe that the permutation $P \cdot (x_i)$ is nothing but a renaming of the indexes of the unknowns. Apart from this renaming, the remaining problem is an equation $(z_i) = U \cdot (x_i)$, which is solved by the above backward substitution.

The algorithm for the LUP-decomposition runs as follows: First, we rewrite A as $A = A \cdot E_n = A \cdot P^2$ by use of a permutation matrix P, which permutes two columns 1 and k of A as described above in 23.1, such that $A \cdot P$ has its $1, 1$-coefficient $\neq 0$. One then writes the new matrix A as a block-configuration of four submatrixes:

$$A = \begin{pmatrix} a_{11} & w \\ v & A' \end{pmatrix}$$

where $v \in \mathbb{M}_{n-1,1}(R)$, $w \in \mathbb{M}_{1,n-1}(R)$, and $A' \in \mathbb{M}_{n-1,n-1}(R)$. The two matrixes v and w define their product matrix $v \cdot w \in \mathbb{M}_{n-1,n-1}(R)$. Supposing that $a_{11} \neq 0$, we now have this equation:

$$A = \begin{pmatrix} a_{11} & w \\ v & A' \end{pmatrix} = \begin{pmatrix} 1 & 0 \\ \frac{1}{a_{11}} \cdot v & E_{n-1} \end{pmatrix} \cdot \begin{pmatrix} a_{11} & w \\ 0 & A' - \frac{1}{a_{11}} \cdot v \cdot w \end{pmatrix} \quad (23.1)$$

where the regular submatrix $A' - \frac{1}{a_{11}} \cdot v \cdot w$ is called the *Schur complement of A with respect to the pivot element a_{11}*. By induction, we assume that this complement has a LUP-decomposition

$$A' - \frac{1}{a_{11}} \cdot v \cdot w = L' \cdot U' \cdot P'$$

which we now use to obtain the desired LUP-decomposition

$$A = \begin{pmatrix} 1 & 0 \\ \frac{1}{a_{11}} \cdot v & L' \end{pmatrix} \cdot \begin{pmatrix} a_{11} & w \cdot (P')^{-1} \\ 0 & U' \end{pmatrix} \cdot \begin{pmatrix} 1 & 0 \\ 0 & P' \end{pmatrix}$$

of A.

Example 92 The goal is to calculate the LUP-decomposition of the matrix $A \in \mathbb{M}_{3,3}(\mathbb{Q})$:

$$A = \begin{pmatrix} 2 & -3 & 1 \\ 1 & -2 & -3 \\ 1 & 4 & 1 \end{pmatrix}.$$

Equation 23.1 yields the following values:

$$a_{11} = 2 \qquad A' = \begin{pmatrix} -2 & -3 \\ 4 & 1 \end{pmatrix} \qquad v = \begin{pmatrix} 1 \\ 1 \end{pmatrix} \qquad w = (-3, 1)$$

The Schur complement B is then computed as

$$\begin{aligned} B &= A' - \frac{1}{a_{11}} \cdot v \cdot w \\ &= \begin{pmatrix} -2 & -3 \\ 4 & 1 \end{pmatrix} - \frac{1}{2} \cdot \begin{pmatrix} 1 \\ 1 \end{pmatrix} \cdot (-3, 1) \\ &= \begin{pmatrix} -\frac{1}{2} & -\frac{7}{2} \\ \frac{11}{2} & \frac{1}{2} \end{pmatrix} \end{aligned}$$

thus

$$A = L \cdot U = \begin{pmatrix} 1 & 0 & 0 \\ \frac{1}{2} & 1 & 0 \\ \frac{1}{2} & 0 & 1 \end{pmatrix} \cdot \begin{pmatrix} 2 & -3 & 1 \\ 0 & -\frac{1}{2} & -\frac{7}{2} \\ 0 & \frac{11}{2} & \frac{1}{2} \end{pmatrix}.$$

The next step is to recursively construct the LUP-decomposition of the Schur complement B. First the required parts are extracted:

$$b_{11} = -\frac{1}{2} \qquad B' = \frac{1}{2} \qquad v' = \frac{11}{2} \qquad w' = -\frac{7}{2}$$

The Schur complement C at this (last) stage is very simple, a 1×1-matrix:

$$\begin{aligned} C &= B' - \frac{1}{b_{11}} \cdot v' \cdot w' \\ &= \frac{1}{2} + 2 \cdot \frac{11}{2} \cdot -\frac{7}{2} \\ &= -38 \end{aligned}$$

therefore

$$B = L' \cdot U' = \begin{pmatrix} 1 & 0 \\ -11 & 1 \end{pmatrix} \cdot \begin{pmatrix} -\frac{1}{2} & -\frac{7}{2} \\ 0 & -38 \end{pmatrix}.$$

Finally, the LUP-decomposition is built up using the components just determined. Luckily, during the entire procedure there has never been a need for an exchange of columns; all permutation matrixes are therefore unit matrixes and can be omitted:

$$A = \begin{pmatrix} 1 & 0 & 0 \\ \frac{1}{2} & 1 & 0 \\ \frac{1}{2} & -11 & 1 \end{pmatrix} \cdot \begin{pmatrix} 2 & -3 & 1 \\ 0 & -\frac{1}{2} & -\frac{7}{2} \\ 0 & 0 & -38 \end{pmatrix}$$

It is now easy to calculate the determinant of A:

$$\det(A) = \det(L) \cdot \det(U) = 1 \cdot 1 \cdot 1 \cdot 2 \cdot -\frac{1}{2} \cdot -38 = 38$$

Of course, in this case we could have applied the formula for determinants of 3×3-matrixes, but in larger sized matrixes, the LUP-decomposition provides a much more efficient procedure than using the definition of determinants directly.

CHAPTER 24

Linear Geometry

The previous mathematical development has covered a considerable number of familiar objects and relations, such as sets, numbers, graphs, grammars, or rectangular tables, which are abstractly recast in the matrix calculus. The axiomatic treatment of modules and, more specifically, vector spaces, has also allowed us to rebuild what is commonly known as coordinate systems. However, one very important aspect of everyday's occupation with geometric objects has not been even alluded to: distance between objects, angles between straight lines, lengths of straight lines connecting two points in space. Even more radically, the concept of a neighborhood has not been thematized, although it is a central concept in the comparison of positions of objects in a sensorial space, such as the visual, tactile, gestural, or auditive space-time. The following chapter is devoted to the very first steps towards the concept of a mathematical model of geometric reality (its Greek etymology being "to measure the earth"). *In this spirit, we shall exclusively deal with real vector spaces in the last two chapters of this part, i.e., the coefficient set is* $\mathbb{R}$*. We shall also always assume that the vector spaces are finite dimensional—unless explicitly stated.*

24.1 Euclidean Vector Spaces

We begin with some preliminary definitions. For a real vector space V, the vector space of linear homomorphisms $Lin_{\mathbb{R}}(V, \mathbb{R})$ is called the *dual space of V* and denoted by V^*. If V is finite-dimensional of dimension

$dim(V) = n$, then we know that $V^* \overset{\sim}{\to} \mathbb{M}_{1,n}(\mathbb{R}) \overset{\sim}{\to} \mathbb{R}^n \overset{\sim}{\to} V$. A linear map $l \in V^*$ is called an $\mathbb{R}$-*linear form on V*.

In order to generate the basic metric structures, one first needs bilinear forms:

Definition 169 *Given a real vector space V, a map* $b : V \times V \to \mathbb{R}$ *is called* $\mathbb{R}$-*bilinear iff for each* $v \in V$, *both maps* $b(v,?) : V \to \mathbb{R} : x \mapsto b(v,x)$ *and* $b(?,v) : V \to \mathbb{R} : x \mapsto b(x,v)$ *are* $\mathbb{R}$-*linear forms. A bilinear form is called* symmetric *iff* $b(x,y) = b(y,x)$ *for all* $(x,y) \in V \times V$. *It is called* positive definite *iff* $b(x,x) > 0$ *for all* $x \neq 0$.

Given a symmetric, positive definite bilinear form b, the pair (V,b) *is called a* Euclidean vector space. *An* isometry $f : (V,b) \to (W,c)$ *between Euclidean spaces is a linear map* $f \in Lin_{\mathbb{R}}(V,W)$ *such that for all* $(x,y) \in V \times V$, *we have* $c(f(x),f(y)) = b(x,y)$. *The set of isometries* $f : (V,b) \to (W,c)$ *is denoted by* $O_{b,c}(V,W)$ *or* $O(V,W)$ *if the bilinear forms are clear. If* $(V,b) = (W,c)$, *one writes* $O(V)$ *instead*.

For a vector x in a Euclidean vector space (V,b), *the* norm *of x is the non-negative real number* $\|x\| = \sqrt{b(x,x)}$.

Lemma 200 *For a Euclidean space* (V,b), *the norm has this property for any vectors* $x, y \in V$:

$$\|x + y\|^2 = \|x\|^2 + 2 \cdot b(x,y) + \|y\|^2,$$

in particular, the form b is determined by the associated norm with the formula

$$b(x,y) = \frac{1}{2}(\|x + y\|^2 - \|x\|^2 - \|y\|^2).$$

Proof We have $\|x + y\|^2 = b(x + y, x + y) = b(x,x) + b(x,y) + b(y,x) + b(y,y) = \|x\|^2 + 2 \cdot b(x,y) + \|y\|^2$. $\square$

Lemma 201 *For a finite-dimensional Euclidean space* (V,b), *the map* $^*b : V \to V^* : v \mapsto b(v,?)$ *is a linear isomorphism, equal to the map* $b^* : V \to V^* : v \mapsto b(?,v)$.

Proof The map $v \mapsto b(v,?)$, where $b(v,?) : V \to \mathbb{R} : w \mapsto b(v,w)$, maps into V^*, where $dim(V^*) = dim(V)$ according to the remark at the beginning of this section. So it is sufficient to show that $^*b$ is a linear injection. If $b(v,?) = 0$, then also $b(v,v) = 0$, but then, $v = 0$, since b is positive definite. Further $b(v_1 + v_2, w) = b(v_1, w) + b(v_2, w)$, and $b(\lambda \cdot v, w) = \lambda \cdot b(v,w)$, so $^*b$ is linear. By symmetry of b, we also have $b^* = {^*b}$. $\square$

Exercise 121 For a Euclidean space (V, b), if $f : V \to V$ is a linear endomorphism, prove that for any $x \in V$, the map $y \mapsto b(x, f(y))$ is a linear form. By lemma 201, there is a vector ${}^\tau f(x) \in V$ such that $b(x, f(y)) = b({}^\tau f(x), y)$ for all y. Show that ${}^\tau f$ is a linear map. It is called the *adjoint of f*.

Proposition 202 *An isometry $f \in O(V, W)$ is always injective, and it is an isomorphism, whose inverse is also an isometry, if V and W have the same finite dimension n. The composition $g \circ f$ of two isometries $f : V \to W$ and $g : W \to X$ is an isometry, and $O(V)$ is a subgroup of $GL(V)$, called the orthogonal group of V.*

Proof For an isometry $f : (V, b) \to (W, c)$ and $v \in V$, $\|f(v)\| = \|v\|$ in the respective norms, i.e., f conserves norms. But then, if $v \neq 0$, $f(v) \neq 0$, so f is injective. If both spaces have the same finite dimension, f must also be surjective, and the inverses of such isometries also conserve norms, and norms define the bilinear forms. So they are also isometries. Further, the composition of isometries is an isometry, since conservation of norms is a transitive relation. Hence $O(V)$ is a subgroup of $GL(V)$. □

Exercise 122 For $V = \mathbb{R}^n, n > 0$, we have the standard bilinear form $(?, ?) : \mathbb{R}^n \times \mathbb{R}^n \to \mathbb{R}$, or *scalar product* with

$$((x_1, \ldots x_n), (y_1, \ldots y_n)) = (x_1, \ldots x_n) \cdot (y_1, \ldots y_n)^\tau = \sum_{i=1,\ldots n} x_i y_i,$$

the product of a row and a column matrix, where we omit the parentheses for the resulting number. Show that the standard bilinear form is symmetric and positive definite.

The form of any non-zero Euclidean space can be calculated by means of matrix products as follows: Let $(e_i)_{i=1\ldots n}$ be a basis of V, and define the *associated matrix of the bilinear form* by $B = (B_{ij}) \in \mathbb{M}_{n,n}(\mathbb{R})$ with $B_{ij} = b(e_i, e_j)$. Then bilinearity of the bilinear form b implies the following formula for the representations $x = \sum_i \xi_i e_i$ and $y = \sum_i \eta_i e_i$ of the vectors x and y by their n-tuples in $\mathbb{R}^n$:

$$b(x, y) = (\xi_i) \cdot (B_{ij}) \cdot (\eta_i)^\tau$$

One recognizes that the scalar product defined above is the special case where $(B_{ij}) = (\delta_{ij}) = E_n$. The question, whether one may find a basis for every Euclidean space such that its associated matrix becomes so simple, can be answered by "yes!", but we need some auxiliary theory which will also justify the hitherto mysterious wording "orthogonal group".

Definition 170 *In a Euclidean space* (V, b), *a vector* x *is said to be* orthogonal *to a vector* y *iff* $b(x, y) = 0$, *in signs:* $x \perp y$. *Since* b *is symmetric, orthogonality is a symmetric relation. A sequence* $(x_1, \ldots x_k)$ *of vectors in* V *is called* orthogonal *iff* $x_i \perp x_j$ *for all* $i \neq j$. *It is called* orthonormal *iff it is orthogonal and we have* $\|x_i\| = 1$ *for all* i.

Two subspaces $U, W \subset V$ *are called* orthogonal to each other, *in signs* $U \perp W$ *iff* $u \perp w$ *for all* $u \in U, w \in W$. *The subspace of all vectors which are orthogonal to a subspace* U *is the largest subspace orthogonal to* U *and is denoted by* $U^{\perp}$.

Proposition 203 (Gram-Schmidt Orthonormalization) *If* $(x_1 \ldots x_k)$ *with* $k > 0$ *is a sequence of linearly independent vectors in a Euclidean space* (V, b), *then there is an orthonormal sequence* $(e_1 \ldots e_k)$ *of linearly independent vectors such that for every index* $i = 1, \ldots k$, $(x_1 \ldots x_i)$ *and* $(e_1 \ldots e_i)$ *generate the same subspace. In particular, if* $(x_1 \ldots x_n)$ *is a basis of* V, *then there is a orthonormal basis* $(e_1, \ldots e_n)$ *such that* $(x_1 \ldots x_i)$ *and* $(e_1 \ldots e_i)$ *generate the same subspaces for all* $i = 1, \ldots n$.

Proof The construction is by induction on the length k of the sequence. For $k = 1$, $e_1 = \frac{1}{\|x_1\|} x_1$ does the job. Suppose that all $x_1, x_2, \ldots x_i$ are represented by an orthonormal sequence $e_1, e_2, \ldots e_i$ in the required way. Setting $e_{i+1} = x_{i+1} + \sum_{j=1,\ldots i} \lambda_j e_j$, if we find a solution, then clearly the space generated by $x_1, \ldots x_{i+1}$ coindexes with the space generated by $e_1, \ldots e_{i+1}$. But the condition that $e_{i+1} \perp e_j$, for all $j = 1, \ldots i$, means that $\lambda_j \cdot \|e_j\|^2 + b(e_j, x_{i+1}) = 0$, which yields $\lambda_j = -b(e_j, x_{i+1})$, since $\|e_j\| = 1$. Now, the resulting e_{i+1} is orthogonal to all previous e_j and cannot vanish, because of the dimension $i + 1$ of the subspace generated by $e_1, \ldots e_{i+1}$. So, to obtain the correct norm 1 of e_{i+1}, replace it by $\frac{1}{\|e_{i+1}\|} e_{i+1}$, and everything is perfect. $\qquad\square$

Observe that the proof of proposition 203 is constructive, i.e., algorithmic, and should be kept in mind together with the result, because in all computer-oriented applications, especially in computer graphics, this construction plays a dominant role.

Example 93 Let $x_1 = (2, 2, 0)$, $x_2 = (1, 0, 2)$ and $x_3 = (0, 2, 1)$ be a basis of $\mathbb{R}^3$. We compute an orthonormal basis $\{e_1, e_2, e_3\}$ using the Gram-Schmidt procedure. For the linear form b we use the ordinary scalar product.

The computation of e_1 is simple. It consists in normalizing x_1:

$$
\begin{aligned}
e_1 &= \frac{1}{\|x_1\|} x_1 \\
&= \frac{1}{\sqrt{2^2 + 2^2 + 0^2}} (2, 2, 0) \\
&= \frac{1}{2\sqrt{2}} (2, 2, 0) \\
&= \left(\frac{\sqrt{2}}{2}, \frac{\sqrt{2}}{2}, 0 \right)
\end{aligned}
$$

For the second vector e_2 we first compute an intermediate value e_2' using the formula from the proof of proposition 203:

$$
\begin{aligned}
e_2' &= x_2 - (e_1, x_2) \cdot e_1 \\
&= (1, 0, 2) - \left(\left(\frac{\sqrt{2}}{2}, \frac{\sqrt{2}}{2}, 0 \right), (1, 0, 2) \right) \cdot \left(\frac{\sqrt{2}}{2}, \frac{\sqrt{2}}{2}, 0 \right) \\
&= (1, 0, 2) - \frac{\sqrt{2}}{2} \cdot \left(\frac{\sqrt{2}}{2}, \frac{\sqrt{2}}{2}, 0 \right) \\
&= \left(1 - \frac{\sqrt{2}}{2} \cdot \frac{\sqrt{2}}{2}, -\frac{\sqrt{2}}{2} \cdot \frac{\sqrt{2}}{2}, 2 \right) \\
&= \left(\frac{1}{2}, -\frac{1}{2}, 2 \right)
\end{aligned}
$$

and then normalize to get e_2:

$$
\begin{aligned}
e_2 &= \frac{1}{\|e_2'\|} e_2' \\
&= \frac{1}{\sqrt{1/2^2 + 1/2^2 + 2^2}} \left(\frac{1}{2}, -\frac{1}{2}, 2 \right) = \frac{\sqrt{2}}{3} \left(\frac{1}{2}, -\frac{1}{2}, 2 \right) \\
&= \left(\frac{\sqrt{2}}{6}, -\frac{\sqrt{2}}{6}, \frac{2\sqrt{2}}{3} \right)
\end{aligned}
$$

The formula for the last vector gets more complex:

$$
\begin{aligned}
e_3' &= x_3 - (e_1, x_3) \cdot e_1 - (e_2, x_3) \cdot e_2 \\
&= x_3 - \left(\left(\frac{\sqrt{2}}{2}, \frac{\sqrt{2}}{2}, 0 \right), (0, 2, 1) \right) \cdot e_1 - \left(\left(\frac{\sqrt{2}}{6}, -\frac{\sqrt{2}}{6}, \frac{2\sqrt{2}}{3} \right), (0, 2, 1) \right) \cdot e_2 \\
&= x_3 - \sqrt{2} \cdot \left(\frac{\sqrt{2}}{2}, \frac{\sqrt{2}}{2}, 0 \right) - \frac{\sqrt{2}}{3} \cdot \left(\frac{\sqrt{2}}{6}, -\frac{\sqrt{2}}{6}, \frac{2\sqrt{2}}{3} \right) \\
&= (0, 2, 1) - (1, 1, 0) - \left(\frac{1}{9}, -\frac{1}{9}, \frac{4}{9} \right) \\
&= \left(-\frac{10}{9}, \frac{10}{9}, \frac{5}{9} \right)
\end{aligned}
$$

Normalizing e_3' finally yields e_3:

$$e_3 = \frac{1}{\|e_3'\|} e_3'$$

$$= \frac{1}{\sqrt{(-\frac{10}{9})^2 + (\frac{10}{9})^2 + (\frac{5}{9})^2}} e_3'$$

$$= \frac{3}{5} \cdot \left(-\frac{10}{9}, \frac{10}{9}, \frac{5}{9}\right)$$

$$= \left(-\frac{2}{3}, \frac{2}{3}, \frac{1}{3}\right)$$

Summarizing, the orthonormal basis is:

$$e_1 = \left(\frac{\sqrt{2}}{2}, \frac{\sqrt{2}}{2}, 0\right), e_2 = \left(\frac{\sqrt{2}}{6}, -\frac{\sqrt{2}}{6}, \frac{2\sqrt{2}}{3}\right), e_3 = \left(-\frac{2}{3}, \frac{2}{3}, \frac{1}{3}\right)$$

Figure 24.1 shows both bases. It is left to the reader to check that the e_i are indeed pairwise orthogonal.

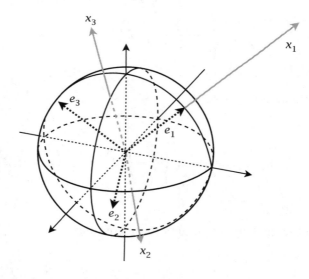

Fig. 24.1. The base x_i and its orthonormalization e_i from example 93.

Exercise 123 Show that, for an endomorphism $f : V \to V$ on a Euclidean space, if A is the matrix of f with respect to an orthonormal basis, the adjoint endomorphism $^\top f$ has the transpose $A^\top$ as its matrix with respect to this basis.

Corollary 204 *For an n-dimensional Euclidean space (V, b), if (e_i) is an orthonormal basis (which exists according to Gram-Schmidt), then the group $O(V)$ identifies with the subgroup $O_n(\mathbb{R}) \subset GL_n(\mathbb{R})$ consisting of all matrixes A with $A^\tau \cdot A = E_n$. In particular, $\det(f) = \pm 1$ for $f \in O(V)$. The orthogonal group is the disjoint union of the normal subgroup $SO(V) \subset O(V)$ of the isometries f with $\det(f) = 1$, called rotations, and the coset $O^-(V)$ of the isometries f with $\det(f) = -1$. $SO(V)$ is called the* special orthogonal group of V.

Proof Given a orthonormal basis $(e_1, \ldots e_n)$ of (V, b), if $f \in O(V)$ is represented by the matrix A relative to this basis, then $\delta_{ij} = b(e_i, e_j) = b(f(e_i), f(e_j)) = \sum_{t=1,\ldots n} A_{ti} A_{tj} = \sum_{t=1,\ldots n} A^\tau_{it} A_{tj} = (A^\tau \cdot A)_{ij}$. This means that $A^\tau \cdot A = E_n$. Conversely, if the latter equation holds, then reading these equalities in the other direction, we have $\delta_{ij} = b(e_i, e_j) = b(f(e_i), f(e_j))$, and therefore f conserves the bilinear form's values for the orthonormal basis $(e_1, \ldots e_n)$. But then, by bilinearity, it conserves bilinear form values $b(x, y)$ for any x and y. The rest is straightforward. $\square$

Corollary 205 *In a Euclidean space (V, b), if U is a subspace, then $U^\perp$ is a complement of U in V, i.e., we have an isomorphism $U \oplus U^\perp \overset{\sim}{\to} V$. In particular,*
$$dim(U) + dim(U^\perp) = dim(V).$$

Exercise 124 Give a proof of the corollary 205 by using proposition 203 and the Steinitz theorem 193.

We are now ready to describe hyperplanes in a Euclidean space.

Definition 171 *In a Euclidean space (V, b) of dimension $n > 0$, a* hyperplane *is the translate $H = T^v(W)$ of a sub-vector space $W \subset V$ of codimension 1, i.e., $dim(W) = n - 1$.*

Proposition 206 *In a Euclidean space (V, b), a hyperplane H can be defined by an equation*
$$H = \{h \mid h \in V, k \perp (h - v)\}$$
where v and k are vectors with $k \neq 0$.

Exercise 125 Give a proof of proposition 206. Use this idea: We know that $H = T^v(W)$ for a subspace W of codimension 1. Since $h \in H$ means

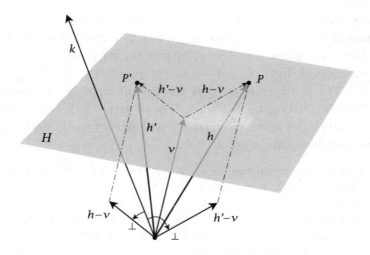

Fig. 24.2. The construction of a 2-dimensional hyperplane H in a 3-dimensional Euclidean space, according to proposition 206.

that $h - v \in W$, and since $W = (W^{\perp})^{\perp}$ with $dim(W^{\perp}) = 1$, we have $h - v \perp k$ for any generator of $W^{\perp}$.

Exercise 126 Rewrite the equation in proposition 206 in terms of an equation using coordinates for an orthonormal basis.

A special type of isometries are the reflections at a hyperplane H of a non-zero Euclidean space (V, b). By definition, a hyperplane is a sub-vector space $H \subset V$ of dimension $dim(H) = dim(V) - 1$. By corollary 205, we have a 1-dimensional complement $H^{\perp} = \mathbb{R} \cdot x$, where $x \neq 0$ is any vector in $H^{\perp}$, and $V = H \oplus H^{\perp}$. This defines a linear map ρ_H on V by

$$\rho_H(y) = y - 2 \cdot \frac{b(y, x)}{b(x, x)} \cdot x$$

where any other generator $x' \in H^{\perp}$ is a scaling of x, $x' = \lambda x$, and therefore yields the same map, so the map only depends on H. In fact, $\rho_H|_H = Id_H$ and $\rho_H|_{H^{\perp}} = -Id_{H^{\perp}}$. Therefore $\rho_H \in O^-(V)$ and $\rho_H^2 = Id_V$. The isometry ρ_H is therefore called the *reflection at H*.

Exercise 127 If $x \neq y$ are two different vectors of equal norm $\|x\| = \|y\|$, then the reflection ρ_H at $H = (x - y)^{\perp}$ exchanges x and y, i.e., $\rho_H(x) = y$.

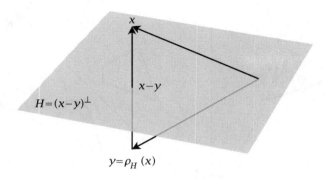

Fig. 24.3. If $\|x\| = \|y\|$, $x \neq y$, and $H = (x - y)^\perp$, then $\rho_H(x) = y$.

This exercise entails the theorem about the fundamental role of reflections:

Proposition 207 *Every $f \in O(V)$ for a non-zero Euclidean space V is the product of at most $dim(V)$ reflections (the identity for zero reflections). Every rotation is the product of an even number of reflections. In particular, for $dim(V) = 2$, a rotation is a product of two reflections.*

Proof Suppose that $f \neq Id_V$. Then there is x such that $f(x) = y \neq x$. Following exercise 127, the reflection ρ_H at $H = (x - y)^\perp$ exchanges x and y and fixes the orthogonal space H pointwise. Therefore $\rho_H \circ f$ fixes the line $\Delta = \mathbb{R}(x - y)$ pointwise, and, since it is an isometry, also H, but not necessarily pointwise. Then the restriction of $g = \rho_H \circ f$ to H is an isometry of H, which has dimension $dim(V) - 1$. So, by recursion, we have $g = Id_\Delta \times g_H$ with $g_H \in O(H)$. If $g_H = Id_H$, we have $f = \rho_H$, and we are finished. Else, we have $g_H = \rho_{H_1} \times \ldots \rho_{H_k}$, $k \leq dim(V) - 1$, for hyperplanes $H_j \subset H$. But each reflection ρ_{H_i} extends to a reflection at $H_i \oplus \Delta$, leaving Δ pointwise fixed, since $\Delta \perp H$, and therefore $\Delta \perp H_i^\perp$, where $H_i^\perp$ is the line orthogonal to H_i in H. So we are done, since $f = \rho_H \circ g$. Finally, since a rotation in a 2-dimensional space cannot be one single reflection, it must be the product of two of them, since it is the product of at most two of them. □

In figure 24.4, the geometrical object x is first reflected through the axis R, then through S. This corresponds to a rotation by an angle a.

Among the orthonormal bases (e_i) and (d_i) of a Euclidean space (V, b) we have an equivalence relation $(e_i) \sim (d_i)$ iff the transition isometry $e_i \mapsto d_i$ has determinant 1. Bases in one of the two equivalence classes

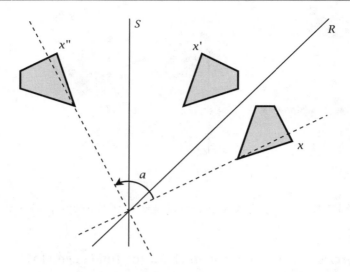

Fig. 24.4. A rotation in $\mathbb{R}^2$ is a product of two reflections.

are said to have the *same orientation*, i.e., each of these two equivalence classes defines an orientation.

24.2 Trigonometric Functions from Two-Dimensional Rotations

In this section we deal exclusively with the case $dim(V) = 2$, i.e., the plane geometry, and the structure of the group $SO(V)$ of rotations.

Proposition 208 *Given an orthonormal basis* (e_i) *of* V, *let* $M_f \in \mathbb{M}_{2,2}(\mathbb{R})$ *be the associated matrix of an isometry* $f \in GL(V)$. *Then*

(i) *we have* $f \in SO(V)$ *(a rotation) iff* $M_f = \begin{pmatrix} a & -b \\ b & a \end{pmatrix}$ *and* $a^2 + b^2 = 1$,

(ii) *we have* $f \in O^-(V)$ *(a reflection) iff* $M_f = \begin{pmatrix} a & b \\ b & -a \end{pmatrix}$ *and* $a^2 + b^2 = 1$.

(iii) *The group* $SO(V)$ *is abelian and the product matrix for two rotations* f *and* g *is*

$$M_g \cdot M_f = \begin{pmatrix} u & -v \\ v & u \end{pmatrix} \cdot \begin{pmatrix} a & -b \\ b & a \end{pmatrix} = \begin{pmatrix} au - bv & -(av + bu) \\ av + bu & au - bv \end{pmatrix}.$$

(iv) *The number a is independent of the chosen orthonormal basis, and so is $|b|$. If another orthonormal basis (e'_i) is chosen with the same orientation, then b does not change.*

(v) *For any two vectors $x, y \in S(V) = \{z \mid z \in V, \|z\| = 1\}$, there is exactly one $f \in SO(V)$ such that $f(x) = y$.*

Proof Clearly, concerning points (i) and (ii), according to corollary 204 the matrixes in the described form define isometries f which are rotations or reflections, respectively. Conversely, if $\begin{pmatrix} a \\ b \end{pmatrix}$ is the first column of M_f, then we must have $a^2 + b^2 = 1$, since the norm must be 1. On the other hand, the second column must be orthogonal to the first, and $\begin{pmatrix} -b \\ a \end{pmatrix}$ is orthogonal to the first and has norm 1. So $\begin{pmatrix} -b \\ a \end{pmatrix}$ must be $\pm f(e_2)$. But the determinant must be 1 if f is a rotation, so the second column must be $\begin{pmatrix} -b \\ a \end{pmatrix}$. Since the determinant must be -1, if f is a reflection, the second column must be the negative of the first, so (i) and (ii) are settled.

For point (iii), knowing that rotation matrixes have the shape described in (i), the formula in (iii) shows that the product of rotations is commutative.

For (iv), we shall see in proposition 214 that the characteristic polynomial $\chi_{M_f}(X)$ of f is independent of its matrix M_f (that result does not presuppose the present one). But the coefficient of X in $\chi_f(X)$ is $-2a$ with the notation described in (i) and (ii). So a is uniquely determined, and therefore also $|b|$. If one changes the base by a rotation, then the new matrix of f is the conjugate of the old matrix by a rotation matrix. A straight calculation shows that b is also invariant.

Statement (v) follows from the fact that the orthogonal spaces $x^{\perp}$ and $y^{\perp}$ are 1-dimensional. So there are two possibilities to map a unit vector in $x^{\perp}$ to a unit vector in $y^{\perp}$. Exactly one of them has positive determinant, and this is our candidate from $SO(V)$. □

Exercise 128 In the case (ii) of a reflection in proposition 208, calculate the reflection formula ρ_H.

We now have to justify the word "rotation" and want to define the cosine and sine functions, together with the associated angles. To this end, recall that U is the unit circle $S(\mathbb{C})$, i.e., the multiplicative group of complex numbers z with norm $|z| = 1$.

Proposition 209 *Suppose that we are given an orthonormal basis of V, $dim(V) = 2$. Consider the maps $\cos : SO(V) \to \mathbb{R} : f \mapsto a = (M_f)_{11}$ and*

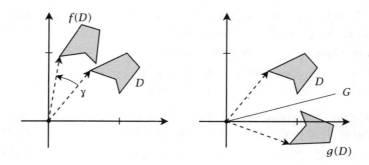

Fig. 24.5. Using $a = \frac{\sqrt{3}}{2}$ and $b = \frac{1}{2}$ in the matrixes of proposition 24.5, on the left is a rotation $f \in SO(\mathbb{R}^2)$ by the angle y, and on the right a reflection $g \in O^-(\mathbb{R}^2)$ about the axis G.

$\sin : SO(V) \to \mathbb{R} : f \mapsto b = (M_f)_{21}$, *then the map*

$$\text{cis} : SO(V) \to U : f \mapsto \cos(f) + i \cdot \sin(f)$$

is an isomorphism of groups. The isomorphism remains unchanged if we select another orthonormal basis of the same orientation.

Proof The only point is the multiplication of complex numbers, which corresponds to the product of rotations, but this is evident from the product formula in proposition 208, (iii). □

Now we know that rotations identify with complex numbers on the unit circle group U in $\mathbb{C}$, but we would like to have the concept of an angle which gives rise to a complex number in U. In fact, we have this result, which will be proved in the advanced topic of part three (volume II of this book):

Proposition 210 *There is a surjective group homomorphism $A : \mathbb{R} \to U$ whose kernel is $Ker(A) = 2\pi\mathbb{Z}$, where π is the positive number 3.1415926... which will be discussed in the advanced topic of part three. So with $f \in SO(V)$ a coset of numbers $\theta \in \mathbb{R}$ is associated such that $\cos(f) + i\sin(f) = A(\theta)$. We also write $\cos(\theta)$ and $\sin(\theta)$ instead of $\cos(f)$ and $\sin(f)$ respectively for the rotation associated with θ. Any such θ is called* angle *of f. The matrix of f is:*

$$M_f = \begin{pmatrix} \cos(\theta) & -\sin(\theta) \\ \sin(\theta) & \cos(\theta) \end{pmatrix}.$$

So the rotation angle θ is determined up to multiples of 2π. The product formula in statement (iii) of proposition 208 translates into the classical goniometric formulas for $\cos(\theta \pm \eta)$ and $\sin(\theta \pm \eta)$:

$$\cos(\theta \pm \eta) = \cos(\theta)\cos(\eta) \mp \sin(\theta)\sin(\eta),$$

$$\sin(\theta \pm \eta) = \sin(\theta)\cos(\eta) \pm \cos(\theta)\sin(\eta).$$

Proof The only point to be proved here are the formulas for $\cos(\theta \pm \eta)$ and $\sin(\theta \pm \eta)$. But both relate to the cos and sin functions of sums/differences of angles, and this means that one looks for the product of the rotations (or the product of one with the inverse of the other) associated with these angles, i.e., $\cos(f \circ g^{\pm 1})$ and $\sin(f \circ g^{\pm 1})$. Then we know the formulas from the product formula in proposition 208, (iii). □

24.3 Gram's Determinant and the Schwarz Inequality

Given an n-dimensional Euclidean space (V, b) and an orthonormal basis (e_i), we may consider the determinant of the linear map associated with a sequence $x_\bullet = (x_i)$ of length n by $f : e_i \mapsto x_i$. On the other hand, we also have the *Gram determinant Gram(x_i) of (x_i)* defined by

$$Gram(x_i) = \det \begin{pmatrix} b(x_1, x_1) & b(x_1, x_2) & \ldots & b(x_1, x_n) \\ b(x_2, x_1) & b(x_2, x_2) & \ldots & b(x_2, x_n) \\ \vdots & \vdots & & \vdots \\ b(x_n, x_1) & b(x_n, x_2) & \ldots & b(x_n, x_n) \end{pmatrix}$$

But the Gram matrix $(b(x_i, x_j))$ clearly equals $M_f^T \cdot M_f$. Therefore we have the Gram equation

$$Gram(x_i) = \det(f)^2$$

For $n = 2$ we immediately deduce the Schwarz inequality:

Proposition 211 (Schwarz Inequality) *If x and y are two vectors in a Euclidean space (V, b), then*

$$|b(x, y)| \leq \|x\| \|y\|,$$

equality holding iff x and y are linearly dependent.

This result may be reinterpreted in terms of the cosine function. If $x, y \neq 0$ choose a 2-dimensional subspace W of V containing x and y, and carrying the induced bilinear form $b|_{W \times W}$. Then we have $\left| \frac{b(x,y)}{\|x\|\|y\|} \right| \leq 1$, and defining $c(x, y) = \frac{b(x,y)}{\|x\|\|y\|}$, we have $b(x, y) = c(x, y) \cdot \|x\| \cdot \|y\|$ with $|c(x, y)| \leq 1$. If $f \in SO(W)$ is the unique rotation with $f(x/\|x\|) = y/\|y\|$, then $c(x, y) = \cos(f) = \cos(\theta(x, y))$, which means that the angle is determined up to integer multiples of 2π by the unit vectors $x/\|x\|$ and $y/\|y\|$ or equivalently the half lines $\mathbb{R}_+ x$ and $\mathbb{R}_+ y$ through x and y. This gives us the famous cosine formula for the bilinear form, where one has chosen an orthogonal basis on a plane containing x and y:

$$b(x, y) = \cos(\theta(x, y)) \cdot \|x\| \cdot \|y\|.$$

We also obtain the following intuitive fact: the triangle inequality for norms.

Corollary 212 (Triangle Inequality) *If x and y are two vectors in a Euclidean space (V, b), then*

$$\|x + y\| \leq \|x\| + \|y\|.$$

Proof Since both sides are non-negative numbers, we may prove that the squares of these numbers fulfill the inequality. But by the Schwarz inequality from proposition 211, we have $\|x + y\|^2 = \|x\|^2 + 2b(x, y) + \|y\|^2 \leq \|x\|^2 + 2\|x\|\|y\| + \|y\|^2 = (\|x\| + \|y\|)^2$. $\qquad \square$

Defining the *distance* between two vectors x and y in a Euclidean space by

$$d(x, y) = \|x - y\|,$$

we obtain these characteristic properties of a distance function:

Proposition 213 *Given a Euclidean space (V, b), the distance function $d(x, y) = \|x - y\|$ has these properties for all $x, y, z \in V$:*

   (i) $d(x, y) \geq 0$, *and* $d(x, y) = 0$ *iff* $x = y$,

   (ii) *(Symmetry)* $d(x, y) = d(y, x)$,

   (iii) *(Triangle Inequality)* $d(x, z) \leq d(x, y) + d(y, z)$.

Proof Claim (i) is true because the norm is always non-negative, and strictly positive if the argument is not zero (b is positive definite). For (ii), we have $d(x, y)^2 = \|x - y\|^2 = b(x - y, x - y) = (-1)^2 b(y - x, y - x) = \|y - x\|^2 = d(y, x)^2$. For (iii), again by the Schwarz inequality from proposition 211, we have $d(x, z) = \|x - z\| = \|(x - y) + (y - z)\| \leq \|x - y\| + \|y - z\|$. $\qquad \square$

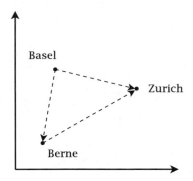

Fig. 24.6. Triangle Inequality: The beeline between Basel and Zurich is shorter than a detour over Berne.

These three properties are those which will later define a metric in our topological considerations. So proposition 213 guarantees that a Euclidean space has a metric which is induced by the norm derived from the given positive definite bilinear form.

... ...

... ...
... ...
... ...

CHAPTER 25

Eigenvalues, the Vector Product, and Quaternions

This chapter deals with the geometry in three dimensions, the most important case since we all live in 3D space, and therefore, computer graphics must completely control this space and its transformations. *Again, in this chapter we shall only deal with finite-dimensional $\mathbb{R}$-vector spaces.*

25.1 Eigenvalues and Rotations

We begin with an analysis of the special orthogonal group $SO(V)$ with $dim(V) = 3$; recall that for the standard scalar product $(?,?)$ on $\mathbb{R}^3$ this group is also denoted by $SO_3(\mathbb{R})$.

First, we want to show that every $f \in SO(V)$ is really what one would call a rotation: it has a rotation axis and turns the space around this axis by a specific angle as introduced in the last chapter. We are looking for a *rotation axis*, i.e., a 1-dimensional sub-vector space $R = \mathbb{R}x$ which remains fixed under f, i.e., $f(x) = x$. Rewriting this equation as $f(x) - 1 \cdot x = 0$ means that there is a solution of the linear equation $(f - 1 \cdot Id_V)(x) = 0$, i.e., the linear map $f - 1 \cdot Id_V$ has a non-trivial kernel, namely $x \in Ker(f - 1 \cdot Id_V) - \{0\}$. We know from the theory of linear equations that the condition $Ker(f - 1 \cdot Id_V) \neq 0$ is equivalent to the vanishing of the determinant $\det(f - 1 \cdot Id_V)$. This means that we have to look for solutions of the equation $\det(f - X \cdot Id_V) = 0$. Let us make this equation more precise.

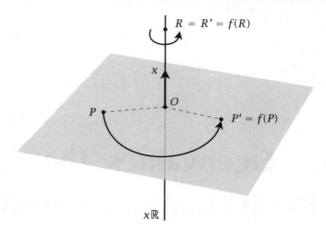

Fig. 25.1. Points on the rotation axis $x\mathbb{R}$ like R are not affected by the rotation.

Lemma 214 *If* V *is a real vector space of finite dimension* n, *if* $f \in$ *End*(V), *and if* $M \in \mathbb{M}_{n,n}(\mathbb{R})$ *is the matrix representation of* f *with respect to a basis, then the characteristic polynomial* $\chi_M = \det(M - X \cdot E_n)$ *is independent of the chosen basis. We therefore also write* χ_f *instead of* χ_M. *We have*

$$\chi_f = \sum_{i=0,\ldots n} t_i X^i = (-1)^n X^n + (-1)^{n-1} tr(f) X^{n-1} \pm \ldots + \det(f)$$

where the second coefficient $tr(f) = \sum_{i=1,\ldots n} M_{ii}$ *is called the* trace *of* f *(or of the matrix which represents* f).

Proof If we change the basis of V, the new matrix M' of f is the conjugate of M by the base change matrix X, this is corollary 195. But we know from corollary 185 that the characteristic polynomial does not change under conjugation. □

So our problem is this: to find special solutions of the characteristic polynomial equation $\chi_f(X) = 0$.

Definition 172 *If* V *is a real vector space of finite dimension* n, *and if* $f \in$ *End*(V), *the zeros* λ *of the characteristic polynomial* χ_f

$$\chi_f(\lambda) = 0$$

are called the eigenvalues *of* f. *The non-zero elements* x *of the non-trivial kernel* $Ker(f - \lambda \cdot Id_V)$ *for an eigenvalue* λ *are called* eigenvectors *of* f *(corresponding to* λ). *They are characterized by* $f(x) = \lambda x$.

We are more concretely interested in the case $dim(V) = 3$, where we have the polynomial

$$\chi_f = -X^3 + tr(f)X^2 + t_1 X + \det(f),$$

and we are looking for solutions $-\lambda^3 + tr(f)\lambda^2 + t_1\lambda + \det(f) = 0$ thereof. It will be shown in the second volume of this book that every polynomial $P \in \mathbb{R}[X]$ of odd degree has at least one root in $\mathbb{R}$. Therefore, the characteristic polynomial has a real eigenvalue λ_0. Let us show that in the case of $f \in SO(V)$, there is a positive eigenvalue. The equation factorizes to

$$
\begin{aligned}
-X^3 + tr(f)X^2 + t_1 X + \det(f) &= -X^3 + tr(f)X^2 + t_1 X + 1 \\
&= -(X - \lambda_0)(X^2 + bX + c)
\end{aligned}
$$

and therefore $\lambda_0 \cdot c = 1$. If $\lambda_0 < 0$, then $c < 0$, and then we have two more real solutions $\lambda_{1,2} = \frac{-b \pm \sqrt{b^2 - 4c}}{2}$ since $b^2 - 4c \geq 0$, and $\mathbb{R}$ has square roots of non-negative elements (see corollary 90 in chapter 9). Therefore $1 = \lambda_0\lambda_1\lambda_2$, which means that one of these solutions must be positive. So, after renaming the roots, we take a positive eigenvalue λ_0 and look at a corresponding eigenvector x, i.e., $f(x) = \lambda_0 x$. Since f is an isometry, we have $\|f(x)\| = |\lambda_0|\|x\| = \|x\|$, whence $\lambda_0 = 1$, and we have a 1-dimensional subspace $\mathbb{R}x$ which is left pointwise fixed under f. This shows:

Proposition 215 *Every rotation in* $SO(V)$ *for an Euclidean space* (V, b) *with* $dim(V) = 3$, *has a rotation axis* $A_f = \mathbb{R}x$, *i.e.,* $f|_{A_f} = Id_{A_f}$. *If* $f \neq Id_V$, *the rotation axis is unique.*

Proof The only open point here is uniqueness of a rotation axis. But if we had two rotation axes, the plane H generated by these axes would be fixed pointwise, so $H^\perp$ would have to be reflected or to be fixed, in both cases, this would not yield a non-trivial transformation in $SO(V)$. □

Since $A_f \oplus A_f^\perp \overset{\sim}{\to} V$, the isometry f which leaves the rotation axis pointwise invariant also leaves invariant the 2-dimensional orthogonal plane $A_f^\perp$, i.e., each point of $A_f^\perp$ is mapped to another point of $A_f^\perp$. Taking an orthonormal basis (a_1, a_2) of $A_f^\perp$, we may rewrite f in terms of the orthonormal basis $(a_0, a_1, a_2), a_0 = x/\|x\|$ as a matrix

$$
M_f = \begin{pmatrix} 1 & 0 & 0 \\ 0 & \cos(\theta) & -\sin(\theta) \\ 0 & \sin(\theta) & \cos(\theta) \end{pmatrix}
$$

which makes explicit the rotation in the plane $A_f^\perp$ orthogonal to the rotation axis A_f. This representation is unique for all orthonormal bases (a'_0, a'_1, a'_2) of same orientation, which represent the orthogonal decomposition defined by the rotation axis, i.e., $A_f = \mathbb{R}a'_0$ and $A_f^\perp = \mathbb{R}a'_1 + \mathbb{R}a'_2$.

The above formula and the invariance of the trace $tr(f)$ implies this uniqueness result:

Corollary 216 *Every rotation in* $SO(V)$ *for an Euclidean space* (V, b) *with* $dim(V) = 3$, *has a rotation angle* θ *whose* $\cos(\theta)$ *is uniquely determined, i.e.,*

$$\cos(\theta) = \frac{1}{2}(tr(f) - 1).$$

So we are left with the elements $f \in O^-(V)$ for $dim(V) = 3$. There are two cases for the characteristic polynomial:

1. Either it has a positive solution $\lambda_0 = 1$, then we have again a rotation axis A_f and the orthogonal complement is left invariant not by a rotation, but by a reflection at a line $B_f \in A_f^\perp$. This line and the axis A_f are left point-wise invariant, whereas the orthogonal complement line $C = B_f^\perp$ in $A_f^\perp$ is inverted by -1. This means that f is a *reflection* ρ_H at the plane H generated by A_f and B_f.

2. Or else, there is no positive eigenvalue. Then we have the eigenvalue -1 and an eigenvector x with $f(x) = -x$. Then the reflection ρ_H at the plane $H = x^\perp$ yields an isometry $g = \rho_H \circ f \in SO(V)$, with rotation axis $A_g = \mathbb{R}x$. Therefore, $f = \rho_H \circ g$ is the *composition of a rotation* g *and a reflection* ρ_H *orthogonal to the rotation axis of* g.

So this classification covers all cases of isometries in $O(V)$, $dim(V) = 3$.

25.2 The Vector Product

If we are given a triple (x_i) of vectors $x_i \in V$ in a Euclidean space (V, b) of dimension 3, and if (e_i) is an orthonormal basis of V, then the linear map $f : e_i \mapsto x_i$ has a determinant $det(f)$, which we also write as $det_{(e_i)}(x_1, x_2, x_3)$. We know that this determinant does only depend on the orientation defined by (e_i). So if we fix the orientation ω of (e_i), we have a function $det_\omega(x_1, x_2, x_3)$. Rewriting this function as a function on the matrix M_f in the basis (e_i), we deduce the following properties from the general properties of the determinant described in theorem 180 of chapter 20:

1. The function $\det_\omega(x_1, x_2, x_3)$ is linear in each argument.

2. The function is skew-symmetric, i.e., $\det_\omega(x_1, x_2, x_3) = 0$ if two of the three x_i are equal. This entails that $\det_\omega(x_{\pi(1)}, x_{\pi(2)}, x_{\pi(3)}) = sig(\pi)\det_\omega(x_1, x_2, x_3)$ for a permutation $\pi \in S_3$.

3. $\det_\omega(x_1, x_2, x_3) = 1$ for any orthonormal basis $(x_1, x_2, x_3) \in \omega$.

Therefore we may introduce this construction of the vector product:

Definition 173 *Given a 3-dimensional Euclidean space (V, b), with the previous notations, fix a pair (x_1, x_2) of vectors in V. Then under the isomorphism $*b : V \xrightarrow{\sim} V^*$, the linear form*

$$d_{(x_1, x_2)} \in V^* \text{ and } d_{(x_1, x_2)}(x) = \det_\omega(x_1, x_2, x)$$

corresponds to a vector $*b^{-1}d_{(x_1, x_2)} \in V$ which we denote by $x_1 \wedge x_2$, and which is characterized by the equation

$$b(x_1 \wedge x_2, x) = \det_\omega(x_1, x_2, x),$$

and which is called the vector product *of x_1 and x_2 (in this order).[1] The orientation ω is not explicitly denoted in the vector product expression, but should be fixed in advance, otherwise the product is only defined up to sign.*

Given the representations $x_1 = \sum_i x_{i1}e_i$ and $x_2 = \sum_i x_{i2}e_i$, the vector product has these coordinates in terms of the basis (e_i):

$$(x_1 \wedge x_2)_1 = x_{21}x_{32} - x_{31}x_{22},$$
$$(x_1 \wedge x_2)_2 = x_{31}x_{12} - x_{11}x_{32},$$
$$(x_1 \wedge x_2)_3 = x_{11}x_{22} - x_{21}x_{12}.$$

We recall that $x_1 \wedge x_2$ is linear in each argument and skew-symmetric, i.e., $x_1 \wedge x_2 = -x_2 \wedge x_1$.

Exercise 129 Show that $(x \wedge y) \perp x$ and $(x \wedge y) \perp y$. Calculate $e_2 \wedge e_1$ for two basis vectors of an orthonormal basis (e_1, e_2, e_3) of the given orientation.

---

[1] Some texts also use the notation $x_1 \times x_2$ instead of $x_1 \wedge x_2$.

Exercise 130 Calculate the vector product $(1, -12, 3) \wedge (0, 3, 6)$ of two vectors in $\mathbb{R}^3$ with the standard scalar product $b = (?, ?)$ and the orientation of the standard basis $(e_1 = (1, 0, 0), e_1 = (0, 1, 0), e_1 = (0, 0, 1))$.

Proposition 217 *Given a 3-dimensional Euclidean space (V, b) with a fixed orientation, the vector product satisfies these identities for any three vectors u, v and w:*

(i) $u \wedge (v \wedge w) = b(u, w)v - b(u, v)w$,

(ii) *(Jacobi Identity)* $u \wedge (v \wedge w) + v \wedge (w \wedge u) + w \wedge (u \wedge v) = 0$.

Proof Since the expressions in question are all linear in each argument, it suffices to verify them for $u = e_i, v = e_j, w = e_k, i, j, k = 1, 2, 3$, where e_1, e_2, e_3 is an orthonormal basis. Further, (i) is skew-symmetric in v and w whence some cases can be omitted. We leave to the reader the detailed verification, which follows from the formulas given in this section for the coordinatewise definition of the vector product. ☐

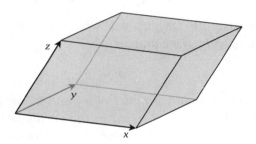

Fig. 25.2. The parallelepiped spanned by the vectors x, y and z.

Remark 28 We should add a remark on surfaces of parallelograms and volumes of parallelepipeds which have not yet been defined and which belong to the chapters on calculus in the second volume of this book. However, from high school the reader may temporarily recall the definitions of surfaces and volumes and read these remarks which will be fully justified later. If we are given $x, y, z \in V$, a Euclidean space of dimension 3, then we can consider the parallelogram spanned by x and y, i.e., the set $Parallel(x, y) = \{t \mid t = \lambda x + \mu y, 0 \le \lambda, \mu \le 1\}$ as well as the parallelepiped spanned by x, y and z, i.e., the set $Parallel(x, y, z) = \{t \mid t = \lambda x + \mu y + \nu z, 0 \le \lambda, \mu, \nu \le 1\}$. The surface of $Parallel(x, y)$ is

$|x \wedge y|$, whereas the volume of $Parallel(x, y, z)$ is $|b(x \wedge y, z)|$, and these numbers are independent of the orientation.

25.3 Quaternions

The flavor of the vector product suggests that it should play a major role in the description of rotations in 3-dimensional Euclidean spaces. We have learned that 2-dimensional rotations are essentially products of complex numbers of norm 1. We shall see in this section that in three dimensions, we also have an algebraic structure, the quaternions, which mimic rotations in three dimensions, and that quaternions are effectively multiplied by use of the vector product. It seems that there is a deep relation between algebra and geometry, a relation which is classical in algebraic geometry, but which turns out to get more and more intense for the groups of transformations in linear geometry. It is not by chance that this field of research is named geometric algebra.

The quaternions were invented by the English mathematician William Rowan Hamilton, who was in search for a correspondence to the geometric interpretation of complex numbers for three dimensions. Although quaternions were overrated by some fanatics (the "quaternionists", see [12]) as being a kind of magic if not divine objects, they are gaining increasing signification in computer graphics and also in aerospace science (see [35]).

Let us first motivate how quaternions relate to the vector product. As is shown by the Jacoby identity in proposition 217, the anticommutative ($x \wedge y = -y \wedge x$) vector product is not associative. In particular, $0 = (x \wedge x) \wedge y$, while $x \wedge (x \wedge y)$ may be different from 0. Quaternions were invented while looking for the construction of an algebraic multiplication structure which was associative, such that the square of an element x cannot be annihilated as it happens with the vector product. As already seen with the complex numbers, such a desideratum may be met by adding supplementary dimensions to the given space. Hamilton's solution was to add one dimension to the three dimensions of $\mathbb{R}^3$ and to delegate the non-vanishing part of the square to that new dimension. His philosophical justification of this procedure was that the three space coordinates must be supplemented by one time coordinate to describe the comprising four-dimensional space-time. It is not by chance (recall the definition of the vector product, which is intimately related to the bilinear form of the Euclidean space) that the new component of the Hamilton product of $u, v \in \mathbb{R}^3$ was just the negative of the scalar product $-(u, v)$. There is in fact nothing arbitrary in Hamilton's construction. It can be shown that not only $\mathbb{C}$ is the only field of dimension two over the reals, but also $\mathbb{H}$ is definitely the only skew field of dimension four over the reals! Here is the formal definition.

Definition 174 *The* quaternions *are a skew field* $\mathbb{H}$ *(for Hamilton), whose additive group is* $\mathbb{R}^4$, *and whose multiplication is defined as follows. One identifies* $\mathbb{R}^4$ *with the direct sum* $\mathbb{R} \oplus \mathbb{R}^3$ *defined by the projections* $R : \mathbb{H} \to \mathbb{R} : (r, x, y, z) \mapsto r$ *and* $P : \mathbb{H} \to \mathbb{R}^3 : (r, x, y, z) \mapsto (x, y, z)$, *such that every* $q \in \mathbb{H}$ *is uniquely decomposed as* $q = r + p$ *with* $r = R(q) \in R(\mathbb{H}) \overset{\sim}{\to} \mathbb{R}$ *and* $p = P(q) \in P(\mathbb{H}) \overset{\sim}{\to} \mathbb{R}^3$. *The summand* p *is called the* pure part *of* q, *and if* $r = 0$, q *is called a* pure quaternion. *The summand* r *is called the* real *part of* q, *and if* $p = 0$, q *is called a* real quaternion. *If the context is clear, the additive notation* $q = r + p$ *is always meant in this understanding.[2] On the pure part, we take the standard scalar product* $(?, ?)$, *together with the orientation given by the canonical basis* $((1, 0, 0), (0, 1, 0), (0, 0, 1))$. *The quaternion product is now defined by*

$$(r + p) \cdot (r' + p') = (r \cdot r' - (p, p')) + (r \cdot p' + r' \cdot p + p \wedge p'),$$

i.e.,

$$R((r + p) \cdot (r' + p')) = r \cdot r' - (p, p'),$$

and

$$P((r + p) \cdot (r' + p')) = r \cdot p' + r' \cdot p + p \wedge p'.$$

On $\mathbb{H}$, conjugation *is defined by* $\overline{?} : \mathbb{H} \to \mathbb{H} : q = r + p \mapsto \overline{q} = r - p$. *The* norm $\|q\|$ *of a quaternion* q *is defined by*

$$\|q\| = \sqrt{q \cdot \overline{q}},$$

which makes sense since $q \cdot \overline{q} = r^2 + (p, p)$ *is a non-negative real quaternion, which coincides with the square norm of* q *when interpreted in the standard Euclidean space* $(\mathbb{R}^4, (?, ?))$.

The immediate properties of $\mathbb{H}$, in particular the skew field properties, are summarized in the following sorite:

---

[2] Remember remark 24 concerning addition in direct sums.

Sorite 218 *Let $q, q' \in \mathbb{H}$ be quaternions. Then*

(i) *Conjugation* $\overline{?} : \mathbb{H} \to \mathbb{H}$ *is a linear anti-involution, i.e.,* $\overline{\overline{q}} = q$ *and* $\overline{q \cdot q'} = \overline{q'} \cdot \overline{q}$.

(ii) $\overline{q} = q$ *iff q is real, i.e., $q = R(q)$.*

(iii) $\overline{q} = -q$ *iff q is pure, i.e., $q = P(q)$.*

(iv) *q is pure iff q^2 is a real number ≤ 0.*

(v) *q is real iff q^2 is a real number ≥ 0.*

(vi) *We have $\|q \cdot q'\| = \|q\| \cdot \|q'\|$.*

(vii) *With the quaternion product, $\mathbb{H}$ becomes a skew field with $1_{\mathbb{H}} = 1+0$. The inverse of $q \neq 0$ is the quaternion $q^{-1} = \frac{1}{\|q\|^2}\overline{q}$.*

(viii) *The injection $\mathbb{R} \to \mathbb{H} : r \mapsto r \cdot 1_{\mathbb{H}}$ identifies the subfield $\mathbb{R}$ of real quaternions, which commutes with all quaternions, i.e., $r \cdot q = q \cdot r$ for all $r \in \mathbb{R}, q \in \mathbb{H}$. Conversely, if $q' \cdot q = q \cdot q'$ for all q, then q' is real.*

(ix) *By linearity in each factor, the multiplication $x \cdot y$ on $\mathbb{H}$ is entirely determined by the following multiplication rules for the four basis vectors $1_{\mathbb{H}}, i, j, k$. Observe that i, j, k are pure and, therefore, their vector product is defined.*

$$1_{\mathbb{H}} \cdot t = t \cdot 1_{\mathbb{H}} = t \text{ for all } t = 1_{\mathbb{H}}, i, j, k,$$
$$i^2 = j^2 = k^2 = -1_{\mathbb{H}},$$
$$ij = -ji = i \wedge j = k,$$
$$jk = -kj = j \wedge k = i,$$
$$ki = -ik = k \wedge i = j.$$

The permutation (i, j, k) induces an automorphism of $\mathbb{H}$.

(x) *Setting $i = (0, 1, 0, 0), j = (0, 0, 1, 0), k = (0, 0, 0, 1)$, the three injections*

$$(i): \mathbb{C} \to \mathbb{H} : a + ib \mapsto a + ib,$$
$$(j): \mathbb{C} \to \mathbb{H} : a + ib \mapsto a + jb,$$
$$(k): \mathbb{C} \to \mathbb{H} : a + ib \mapsto a + kb$$

define identifications of the field $\mathbb{C}$ with the subfields $\mathbb{R} + i\mathbb{R}, \mathbb{R} + j\mathbb{R}, \mathbb{R} + k\mathbb{R}$, respectively, which are related with each other by the automorphism of $\mathbb{H}$ induced by the permutation (i, j, k).

(xi) *We have $(q, q') = \frac{1}{2}(\overline{q}q' + \overline{q'}q)$.*

Proof For (i) and a quaternion $q = r + p$, we have $\overline{\overline{q}} = \overline{(r - p)} = r - (-p) = r + p = q$. And for $q' = r' + p'$, we have

$$
\begin{aligned}
\overline{q \cdot q'} &= \overline{(r \cdot r' - (p, p')) + (r \cdot p' + r' \cdot p + p \wedge p')} \\
&= (r \cdot r' - (p, p')) - (r \cdot p' + r' \cdot p + p \wedge p') \\
&= (r' \cdot r - (p', p)) + (r \cdot (-p) + r' \cdot (-p') + p' \wedge p) \\
&= \overline{q'} \cdot \overline{q}
\end{aligned}
$$

Points (ii) and (iii) are obvious.

For (iv), we have this general formula: $q^2 = (r^2 - (p, p)) + 2rp$. If $q = p$, then $q \cdot q = -\|p\|^2 \leq 0$. Conversely, if $r \neq 0$, then if $p \neq 0$, $P(q^2) = 2rp \neq 0$. If q is real, then $q^2 = r^2 > 0$.

The proof of point (v) works along the same straightforward calculation, we therefore leave the proof as an exercise for the reader.

Next, we prove (viii). The commutativity of real quaternions with all quaternions is immediate from the definition of the quaternion product. Conversely, if $q \cdot q' = q' \cdot q$ for all q', then $p \wedge p' = p' \wedge p = -p \wedge p' = 0$ for all p'. But for $p \neq 0$, taking p, p' linearly independent yields a contradiction, so $p = 0$.

As to (vi), $\|q \cdot q'\| = \sqrt{q \cdot q' \cdot \overline{q \cdot q'}} = \sqrt{q \cdot q' \cdot \overline{q'} \cdot \overline{q}} = \sqrt{q \cdot \|q'\|^2 \cdot \overline{q}}$, but real quaternions commute with all quaternions by (viii), so $\sqrt{q \cdot \|q'\|^2 \cdot \overline{q}} = \sqrt{\|q'\|^2 \cdot q \cdot \overline{q}} = \sqrt{\|q'\|^2 \cdot \|q\|^2} = \|q'\| \cdot \|q\|$, and we are done.

Next, we prove (ix). The first point is $\mathbb{R}$-linearity in each argument. But in view of the defining formula of multiplication, this follows from linearity of the scalar product $(?, ?)$ and of the vector product $\wedge$. The permutation (i, j, k) of the basis ($1_{\mathbb{H}}$ is fixed) transforms the equation $i \cdot j = k$ into $j \cdot k = i$ and this into $k \cdot i = j$, and this latter into the first, further $i^2 = j^2 = k^2 = -1$ is invariant under all permutations of i, j, k, and so are the products with $1_{\mathbb{H}}$.

Point (vii) is straightforward, except that we must verify associativity. This follows from the associativity of multiplication rules for the basis $(1_{\mathbb{H}}, i, j, k)$ in (ix), since multiplication is $\mathbb{R}$-linear in each factor. The reader should verify all equalities $x \cdot (y \cdot z) = (x \cdot y) \cdot z$ for three arbitrary basis elements, but should also use the permutation (i, j, k) to minimize the cases to be studied.

Point (x) is obvious.

Point (xi) is an immediate consequence of the definition of the product. $\square$

We have now established all the necessary algebraic ingredients to tackle the promised geometric features. To this end, recall that we have a group isomorphism of $SO_2(\mathbb{R})$ and the multiplicative group $U \subset \mathbb{C}$ of complex numbers of norm 1, i.e., the *unit circle* $U = S(\mathbb{C}) = \{z \mid \|z\| = 1\}$. Therefore we suggest to consider the group $S(\mathbb{H}) = \{q \mid \|q\| = 1\}$ of unit quaternions (in a Euclidean space, this subset is called the *unit sphere*). This is a group because of point (vi) in sorite 218. The relation between quaternion multiplication and 3-dimensional isometries is this:

Proposition 219 *Let $s, s' \in \mathbb{H}^*$ be non-zero quaternions.*

(i) *If $s \in P(\mathbb{H})$ is pure, then the map $q \mapsto -s \cdot q \cdot s^{-1}$ leaves the pure quaternion space $P(\mathbb{H})$ invariant and its restriction to $P(\mathbb{H})$ is the reflection $\rho_{s^\perp}$ in $P(\mathbb{H})$ at the plane $s^\perp$ orthogonal to s.*

(ii) *The map $q \mapsto Int_s(q) = s \cdot q \cdot s^{-1}$ leaves the pure quaternion space $P(\mathbb{H})$ invariant, and its restriction Int_s^P to $P(\mathbb{H})$ is a rotation in $SO_3(\mathbb{R})$.*

(iii) *We have $Int_s^P = Int_{s'}^P$ iff $\mathbb{R}s = \mathbb{R}s'$.*

(iv) *The restriction Int_s^{PS} of the map Int_s^P to arguments s in the unit sphere $S(\mathbb{H})$ is a surjective group homomorphism*

$$Int^{PS} : S(\mathbb{H}) \to SO_3(\mathbb{R})$$

with kernel $Ker(Int^{PS}) = \{\pm 1\}$.

Proof By criterion (iii) in sorite 218, we suppose q^2 is real and ≤ 0. So $(\pm s \cdot q \cdot s^{-1})^2 = \pm s \cdot q^2 \cdot s^{-1} = q^2$ is also real and ≤ 0, i.e., $\pm s \cdot q \cdot s^{-1}$ is pure. These maps are evidently $\mathbb{R}$-linear. Since $\| \pm s \cdot q \cdot s^{-1} \| = \|s\| \cdot \|q\| \cdot \|s\|^{-1} = \|q\|$, the maps $q \mapsto \pm s \cdot q \cdot s^{-1}$ conserve norms and therefore their restrictions to $P(\mathbb{H})$ are in $O_3(\mathbb{R})$. Now, if s is pure, then $-s \cdot s \cdot s^{-1} = -s$, whereas for a pure $q \perp s$, we have $0 = (q, s) = \frac{1}{2}(\overline{q} \cdot s + \overline{s} \cdot q) = \frac{-1}{2}(q \cdot s + s \cdot q)$, i.e., $-q \cdot s = s \cdot q$, whence $-s \cdot q \cdot s^{-1} = q \cdot s \cdot s^{-1} = q$, and therefore the plane $s^\perp$ orthogonal to s remains fixed, i.e., $q \mapsto -s \cdot q \cdot s^{-1}$ on $P(\mathbb{H})$ is a reflection at $s^\perp$.

The map Int_s for a real s is the identity, so suppose s is not real. Consider for $\mu \in [0, 1]$, the real unit interval, the quaternion $s_\mu = (1 - \mu) + \mu \cdot s$, so $s_0 = 1, s_1 = s$, and $s_\mu \neq 0$ for all $\mu \in [0, 1]$. So $Int_{s_\mu}^P \in O_3(\mathbb{R})$ for all $\mu \in [0, 1]$, i.e., $f(\mu) = det(Int_{s_\mu}^P)$ is never zero, and it is always ± 1. We shall show in the chapter on topology in volume II of this book, that f has the property of being continuous (a concept to be introduced in that chapter). Evidently, $f(0) = 1$. On the other hand, if we had $f(1) = -1$, then by continuity of f, there would exist $\mu \in [0, 1]$ with $f(\mu) = 0$, a contradiction, so $Int_s^P \in SO_3(\mathbb{R})$.

If $s' = \lambda s, \lambda \in \mathbb{R}^*$, then $s' \cdot q \cdot s'^{-1} = \lambda s \cdot q \cdot \lambda^{-1} s^{-1} = s \cdot q \cdot s^{-1}$. Conversely, if $Int_{s'}^P = Int_s^P$, then $-Int_{s'}^P = -Int_s^P$, so the reflection axis $\mathbb{R}s = \mathbb{R}s'$ is uniquely determined. As to (iv), the map Int^{PS} is a group homomorphism, since $(s' \cdot s) \cdot q \cdot (s' \cdot s)^{-1} = s' \cdot s \cdot q \cdot s^{-1}s'^{-1}$. it is surjective, since every rotation is the product of an even number of reflections at hyperplanes. So since the reflections are represented by the quaternion multiplications $-s \cdot q \cdot s^{-1}$ by (i), rotations are represented by the product of an even number such such have the required map. Since two rotations are equal iff their quaternions s define the same line $\mathbb{R}s$, and since any line has exactly two points $t, -t$ on the sphere $S(\mathbb{H})$, the kernel is $\{\pm 1\}$. □

So we recognize that rotations in $SO_3(\mathbb{R})$ can be described by conjugations Int_s^{PS} by quaternions, and the composition of rotations corresponds to the conjugation of products. This is exactly what was expected from the 2-dimensional case of complex multiplication. Observe that the identification of the quotient group $S(\mathbb{H})/\{\pm 1\}$ and $SO_3(\mathbb{R})$ gives us an interesting geometric interpretation of $SO_3(\mathbb{R})$: Observe that each line $\mathbb{R} \cdot x \subset \mathbb{R}^3$ intersects the unit sphere $S(\mathbb{R}^3)$ exactly in two points $s, -s$ of norm 1. These points are identified via the two-element group $\{\pm 1\}$. So the quotient group identifies with the set $\mathbb{P}_3(\mathbb{R})$ of lines through the origin (1-dimensional subspaces) in $\mathbb{R}^4$. This space is called the three-dimensional *projective space of* $\mathbb{R}^3$. So we have this corollary:

Corollary 220 *The group* $SO_3(\mathbb{R})$ *is isomorphic to the projective space* $\mathbb{P}_3(\mathbb{R})$ *with the group structure induced from the quaternion multiplication.*

The only thing which is not clear at this point is the rotation axis which is induced by a quaternion conjugation. Here is the result:

Proposition 221 *Let* $s = r + p$ *be a non-zero quaternion. If s is real, it induces the identity rotation. So let us suppose $p \neq 0$. Then $\mathbb{R}p$ is the rotation axis of* Int_s^{PS}. *The rotation angle* $\theta \in [0, \pi]$ *is given by*

$$\tan\left(\tfrac{\theta}{2}\right) = \sin\left(\tfrac{\theta}{2}\right) / \cos\left(\tfrac{\theta}{2}\right) = \|p\|/r \ \ if \ r \neq 0,$$

and by

$$\theta = \pi \ \ if \ r = 0,$$

the second case corresponding to a reflection orthogonally through $\mathbb{R}p$.

Proof Wlog, we may suppose that $\|s\| = 1$. If $p \neq 0$, then $s \cdot p \cdot s^{-1} = (r + p) \cdot p \cdot (r - p) = p \cdot (r^2 - p^2) = p(r^2 + (p, p)) = p$. So p generates the rotation axis. We omit the somewhat delicate proof of the tangent formula and refer to [4]. □

Remark 29 Of course, this is a fairly comfortable situation for the parametric description of rotations via quaternions. However, there is an ambiguity by the two-element kernel of the group homomorphism Int^{PS}. But there is no right inverse group homomorphism $f : SO_3(\mathbb{R}) \to S(\mathbb{H})$, i.e., such that $Int^{PS} \circ f = Id_{SO_3(\mathbb{R})}$ is impossible. For details, see [4].

For practical purposes it is advantageous to restate the fact that $s \in S(\mathbb{H})$. If $s = r + p$ is the real-pure decomposition, then $s \in S(\mathbb{H})$ means $r^2 + \|p\|^2 = 1$. Thus, for the rotation angle θ calculated in proposition 221, we have $r = \cos\left(\frac{\theta}{2}\right)$ and $p = \sin\left(\frac{\theta}{2}\right)(ix + jy + kz)$, where $(x, y, z) \in S^2 = S(\mathbb{R}^3)$, the unit sphere in $\mathbb{R}^3$. In other words, writing $s_\theta = \left(\cos\left(\frac{\theta}{2}\right), \sin\left(\frac{\theta}{2}\right)\right)$ and $s_{dir} = (x, y, z)$ we have a representation of s as $s_S = (s_\theta, s_{dir}) \in S^1 \times S^2$, the Cartesian product of the unit circle and the unit sphere.

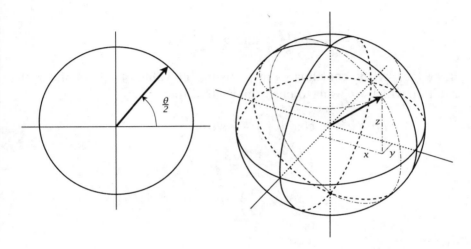

Fig. 25.3. The representation of $s \in S(\mathbb{H})$ as $s_S(s_\theta, s_{dir})$, where s_θ is a vector on the unit circle S^1 and s_{dir} a vector on the unit sphere S^2.

Example 94 Let us now consider a rotation of 120 degrees about the axis defined by $(1, 1, 1)$. This rotation should take the x-axis to the y-axis, the y-axis to the z-axis, and the z-axis to the x-axis.

According to the representation above, the quaternion s describing this rotation can be written as

$$s_S = (s_\theta, s_{dir})$$

where

$$s_\theta = \left(\cos\left(\frac{\theta}{2}\right), \sin\left(\frac{\theta}{2}\right)\right) \text{ and } s_{dir} = \frac{1}{\sqrt{3}}(1, 1, 1).$$

Now to rotate a point $p = (p_x, p_y, p_z)$ we first have to write it as a pure quaternion $\hat{p} = ip_x + jp_y + kp_z$ and then calculate

$$p' = s \cdot \hat{p} \cdot \bar{s},$$

knowing from sorite 218, (vii), that for $s \in S(\mathbb{H})$, $s^{-1} = \bar{s}$. Some calculations using the multiplication rules lined out in sorite 218, (x), (we advise the student to do them as an exercise) lead to the following result:

$$\begin{aligned} p' = \; & i(p_x(u^2 - v^2) + 2p_y(v^2 - uv) + 2p_z(v^2 + uv)) + \\ & j(2p_x(v^2 + uv) + p_y(v^2 - u^2) + 2p_z(v^2 - uv)) + \\ & k(2p_x(v^2 - uv) + 2p_y(v^2 + uv) + 2p_z(u^2 - v^2)), \end{aligned}$$

where

$$u = \cos\left(\frac{\theta}{2}\right) \text{ and } v = \frac{1}{\sqrt{3}}\sin\left(\frac{\theta}{2}\right).$$

If we look at the coordinates of p' using the value of $\frac{2\pi}{3}$ for the 120°-rotation which we want to perform, we find that

$$\begin{aligned} u^2 - v^2 &= \cos^2\left(\frac{\pi}{3}\right) - \frac{1}{3}\sin^2\left(\frac{\pi}{3}\right) \\ &= \left(\frac{1}{2}\right)^2 - \frac{1}{3}\left(\frac{\sqrt{3}}{2}\right)^2 \\ &= \left(\frac{1}{2}\right)^2 - \left(\frac{1}{2}\right)^2 \\ &= 0 \end{aligned}$$

$$\begin{aligned} v^2 - uv &= \frac{1}{3}\sin^2\left(\frac{\pi}{3}\right) - \cos\left(\frac{\pi}{3}\right) \cdot \frac{1}{\sqrt{3}} \cdot \sin\left(\frac{\pi}{3}\right) \\ &= \frac{1}{3}\left(\frac{\sqrt{3}}{2}\right)^2 - \frac{1}{2} \cdot \frac{1}{\sqrt{3}} \cdot \frac{\sqrt{3}}{2} \\ &= 0 \end{aligned}$$

$$\begin{aligned} v^2 + uv &= \frac{1}{3}\sin^2\left(\frac{\pi}{3}\right) + \cos\left(\frac{\pi}{3}\right) \cdot \frac{1}{\sqrt{3}} \cdot \sin\left(\frac{\pi}{3}\right) \\ &= \frac{1}{3}\left(\frac{\sqrt{3}}{2}\right)^2 + \frac{1}{2} \cdot \frac{1}{\sqrt{3}} \cdot \frac{\sqrt{3}}{2} \\ &= \frac{1}{2} \end{aligned}$$

Here we use the trigonometric facts that $\sin(\frac{\pi}{3}) = \frac{\sqrt{3}}{2}$, and $\cos(\frac{\pi}{3}) = \frac{1}{2}$. Putting this together again, we get

$$p' = ip_z + jp_x + kp_y$$

and for $p = (1, 0, 0)$ we do indeed get $p' = (0, 1, 0)$.

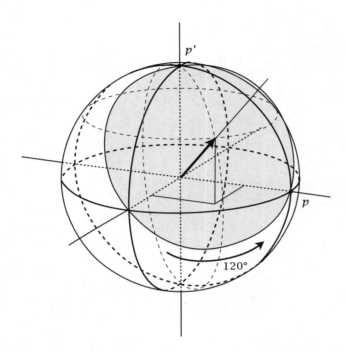

Fig. 25.4. The rotation by $\theta = \frac{2\pi}{3}$ around the axis defined by the vector $(\frac{1}{\sqrt{3}}, \frac{1}{\sqrt{3}}, \frac{1}{\sqrt{3}})$ maps a point p on the x-axis to a point p' on the y-axis.

CHAPTER 26

Second Advanced Topic

In this second advanced chapter, we shall shortly describe the theory of finite commutative skew fields, called Galois fields, and then give two important applications of this theory: The first is Reed's and Solomon's error correction code, which is of great importance in media encoding for CDs, DVDs, ADSL, etc. The second is Rivest's, Shamir's, and Adelman's encryption algorithm, which is of practical use in digital signature generation, for example.

26.1 Galois Fields

Galois fields are by definition commutative fields with finite cardinality. (In field theory, it can be shown that every finite skew field is in fact automatically commutative, so no other finite fields exist.) Their name has been chosen in honor of the French mathematician Evariste Galois (1811–1832), who invented the algebraic theory of polynomial equations and thereby was, among others, able to solve some of the most difficult questions in mathematics: The conjecture of the impossibility of a generally valid construction by compass and straightedge of the trisection of an angle or the Delian problem concerning the duplication of the cube by the same type of construction. Unfortunately, he was killed in a duel at the age of 20. His research had not been understood during his short lifetime. The famous mathematician Siméon-Denis Poisson commented on a Galois paper: "His argument is neither sufficiently clear nor sufficiently developed to allow us to judge its rigor." Only in 1843, the mathematician Joseph Liouville recognized Galois' work as a solution "as correct as it is deep of this lovely problem: Given an irreducible equation of prime degree, decide whether or not it is soluble by radicals." This is in fact the abstract statement implying the solution of the old problems mentioned above.

We shall outline the structure theory of Galois fields, because they play a crucial role in coding theory, the theory dealing with the methods for storing and transmitting digital information through noisy media. In the next section, an application of Galois field theory will be given for the famous Reed-Solomon error correction code.

Finite fields are constructed from the uniquely determined minimal field $P(K) \subset K$ contained in a field K. This is the intersection of all subfields of K, which clearly is a field. Its structure is given as follows:

Lemma 222 *Given a field K, consider the unique homomorphism of rings $p : \mathbb{Z} \to K$ defined by $p(n) = n \cdot 1_K$. Then the principal ideal $Ker(p)$ generated by the non-negative characteristic $char(K)$ of K is $Ker(p) = (char(K))$. $Ker(p)$ is either trivial, i.e., $char(K) = 0$, and then $P(K)$ consists of all fractions $n/m = n \cdot m^{-1}$ and is therefore obviously isomorphic to $\mathbb{Q}$. Or else, $char(K)$ is a prime number since the image is isomorphic to $\mathbb{Z}/(char(K))$ and, as a subset of a field, must be an integral domain. But then the image is a field, i.e., $P(K) \xrightarrow{\sim} \mathbb{Z}/(char(K))$.*

This means that Galois fields K have a positive prime characteristic p and are somehow built from their prime field $\mathbb{Z}_p$. Clearly, a Galois field is a vector space over its prime field, and evidently of finite dimension n. So, by the Steinitz theorem, we have the vector space isomorphism $K \xrightarrow{\sim} \mathbb{Z}_p^n$, and therefore $card(K) = p^n$. Moreover, since the group K^* has $p^n - 1$ elements, these are all roots of the polynomial $X^{p^n-1} - 1$. Together with 0, the elements of a field K with characteristic p and p^n elements are the set of roots of the polynomial $X^{p^n} - X \in \mathbb{Z}_p[X]$. Since this polynomial has at most p^n different roots, this polynomial in K decomposes into a product

$$X^{p^n} - X = \prod_{x \in K} (X - x)$$

of p^n different linear terms. The question here is, whether the characteristic and the polynomial uniquely determine K up to a field isomorphism.

To this end, we sketch a short but comprehensive run through the existence and uniqueness of fields defined by roots of certain polynomials.

In all these discussions, a *field extension* is a pair of fields K, L such that $K \subset L$. L is called an *extension of K*. If $T \subset L$, we denote by $K(T)$ the smallest extension of K in L containing T and call it the *extension of K generated by T*. To begin with, we have this lemma about uniqueness of extensions.

Lemma 223 *If $K \subset L$ is an extension, and if $x \in L$ is a root $f(x) = 0$ of a polynomial $f \in K[X]$, then there is a uniquely determined irreducible polynomial $r = X^m + a_{m-1}X^{m-1} + \ldots a_0 \in K[X]$ such that the extension $K(x) \subset L$ of K is isomorphic to $K[X]/(r)$, which is of dimension $m = \deg(r)$ as a K-vector space, a basis being defined by the sequence $(x^{m-1}, \ldots x^2, x, 1)$. The polynomial r is called the* defining polynomial of x. *In this case, x is said to be* algebraic over K.

An extension $K \subset L$, where every element of L is algebraic over K is called an algebraic extension. *If the dimension of an extension $K \subset L$ as a K-vector space is finite, L is algebraic. In particular the extension $K(x)$ by an algebraic element x is algebraic. If extensions $K \subset L, L \subset M$ are algebraic, then so is $K \subset M$.*

Conversely, given an irreducible polynomial $r = X^m + a_{m-1}X^{m-1} + \ldots a_0 \in K[X]$, there is an algebraic extension $K \subset L = K(x)$ such that the defining polynomial of x is r.

Example 95 A Galois field K of characteristic p is an algebraic extension of its prime field isomorphic to $\mathbb{Z}_p$. In fact, if $x \in K$, then the ring homomorphism $\mathbb{Z}_p[X] \to K : X \mapsto x$ has a kernel which is either trivial or a maximal ideal generated by an irreducible polynomial. If it were trivial, we would have the powers $1, x, x^2, \ldots x^m$ linearly independent over the prime field, for every positive power m, which contradicts the finiteness of K.

Exercise 131 Show that the $\mathbb{R} \subset \mathbb{C}$ is an algebraic extension.

Proposition 224 *Let $K \subset L, K' \subset L'$ be two field extensions, such that there is an isomorphism $k : K \overset{\sim}{\to} K'$ with this property: There is a polynomial $f \in K[X]$ such that $f = (X - x_1)(X - x_2) \ldots (X - x_r)$ in L, and such that $L = K(x_1, x_2, \ldots x_r)$, and the extension $k[X] : K[X] \overset{\sim}{\to} K'[X]$ of k to the polynomial algebras maps f to f' such that $L' = K'(x_1', x_2', \ldots x_r')$ and $f' = (X - x_1')(X - x_2') \ldots (X - x_r')$. Then there is an extension $F : L \overset{\sim}{\to} L'$ of f, i.e., $F|_K = f$.*

The proof of proposition 224 works by induction on the maximal degree of the irreducible factors in the decomposition of f in $K[X]$, and then on the number of such maximal degree factors. The inductive step is in fact the above lemma 223.

Proposition 225 *If K is a field, and if $f \in K[X]$ is a polynomial, there exists a* splitting field extension $K \subset L$ *of f, i.e., $L = K(x_1, \ldots x_r)$ where x_i are roots of f such that in $L[X]$, $f = a(X - x_1)(X - x_2) \ldots (X - x_r)$.*

The proof of this proposition is again an immediate consequence of lemma 223. In fact, for each irreducible factor g of f in $K[X]$, we may embed K in the field $K[X]/(g)$, where g has a root. And so forth until f splits into linear factors.

Corollary 226 *For a given prime characteristic p and a given exponent $n > 0$, there is a splitting field $\mathrm{GF}(p^n)$ of the polynomial $X^{p^n} - X$, and any two such fields are isomorphic. The cardinality of $\mathrm{GF}(p^n)$ is p^n since the roots of $X^{p^n} - X$ are pairwise different.*

The only open point in this corollary is the number of different roots. This is true because the algebraic differential quotient $(p^n - 1)X^{p^n - 2} = -1 \cdot X^{p^n - 2}$ of $X^{p^n - 1} - 1$ is relatively prime to $X^{p^n - 1} - 1$, which is only possible if all roots differ, see [46, page 43] for details.

Summarizing, for every prime p and every positive natural exponent n, there is one, and up to field isomorphisms only one Galois field $\mathrm{GF}(p^n)$ of p^n elements. Let us now show that in fact, $\mathrm{GF}(p^n) = \mathbb{Z}_p(\zeta)$, i.e., an algebraic extension by a single element ζ.

Proposition 227 *The multiplicative group $\mathrm{GF}(p^n)$ of a Galois field is cyclic, i.e., isomorphic to $\mathbb{Z}_{p^n - 1}$. A generator ζ of this group is called a primitive root of unity.*

The proof is easy, see [46, page 42] for details.

Therefore a Galois field $\mathrm{GF}(p^n)$ is an algebraic extension $\mathrm{GF}(p^n) = \mathbb{Z}_p(\zeta)$ by a primitive root of unity ζ, whose defining polynomial $Z \in \mathbb{Z}_p[X]$ is of degree n, the powers $1, \zeta, \zeta^2, \ldots \zeta^{n-1}$ building a basis over the prime field. But observe that there may be different defining polynomials for the same Galois field.

26.1.1 Implementation

We shall discuss the Reed-Solomon error correction code for the special prime field $\mathbb{Z}_2$ which may be identified with the bit set $Bit = \{0,1\}$ for computerized implementation. In this case, we need to control the arithmetics on $GF(2^n)$ on the basis of bit-wise encoding.

Identifying $GF(2^n)$ with $\mathbb{Z}_2[X]/(Z)$, where Z is the defining polynomial of a primitive $(2^n - 1)$-th root of unity ζ, we know that elements $u \in GF(2^n)$ are uniquely represented by the basis $(x^{n-1}, \ldots x^2, x^1, 1)$, according to the vector space identification $GF(2^n) \xrightarrow{\sim} \mathbb{Z}_2^n$, where x is the image of X in $\mathbb{Z}_2[X]/(Z)$. So we identify $u(x) \in \mathbb{Z}_2[X]/(Z)$ with the vector $u = (u_{n-1}, \ldots u_2, u_1, u_0) \in \mathbb{Z}_2^n$, i.e., with a bit sequence of length n, which encodes the class of the polynomial residue $u(x) = u_{n-1}x^{n-1} + \ldots u_2x^2 + u_1x + u_0$ in $\mathbb{Z}_2[X]/(Z)$. Arithmetic on this representation is as follows:

- Addition is the coordinate-wise addition of bits and has nothing to do with the defining polynomial, except for its degree n:

$$(u_{n-1}, \ldots u_1, u_0) + (v_{n-1}, \ldots v_1, v_0) = (u_{n-1} + v_{n-1}, \ldots u_1 + v_1, u_0 + v_0)$$

  This addition may also be viewed as the logical exclusive alternative *xor* on the Boolean algebra $Bit = 2$, via $a + b = \neg(a \Leftrightarrow b) = xor(a, b)$.

- Multiplication $u(x) \cdot v(x)$ is decomposed as follows: multiplication of $u(x)$ with a constant $v \in \mathbb{Z}_2$ is the coordinate-wise multiplication of bits, i.e., corresponding to logical conjunction $a \cdot b = a \& b$. The multiplication $u(x) \cdot x$ decomposes as a sequence of three steps: (1) a shift of $u = (u_{n-1}, \ldots u_2, u_1, u_0)$ to the left, yielding $l(u) = (u_{n-2}, \ldots u_1, u_0, 0)$, (2) the calculation of the residue in $x^n = 1 \cdot Z + r(x)$ (to be done only once for all), (3) the addition $l(u) + r$ in case $u_{n-1} \neq 0$. Now, the full multiplication is written as a succession of these elementary operations:

$$u(x) \cdot v_{n-1}$$
$$(u(x) \cdot v_{n-1})x$$
$$((u(x) \cdot v_{n-1})x) + v_{n-2}$$
$$(((u(x) \cdot v_{n-1})x) + v_{n-2})x$$
$$((((u(x) \cdot v_{n-1})x) + v_{n-2})x) + v_{n-3}$$
$$\vdots$$
$$(\ldots (u(x) \cdot v_{n-1})x) \ldots) + v_0$$

Example 96 We consider the Galois field $GF(2^4)$. It has 16 elements, which are represented as four-bit words, e.g., $u = (1,1,0,1)$ corresponds to $u(x) = x^3 + x^2 + 1$. The primitive root of unity can be defined by the polynomials $Z = X^4 + X + 1$ or $Z = X^4 + X^3 + 1$. Let us take $Z = X^4 + X + 1$. Then we have the remainder formula $X^4 = 1 \cdot Z + (X + 1)$ (observe that $1 = -1$ in $\mathbb{Z}_2$). So $x^4 = x + 1$.

Exercise 132 Calculate the 16 powers $x^i, i = 1, 2, 3, \ldots 16$ of x in the example 96 by means of the multiplication algorithm described above.

The product in $GF(p^n)$ is also easily encoded by observing the fact that $GF(p^n)^* \xrightarrow{\sim} \mathbb{Z}_{p^n-1}$. So the exhibition of a generator, in fact x in the representation $GF(p^n) \xrightarrow{\sim} \mathbb{Z}_p[X]/(Z)$, of this group, together with the multiplication table yielding the different powers of x, enables us to deal with the exponents in $\mathbb{Z}_{p^n-1}$, and to build lookup tables for the transformation between exponents and the powers of x.

26.2 The Reed-Solomon (RS) Error Correction Code

The Reed-Solomon error correction code was invented in 1960 by MIT's Lincoln Laboratory members Irving S. Reed and Gustave Solomon and published in [39]. When it was written, digital technology was not advanced enough to implement the concept. The key to the implementation of Reed-Solomon codes was the invention of an efficient decoding algorithm by Elwyn Berlekamp, a professor of electrical engineering at the University of California, Berkeley (see his paper [5]). The Reed-Solomon code is used in storage devices (including tape, Compact Disk, DVD, barcodes, etc.), wireless or mobile communications (such as cellular telephones or microwave links), satellite communications, digital television, high-speed modems such as ADSL. The encoding of digital pictures sent back by the Voyager space mission in 1977 was the first significant application.

The following development is akin to the elegant exposition in [42]. The Reed-Solomon code creates a redundant information which is very tolerant against changing part of the encoded information. We call the code $RS(s, k, t)$, where s, k, t are natural numbers which define specific choices as follows:

We start from a sequence $S = (b_i)$ of bits, which are subdivided into words of s bits each. So our sequence is interpreted as a sequence (c_j), where $c_j = (b_{i(j)}, b_{i(j)+1}, \ldots b_{i(j)+s-1}) \in GF(2^s)$. This sequence is split into blocks $c = (c_0, c_1, \ldots c_{k-1}) \in GF(2^s)^k$ of k elements $c_j \in GF(2^s)$ each. The encoding is an injective $GF(2^s)$-linear map

$$\epsilon : GF(2^s)^k \to GF(2^s)^{k+2t}$$

with k, t such that $k + 2t \leq 2^s - 1$. To define ϵ, one considers the polynomial $p = (X - \zeta)(X - \zeta^2) \ldots (X - \zeta^{2t}) \in GF(2^s)[X]$ for a primitive $(2^s - 1)$-th root ζ of unity in the cyclic group $GF(2^s)^*$. One then encodes the block vector $c = (c_0, c_1, \ldots c_{k-1}) \in GF(2^s)^k$ as a polynomial $c(X) = \sum_{i=0,\ldots k-1} c_i X^i \in GF(2^s)[X]$. Then, $p \cdot c(X) = \sum_{i=0,\ldots k+2t-1} d_i X^i$, and we set

$$\epsilon(c) = d = (d_0, d_1, \ldots d_{k+2t-1})$$

which is obviously linear in c and injective, since c can be recovered from $p \cdot c(X)$, and the latter from $d = \epsilon(c)$; denote the polynomial $\sum_{i=0,\ldots k+2t-1} d_i X^i$ as a function of d by $d(X)$, i.e., $p \cdot c(X) = d(X)$, whence $c(X) = d(X)/p$.

Proposition 228 (2t-error detection) *If with the above notations, the encoded value d is altered at at most $2t$ positions to the measured value f by noise $e = f - d \in GF(2^s)^{k+2t}$, then $e = 0$ iff $(f(\zeta^i))_{i=1,\ldots 2t} = 0$, where $f(\xi)$ is the evaluation of $f(X)$ at ξ.*

In fact,

$$(f(\zeta^i))^\top = \begin{pmatrix} 1 & \zeta & \zeta^2 & \cdots & \zeta^{k+2t-1} \\ 1 & \zeta^2 & \zeta^4 & \cdots & \zeta^{2(k+2t-1)} \\ \vdots & \vdots & \vdots & & \vdots \\ 1 & \zeta^{2t} & \zeta^{4t} & \cdots & \zeta^{2t(k+2t-1)} \end{pmatrix} \cdot (e_i)^\top$$

since by construction $d(\zeta^i) = 0$. Now, if all $e_i = 0$ except of at most $2t$ indexes $i_1 < i_2 < \ldots i_{2t}$, then the above equation reduces to

$$(f(\zeta^i))^\top = \begin{pmatrix} \zeta^{i_1} & \zeta^{i_2} & \cdots & \zeta^{i_{2t}} \\ \zeta^{2i_1} & \zeta^{2i_2} & \cdots & \zeta^{2i_{2t}} \\ \vdots & \vdots & & \vdots \\ \zeta^{2ti_1} & \zeta^{2ti_2} & \cdots & \zeta^{2ti_{2t}} \end{pmatrix} \cdot (e_{i_j})^\top = Z \cdot (e_{i_j})^\top$$

where (e_{i_j}) is a row vector of length $2t$. But the $2t \times 2t$-matrix Z of the ζ-powers is of rank $2t$, whence the claim, since the map defined by Z is an isomorphism. Let us understand why Z is regular. We have

$$Z = \zeta^{i_1 + i_2 + \ldots i_{2t}} \cdot \begin{pmatrix} 1 & 1 & \ldots & 1 \\ \zeta^{i_1} & \zeta^{i_2} & \ldots & \zeta^{i_{2t}} \\ \zeta^{2i_1} & \zeta^{2i_2} & \ldots & \zeta^{2i_{2t}} \\ \vdots & \vdots & & \vdots \\ \zeta^{(2t-1)i_1} & \zeta^{(2t-1)i_2} & \ldots & \zeta^{(2t-1)i_{2t}} \end{pmatrix}.$$

The matrix to the right of the power of ζ is a Vandermonde matrix, whose determinant is known to be $\prod_{a \leq u < v \leq 2t}(\zeta^{i_u} - \zeta^{i_v})$. See [23] for a proof. Since we suppose $k + 2t \leq 2^s - 1$, no two powers ζ^{i_u}, ζ^{i_v} are equal for $u \neq v$, and the Vandermonde matrix is regular.

Proposition 229 (t-error correction) *If with the above notations, the encoded value d is altered at most at t positions, and if we also know that at most 2t positions can be altered in the measured value f by noise $e = f - d \in \mathrm{GF}(2^s)^{k+2t}$, then the error can be calculated from f, and therefore the original value d and therefore c can be reconstructed.*

We know that in this case the above Vandermonde matrix is regular. So, if the set of indexes $i_1 < i_2 < \ldots i_t$ where the errors $e_{i_1}, \ldots e_{i_t}$ may differ from zero is known, the values of these errors are known from the measured values f by proposition 228. But we do not know for which indexes the t possible values are altered. So we would have to check all t-element sets of indexes. However, we know that at most $2t$ indexes can have altered values. So any two sets of t indexes for altered error values must be contained in a set of $2t$ indexes of a priori possible errors. But the regularity of the Vandermonde matrix for this 2t-index set implies a unique solution, which means that both t-element index sets can only contain a number of non-vanishing errors in their intersection. So the errors can be recalculated and we are done.

Example 97 Let us consider an example from music storage on compact disks. It can be shown that one hour of stereo music with 16-bit resolution and 44100 sampling rate needs roughly $635MB = 635 \times 8 \times 10^6 = 5080000000$ bits. Suppose that we want to be able to correct bursts of up to 200 bit errors. This means that if a block has s bits, such a burst hits up to $\lceil 200/s \rceil$ blocks, where $\lceil x \rceil$ is the least integer $\geq x$. We may reconstruct such errors if $t \geq \lceil 200/s \rceil$. Moreover, the condition $k + 2t \leq 2^s - 1$ implies $k + 2\lceil 200/s \rceil \leq k + 2t \leq 2^s - 1$ and can be met by $k = 2^s - 1 - \lceil 200/s \rceil$, whence also $t = \lceil 200/s \rceil$. So one block of $k \cdot s$ bits makes a total of

$\lceil 5080000000/(k \cdot s) \rceil$ blocks, and each being expanded to $k + 2t$ blocks, we get a number of

$$f(s) = \lceil 5080000000/(k \cdot s) \rceil \cdot (k + 2t)$$
$$= \lceil 5080000000/(2^s - 1 - \lceil 200/s \rceil)s \rceil \cdot (2^s - 1)$$

bits. This yields the following numbers of MB required to meet this task for $s = 6 \ldots 12$:

s	6	7	8	9	10	11	12
$f(s)$	−8001.0	1168.77	789.878	697.818	660.84	647.012	640.319

Observe that below $s \leq 6$ no reasonable bit number is possible, and that for $s = 12$, we get quite close to the original size.

26.3 The Rivest-Shamir-Adelman (RSA) Encryption Algorithm

The RSA algorithm was published in 1978 by Ron Rivest, Adi Shamir, and Leonhard Adleman [40]. It can be used for public key encryption and digital signatures. Its security is based on the difficulty of factoring large integers, and again uses the theory of Galois fields.

The **first step** is the *generation of the public and private key* and runs as follows:

1. Generate two different large primes, p and q, of approximately equal size, such that their product $n = pq$ is of a bit length, e.g., 1024 bits, which is required for the representation of the message as a big number, see below in the next paragraph.

2. One computes $n = pq$ and $\phi = (p - 1)(q - 1)$.

3. One chooses a natural number $e, 1 \leq e \leq \phi$, such that $gcd(e, \phi) = 1$. Then by definition, the couple (e, n) is the *public key*.

4. From the previous choice and the verification that $gcd(e, \phi) = 1$, one computes the inverse $d, 1 \leq d \leq \phi$, i.e. $de = 1 \mod \phi$. By definition, the couple (d, n) is the *private key*.

5. The values p, q, ϕ, and d are kept secret.

The **second step** describes the *encryption* of the message sent from A to B. The private knowledge (p_B, q_B, ϕ_B) is attributed to B who will be able to decipher the encrypted message from A.

1. B's public key (n_B, e_B) is transmitted to A.
2. The message is represented by a (possibly very large) natural number $1 \le m \le n_B$.
3. The encrypted message (the "cyphertext") is defined by

$$c = m^{e_B} \mod n_B.$$

4. The cyphertext c is sent to B.

Since B knows that the message number m is unique in $\mathbb{Z}_{n_B}$, then once we have recalculated $m \mod n_B$, we are done.

The **third step** deals with the *decryption of the original message number m by B*.

1. Receiver B must use his or her private key (d_B, n_B) and calculate the number $c^{d_B} \mod n_B$, where he must use the fact that $m = c^{d_B} \mod n_B$.
2. He then reconstructs A's full message text from the numeric representation of m.

Why is it true that $m = c^{d_B} \mod n_B$? By construction, we have $c^{d_B} = m^{e_B d_B} = m^{1 + s \cdot \phi}$ in $\mathbb{Z}_{n_B}$. Consider now the projection

$$\pi : \mathbb{Z} \to \mathbb{Z}_{p_B} \times \mathbb{Z}_{q_B} : z \mapsto (z \mod p_B, z \mod q_B)$$

of rings. The kernel is the principal ideal $(p_B q_B)$ by the prime factorization theory. So we have an injection of rings $\mathbb{Z}_{n_B} \to \mathbb{Z}_{p_B} \times \mathbb{Z}_{q_B}$, and because both rings have equal cardinality, this is an isomorphism. To show that $m = m^{e_B d_B}$ in $\mathbb{Z}_{n_B}$ is therefore equivalent to show this equation holds in each factor ring $\mathbb{Z}_{p_B}$ and $\mathbb{Z}_{q_B}$. Now, if $m = 0$ in $\mathbb{Z}_{p_B}$ or in $\mathbb{Z}_{q_B}$, the claim is immediate, so let us assume that $m \ne 0 \mod p_B$. Then we have $m = m^{e_B d_B} = m^1 m^{s\phi}$, and it suffices to show that $m^\phi = m^{(p_B - 1)(q_B - 1)} = 1$ in $\mathbb{Z}_{p_B}$. This follows readily from the small Fermat theorem 134 $m^{p_B - 1} = 1$ in $\mathbb{Z}_{p_B}$, for $m \ne 0 \mod p_B$. The same argument holds for q_B, and we are done.

For an in-depth treatment of cryptography, theory and implementation, see [44] and [41].

Appendix

Further Reading

Set theory. Keith Devlin's *The Joy of Sets* [20] is a modern treatment of set theory, including non-well-founded sets. As the title indicates, the style is rather relaxed, but the treatment is nevertheless elaborate.

Graph theory. Harris', Hirst's and Mossinghoff's *Combinatorics and Graph Theory* [26] includes recent results and problems that emphasize the cross-fertilizing power of mathematics. Frank Harary's *Graph theory* [25] is a very classical book including many interesting problems and written by one of the leading graph theorists.

Abstract algebra. Among the wealth of books on algebra, Bhattacharya's, Jain's and Nagpaul's *Basic Abstract Algebra* [7] is very readable. A text with a more practical focus is *Discrete Mathematics* [8] by Norman Biggs.

Number theory. The branch of mathematics that deals with natural numbers and their properties, such as primes, is called *number theory.* Despite its elementary basis, number theory quickly becomes very involved, with many unsolved problems. Andrews' *Number Theory* [2] provides a clear introduction to the subject.

Formal logic. Alonzo Church and Willard Van Orman Quine were pioneers of mathematical logic. Church's *Introduction to Mathematical Logic* [15] and Quine's *Mathematical Logic* [37] deal with with classical propositional and predicate logic, and their relation to set theory, with a slight philosophical flavor. A modern text is Dirk van Dalen's *Logic and Structure* [17], which also provides an exposition of natural deduction and intuitionistic logic.

Languages, grammars and automata. Hopcroft's, Motwani's and Ullman's *Introduction to Automata Theory, Languages, and Computation* [28] provides a comprehensive exposition of formal languages and automata. Although the theory of computation has its origin in the work of Alan Turing and Alonzo Church, the modern treatment is much indebted to Martin Davis' *Computability and Unsolvability* [19] from 1958.

Linear algebra. Werner Greub's *Linear Algebra* is a classical text, which is written in a lucid and precise style, also extending to multilinear algebra in a second volume. The two volumes *Basic Linear Algebra* [9] and *Further Linear Algebra* [10] by Blyth and Robertson are more recent texts for first year students, working from concrete examples towards abstract theorems, via tutorial-type exercises. Marcel Berger's *Geometry* [4] is one of the best introductions to linear geometry, including a large number of examples, figures, and applications from different fields, including the arts and classical synthetic geometry.

Computer mathematics. Mathematics in the context of computers can be roughly divided into three domains: algorithms, numerics and computer algebra.

Donald Knuth's series of *The Art of Computer Programming* [31, 32, 33] has the great merit of introducing to computer science a more rigorous mathematical treatment of algorithms such as sorting and searching or machine arithmetic. His books have become the yardstick for all subsequent literature in this branch of computer science. They are, however, not for the faint of heart. A more recent book is Cormen's, Leiserson's, Rivest's and Stein's *Introduction to Algorithms* [16].

Numerics is probably the oldest application of computers, accordingly the literature is extensive. A recent publication is Didier Besset's *Object-Oriented Implementation of Numerical Methods* [6] which exploits object-oriented techniques.

In contrast to numerics, computer algebra focusses on the symbolic solution of many of the problems presented in this book. A recent work is von zur Gathen's and Gerhard's *Modern Computer Algebra* [47]. Computer algebra is the foundation of such symbolic computation systems as Maple or Mathematica.

Bibliography

[1] Aczel, Peter. *Non-Well-Founded Sets*. CSLI LN 14, Stanford, Cal. 1988.

[2] Andrews, George E. *Number Theory*. Dover, New York 1994.

[3] Barwise, Jon & Moss, Lawrence. *Vicious Circles*. CSLI Publications, Stanford, Cal. 1996.

[4] Berger, Marcel. *Geometry I, II*. Springer, Heidelberg et al. 1987.

[5] Berlekamp, Elwyn. "Bit-Serial Reed-Solomon Encoders." *IEEE Transactions on Information Theory*, IT 28, 1982, pp. 869–874.

[6] Besset, Didier H. *Object-Oriented Implementation of Numerical Methods*. Morgan Kaufmann, San Francisco et al. 2001.

[7] Bhattacharya, P. B., Jain, S. K. & Nagpaul S. R. *Basic Abstract Algebra*. Cambridge University Press, Cambridge 1994.

[8] Biggs, Norman L. *Discrete Mathematics*. Oxford University Press, Oxford 2002.

[9] Blyth, Thomas S. & Robertson, Edmund F. *Basic Linear Algebra*. Springer, Heidelberg et al. 2002.

[10] Blyth, Thomas S. & Robertson, Edmund F. *Further Linear Algebra*. Springer, Heidelberg et al. 2001.

[11] Bornemann, Folkmar. "PRIME Is in P: A Breakthrough for 'Everyman'." *Notices of the AMS*, vol. 50, No. 5, May 2003.

[12] Bourbaki, Nicolas. *Eléments d'histoire des mathématiques*. Hermann, Paris 1969.

[13] Cap, Clemens H. *Theoretische Grundlagen der Informatik.* Springer, Heidelberg et al. 1993.

[14] Chomsky, Noam. "Three models for the description of language." *I.R.E. Transactions on information theory,* volume 2, pp. 113–124, IT, 1956.

[15] Church, Alonzo. *Introduction to Mathematical Logic.* Princeton University Press, Princeton 1996.

[16] Cormen, Thomas H., Leiserson, Charles E., Rivest, Ronald L. & Stein, Clifford. *Introduction to Algorithms.* MIT Press, Cambridge 2001.

[17] van Dalen, Dirk. *Logic and Structure.* Springer, Heidelberg et al. 1994.

[18] Date, C. J. *An Introduction to Database Systems.* Addison-Wesley, Reading 2003.

[19] Davis, Martin. *Computability and Unsolvability.* Dover, New York 1985.

[20] Devlin, Keith J. *The Joy of Sets: Fundamentals of Contemporary Set Theory.* Springer, Heidelberg et al. 1999.

[21] Goldblatt, Robert. *Topoi—The Categorical Analysis of Logic.* North-Holland, Amsterdam, 1984.

[22] Greub, Werner. *Linear Algebra.* Springer, Heidelberg et al. 1975.

[23] Gröbner, Wolfgang. *Matrizenrechnung.* Bibliographisches Institut, Mannheim 1966.

[24] Groff, James R. & Weinberg, Paul N. *SQL: The Complete Reference.* McGraw-Hill Osborne, New York 2002.

[25] Harary, Frank. *Graph Theory.* Addison-Wesley, Reading 1972.

[26] Harris, John M., Hirst, Jeffrey L. & Mossinghoff, Michael J. *Combinatorics and Graph Theory.* Springer, Heidelberg et al. 2000.

[27] Hilbert, David & Ackermann, Wilhelm. *Grundzüge der theoretischen Logik.* Springer, Heidelberg et al. 1967.

[28] Hopcroft, John E., Motwani, Rajeev & Ullman, Jeffrey D. *Introduction to Automata Theory, Languages and Computation.* Addison Wesley, Reading 2000.

[29] Jensen, Kathleen & Wirth, Niklaus. *PASCAL—User Manual and Report ISO Pascal Standard.* Springer, Heidelberg et al. 1974.

[30] Garey, Michael R. & Johnson, David S. *Computers and Intractibility: A Guide to the Theory of NP-Completeness*. W H Freeman & Co., New York 1979.

[31] Knuth, Donald Ervin. *The Art of Computer Programming. Volume I: Fundamental Algorithms*. Addison-Wesley, Reading 1997.

[32] Knuth, Donald Ervin. *The Art of Computer Programming. Volume II: Seminumerical Algorithms*. Addison-Wesley, Reading 1998.

[33] Knuth, Donald Ervin. *The Art of Computer Programming. Volume III: Sorting and Searching*. Addison-Wesley, Reading 1998.

[34] Kruse, Rudolf et al. *Foundations of Fuzzy Systems*. John Wiley & Sons, New York 1996.

[35] Kuipers, Jack B. *Quaternions and Rotation Sequences*. Princeton University Press, Princeton-Oxford 1998.

[36] Mac Lane, Sounders & Moerdijk, Ieke. *Sheaves in Geometry and Logic*. Springer, Heidelberg et al. 1992.

[37] Quine, Willard Van Orman. *Mathematical Logic*. Harvard University Press, Cambridge 1981.

[38] Rasiowa, Helena & Sikorski, Roman. *The Mathematics of Metamathematics*. Polish Scientific Publishers, Warsaw 1963.

[39] Reed, Irving S. & Solomon, Gustave. "Polynomial Codes over Certain Finite Fields." *Journal of the Society for Industrial and Applied Mathematics*, Vol. 8, 1960, pp. 300–304.

[40] Rivest, Ronald L., Shamir, Adi & Adleman, Leonard A. "A method for obtaining digital signatures and public-key cryptosystems." *Communications of the ACM*, Vol. 21, Nr. 2, 1978, pp. 120–126.

[41] Schneier, Bruce. *Applied Cryptography: Protocols, Algorithms, and Source Code in C*. John Wiley & Sons, New York 1996.

[42] Steger, Angelika. *Diskrete Strukturen I*. Springer, Heidelberg et al. 2001.

[43] Stetter, Franz. *Grundbegriffe der Theoretischen Informatik*. Springer, Heidelberg et al. 1988.

[44] Stinson, Douglas R. *Cryptography: Theory and Practice*. CRC Press, Boca Raton 1995.

[45] Surma, Stanislav. *Studies in the History of Mathematical Logic*. Polish Academy of Sciences, 1973.

[46] van der Waerden, Bartel Leendert. *Algebra I*. Springer, Heidelberg et al. 1966.

[47] von zur Gathen, Joachim & Gerhard, Jürgen. *Modern Computer Algebra*. Cambridge University Press, Cambridge 1999.

[48] Zadeh, Lotfi A. "Fuzzy Sets." *Information and Control* 8:338–363, 1965.

Index

Symbols